51129

C000115990

CYCLIC NUCLEOTIDE PHOSPHODIESTERASES: STRUCTURE, REGULATION AND DRUG ACTION

WILEY SERIES ON MOLECULAR PHARMACOLOGY OF CELL REGULATION

Volume 1

G-Proteins as Mediators of Cellular Signalling Processes
  Edited by Miles D. Houslay and Graeme Milligan

Volume 2

Cyclic Nucleotide Phosphodiesterases: Structure, Regulation and Drug Action
  Edited by Joe Beavo and Miles D. Houslay

Further volumes in preparation

CYCLIC NUCLEOTIDE PHOSPHODIESTERASES: STRUCTURE, REGULATION AND DRUG ACTION

Edited by

Joe Beavo

Department of Pharmacology, University of Washington, USA

and

Miles D. Houslay

Institute of Biochemistry, University of Glasgow, UK

A Wiley–Interscience Publication

WILEY

JOHN WILEY & SONS

Chichester • New York • Brisbane • Toronto • Singapore

Other Wiley Editorial Offices

John Wiley & Sons, Inc., 605 Third Avenue,
New York, NY 10158-0012, USA

Jacaranda Wiley Ltd, G.P.O. Box 859, Brisbane,
Queensland 4001, Australia

John Wiley & Sons (Canada) Ltd, 22 Worcester Road,
Rexdale, Ontario M9W 1L1, Canada

John Wiley & Sons (SEA) Pte Ltd, 37 Jalan Pemimpin 05-04,
Block B, Union Industrial Building, Singapore 2057

Library of Congress Cataloging-in-Publication Data:
Cyclic nucleotide phosphodiesterases: structure, regulation and drug
action/edited by Joe Beavo and Miles D. Houslay.
    p. cm.—(Wiley series on molecular pharmacology of cell
regulation; v. 2)
    'A Wiley–Interscience publication.'
    ISBN 0 471 92707 4
    1. Cyclic nucleotide phosphodiesterases.   2. Isoenzymes.
I. Beavo, Joe. II. Houslay, Miles D. III. Series.
    [DNLM: 1. Isoenzymes. 2. Nucleotides, Cyclic.
3. Phosphodiesterases. QU 136 185]
QP609.C92176 1990
612'.01513—dc20
DNLM/DLC
for Library of Congress   90-11982
                                                      CIP

British Library Cataloguing in Publication Data:
Cyclic nucleotide phosphodiesterases:
    structure, regulation and drug action.
    1. Organisms. Isoenzymes
    I. Beavo, Joe II. Houslay, Miles D.
    574.1925

ISBN 0 471 92707 4

Typeset by APS, Salisbury, Wiltshire
Printed and bound by Biddles Ltd, Guildford, Surrey.

CONTRIBUTORS

Joe Beavo, *Department of Pharmacology, University of Washington, Seattle*

Per Belfrage, *Department of Medical and Physiological Chemistry, University of Lund, Lund*

Harry Charbonneau, *Department of Biochemistry, University of Washington, Seattle*

Marco Conti, *The Laboratories for Reproductive Biology, and Departments of Pediatrics and Physiology, University of North Carolina at Chapel Hill, Chapel Hill*

Jackie D. Corbin, *Department of Molecular Physiology and Biophysics and the Howard Hughes Medical Institute, Vanderbilt University, Nashville*

Ronald L. Davis, *Department of Cell Biology, Baylor College of Medicine, Houston*

Eva Degerman, *Department of Medical and Physiological Chemistry, University of Lund, Lund*

Paul J. England, *Department of Cellular Pharmacology, Smith Kline and French Research Ltd, The Frythe, Welwyn*

Paul W. Erhardt, *Berlex Laboratories, Cedarknolls*

Sharron H. Francis, *Department of Molecular Physiology and Biophysics and the Howard Hughes Medical Institute, Vanderbilt University, Nashville*

Peter G. Gillespie, *Department of Cell Biology and Neuroscience, University of Texas Southwestern Center at Dallas, Dallas*

Miles D. Houslay, *Molecular Pharmacology Group, Department of Biochemistry, University of Glasgow, Glasgow*

J. A. Kahn, *Biochemistry Research Group, School of Biological Sciences, University College of Swansea, Swansea*

Elaine Kilgour, *Biochemistry Group, School of Biological Sciences, University College of Swansea, Swansea*

Molecular Pharmacology Group, Department of Biochemistry, University of Glasgow, Glasgow

Vincent C. Manganiello, *Section of Biochemical Physiology, Laboratory of Cellular Metabolism, NHLBI, NIH, Bethesda*

Marilyn J. Mooibroek, *Department of Medical Biochemistry, University of Calgary, Calgary*

Seiko Murashima, *Second Department of Internal Medicine, Mie University School of Medicine, 1515 Kamihama-cho, Tsu-shi, Mie*

R. P. Newton, *Biochemistry Research Group, School of Biological Sciences, University College of Swansea, Swansea*

Martin L. Reeves, *Department of Cellular Pharmacology, Smith Kline and French Research Ltd, The Frythe, Welwyn*

S. G. Salih, *Biochemistry Research Group, School of Biological Sciences, University College of Swansea, Swansea*

Rajendra K. Sharma, *Department of Medical Biochemistry, University of Calgary, 3330 Hospital Drive NW, Calgary*

Carolyn J. Smith, *Section of Biochemical Physiology, Laboratory of Cellular Metabolism, NHLBI, NIH, Bethesda*

Johannes V. Swinnen, *The Laboratories for Reproductive Biology, Departments of Pediatrics and Physiology, University of North Carolina at Chapel Hill, Chapel Hill*

Takyuki Tanaka, *Department of Neurosurgery, University of Nagoya School of Medicine, Furo-cho, Chikusa-ku, Nagoya*

Melissa K. Thomas, *Department of Molecular Physiology and Biophysics and the Howard Hughes Medical Institute, Vanderbilt University, Nashville*

Jerry H. Wang, *Department of Medical Biochemistry, University of Calgary, Calgary*

CONTENTS

SERIES PREFACE

Signal transduction processes play a pivotal role in co-ordinating cellular functioning. At one extreme there is the 'simple' chemotactic behaviour of bacteria and *Dictyostelium* and the other the co-ordinated metabolic, physiological and developmental behaviour of complex multicellular species. For many years now such processes have been recognised and indeed exploited. Thus, we see the development of novel therapeutic agents at one end by mankind and at the other natural subversion, for example through the action of certain bacterial exotoxins.

There is a long and impressive history of the action of natural substances on physiological processes. This has been very rewarding in developing our understanding of a myriad of physiological processes as well as providing us with an impressive pharmacopaeia. However, over the past two decades we have seen an exponential growth in our appreciation not only of the underlying mechanisms which determine how cells communicate with each other but also in the molecular nature of the entities involved.

In the case of receptors, the paucity of purified material has, until recently, bedevilled their structural analysis. Now, however, we see not only receptors being purified by sophisticated affinity matrix strategies but also the genes of receptors which have yet to be purified being isolated. We can expect that the overexpression, crystallisation and consequent structural analyses of both receptors and intracellular targets which are modified during signal transduction processes will allow for the rational development of novel therapeutic agents. This is, of course, true for many other targets for 'drug action', such as enzymes and nucleic acids, and a discussion of a range of processes, linked by a common theme of 'cell regulation', will form the basis of this series.

It is clear that much of our understanding of the nature of the molecular events which underly cell regulation has come from drugs acting on particular targets such as cell receptors, proteins modified by the action of intracellular second messengers and the metabolic processes involved in producing bioactive species. Originally, therapeutic agents arose essentially by serendipidity and were then refined by experimentation. However, the changing face of research has been a drive towards understanding key regulatory systems and then attempting intervention by logical drug design. This is a time consuming, multidisciplinary approach which is currently at the leading edge of our present technological expertise.

In this series it is hoped to provide books which address topical areas of endeavour which are of fundamental importance to our understanding of the control of biological processes. These will provide expert contributions covering a range of endeavour in depth. Thus an overview and historical perspective will serve to preface each volume and analyses will be given to cover a variety of perspectives as might be desired by pharmacologists, biochemists, cell biologists, molecular biologists, and clinicians. Each volume should provide a wealth of detail from the 'whole organism' perspective down to the molecules and genes involved. Whilst the areas chosen will essentially relate to 'leading edge activities', in many subjects to be covered there is either evidence or potential of clinical interest either at the diagnostic or therapeutic level. In this regard, the series will offer a bridge across the basic science—medical science divide.

PREFACE

Information about the molecular mechanisms by which signals are transduced from outside the cell to the inside has increased exponentially in the last decade. Much is now known about the primary messengers, their receptors and the mechanisms by which the receptors are coupled to biochemical responses in the cell. The oldest and probably best understood signal transduction pathway uses cAMP as a second messenger. More recently, great progress has been made in our understanding of cGMP-mediated signal transduction. In particular the mechanisms of cyclic nucleotide synthesis, of their control by different hormones and drugs and of their interaction with their intracellular 'receptors' are now well established.

In contrast to our understanding of cyclic nucleotide formation and functions, our understanding of cyclic nucleotide degradation is much less well developed. This is true even though cyclic nucleotide PDE activity has been known for over 20 years and its activity is a crucial determinant of cyclic nucleotide steady-state levels. A major reason for the relative lack of progress in the area is the fact that it only recently has been appreciated that a large number of different isoenzymes of cyclic nucleotide phosphodiesterase exist and that these isoenzymes are differentially expressed and regulated. However, in the last two years a number of studies have started to elucidate many of our questions about PDE regulation, structure and function. This book brings together in one volume a brief review of several recently developed aspects of our understanding of PDE structure and control. Many of the chapters summarize data about individual isoenzyme families. Others integrate information about hormonal control, inhibitor design and structure–junction relationships. Each chapter is written by an expert in the field and only one chapter on each topic is presented even though in most cases many laboratories have contributed to our knowledge in the area. However, the authors have described not only work from their own laboratories, but also pertinent data from other investigators. We think that this approach has allowed each area to be discussed in some depth without making the book too long. We hope that the volume will prove to be a useful reference resource for years to come. It is the only recent volume of its kind that addresses in such detail questions of PDE regulation and mechanism from multiple viewpoints. Finally, the editors would like to thank each of the authors for presenting such up to date information which in many cases preceded primary publication of the results. We would invite comments about the book from readers and hope that you enjoy reading it.

PART A

INTRODUCTION

1

MULTIPLE PHOSPHODIESTERASE ISOENZYMES
Background, Nomenclature and Implications

Joe Beavo

Department of Pharmacology, SJ-30 University of Washington, Seattle, WA 98195, USA

1.1 BACKGROUND

Enzyme activity that catalyzes the hydrolysis and inactivation of cAMP and cGMP was described almost immediately after the discovery that cyclic nucleotides exist in the cell. In fact the demonstration of enzymatic degradation of cyclic nucleotides by phosphodiesterases (PDEs) formed part of the criteria that these compounds act as second messengers. Within a few years it became apparent that more than one enzyme was able to catalyze the hydrolysis of cAMP and cGMP. We now know that many different PDEs selectively catalyze the hydrolysis of purine 3′, 5′ nucleoside monophosphates. However, most early studies and many current textbooks still describe just two forms of PDE, often termed cAMP-PDE and cGMP-PDE. Occasionally, other forms were acknowledged, usually by the additional description of being a 'high- or low-K_m' PDE. It is unfortunate that this rather simplified perception of PDE diversity has held on so long as it has hindered interpretation of many studies. For example, an investigator might report kinetic, structural, or regulatory properties of a particular PDE preparation that were substantially different than those described by others. Unfortunately, many of these differences were then ascribed to lack of reproducibility between laboratories rather than to the fact that different isoenzymes were being studied. In a similar context, many early studies did not appropriately characterize the PDE activity(s) being described. As a result it is now often quite difficult to determine in retrospect which isoenzyme or mixture

Cyclic Nucleotide Phosphodiesterases: Structure, Regulation and Drug Action
Edited by J. Beavo and M. D. Houslay © 1990 John Wiley and Sons Ltd

of isoenzymes was contributing to the activity being measured.

We still do not know precisely how many different PDE isoenzymes are present in most mammalian tissues. However, recent information obtained from cloning, primary sequence, and drug selectivity studies, makes it clear that a relatively large superfamily of PDEs encompassing at least five distinct PDE isoenzyme families exists. Moreover, many of the families have multiple subfamilies and individual members. It now becomes important for future investigators to determine exactly which individual PDE isoenzyme is contributing to a particular PDE activity being described. It is hoped that the readers of this book will find it contains much of the information about individual isoenzymes necessary to make this determination. If so it should contribute to a more widespread understanding of PDE regulation and function.

1.2 NOMENCLATURE AND ISOENZYME FAMILY ORGANIZATION

This introductory chapter describes the author's current view about the number of distinct purine 3′, 5′ nucleoside monophosphate PDEs clearly demonstrated to be present in mammalian tissues. Because of the newly appreciated complexity, structural relationships and size of the PDE isoenzyme families, most older nomenclature systems are rather inadequate. A somewhat expanded and more inclusive version containing elements of several older nomenclatures is used in this chapter. Table 1.1 lists five major PDE isoenzyme families, all thought to be coded for by related but distinct members of a larger 'supergene' family. A separate descriptive name and Roman numeral is assigned to each family. A capital letter and often a descriptive name is given to closely related subfamilies. Arabic numerals are used to designate individual PDEs thought to be the products of alternative splicing of one of the subfamily genes. Finally a Greek letter is used to designate individual subunits of a particular isoenzyme. It should be pointed out that both the names and Roman numerals are arbitrarily assigned. In particular, the Roman numerals do not necessarily correspond to orders of elution from DEAE–cellulose or any other fractionation process. For example, family V PDEs commonly elute from DEAE at the beginning or middle of the salt gradient, depending on which particular subfamily is being described. Nevertheless, where possible, names and numbers similar or identical to several commonly used older nomenclature systems are used. In general, members of one family are related to members of any other family to the extent of approximately 15–25% sequence identity, much of which occurs in one domain thought to be catalytically active. In most cases, each family includes several subfamilies of more closely related PDEs. Although incomplete, the data available suggest that members of a particular subfamily are coded for by different but highly homologous genes ($\sim 80\%$ identity). In several cases only circumstantial evidence is available to validate this conclusion. As an example of the nomenclature system, the 61 kDa Ca^{2+}–CaM-dependent PDE expressed as a major isoenzyme in bovine brain would be called Ca^{2+}–CaM-dependent PDE, I_{A2}. Similarly, the large subunit of the major, light-activated, cGMP-specific PDE present in rod photoreceptor outer segments (Chapter 7) would be designated as the α subunit of the rod membrane-associated cGMP-specific PDE, $V_{B1\alpha}$. Since sequence information is not yet available for many of the PDEs, some of the classifications and placements are necessarily still tentative. Where published data are strong yet still partially circumstantial, a PDE is referred to as a type or subtype rather than as a specific isoenzyme. Where data are entirely indirect or only kinetic in nature, it is referred to as a PDE form. This list is not meant to be inclusive and is expected to change as more information becomes available. It is likely that additional members will be found and that some of the tentative placements will have to be changed as more cDNA and primary sequence information becomes available. However, it should serve as a working model for discussion and reference.

As should be obvious from Table 1.1, data are incomplete for many (perhaps all) of the isoenzyme families. I hope that it is also obvious that much more information needs to be obtained before we really begin to understand the number of different isoenzymes in this supergene family. Not until that is accomplished will we be able to fully understand the regulation and control of cyclic nucleotide metabolism in various cell types by these enzymes. One of the first orders of business will have to be the development of individual isoenzyme-selective probes for each of the isoenzymes listed in the table as well as any others that are identified in the future. Although currently available drugs and antibodies are able to distinguish between most of the PDE families and a few of the subfamilies, very few are able to distinguish among individual family members within subfamilies. Initially, isoenzyme-specific probes will probably take the form of sequence-specific oligonucleotides and antibodies. Currently, no probes of this type have been used in characterization studies although a few nucleotide sequences that should serve the purpose have been reported. It is hoped that once individual isoenzymes can be unambiguously identified using such probes, isoenzyme-selective drugs that will discriminate among the various different PDEs can be developed. It is expected that any future characterization of unknown PDE activities will not be considered complete until the exact family, subfamily and individual characteristics for the enzyme being studied are determined. It should be noted that the specific sequence information given in Table 1.1 is for the most part arbitrarily taken from the N-terminal regions of the isoenzymes. They are not intended to necessarily be the best ones for developing such probes although many should serve to initiate such studies. In cases where sequence for only one member of a family or subfamily is known, it is, of course, not known whether the indicated sequence will distinguish among family and

Table 1.1 Multiple isoenzymes of cyclic nucleotide phosphodiesterase: nomenclature, family relationships, and characteristics

PDE families	Molecular mass (kDa)	Primary sequence[a]	A unique sequence[b]	Comments	References
I. Ca²⁺–CaM-dependent PDE family					
A. 60 kDa subfamily				Will hydrolyze both cAMP and cGMP. Many inhibitors of CaM activation; few very selective for catalytic site	1
1. 59 kDa heart type	59	yes	DDHVTIR RKHLQRPIF	All hydrolyze both cAMP and cGMP with similar efficiency	2
2. 61 kDa brain type	61	yes	YLIGEQT EKMWQRLKG	Alternative splicing of PDE I$_A$ gene inferred from primary protein sequence	3
B. 63 kDa subfamily				Alternative splicing of PDE I$_A$ gene inferred from primary protein sequence	
				Peptide maps and limited sequence indicates that this subfamily is a different gene product from PDE I$_A$ subfamily	
1. 63 kDa brain type	63	yes, part	NA	Most early preparations of brain Ca²⁺–CaM-dependent PDE contained variable amounts of the 61 kDa and this 63 kDa isoenzyme	4
C. 1. 58 kDa lung form[c]	58	no	none	May have CaM bound as a subunit	5
D. 1. 67 kDa smooth muscle form[c]	67	no	none	Originally described in uterine smooth muscle	6

Table 1.1 (*Cont.*)

PDE families	Molecular mass (kDa)	Primary sequence[a]	A unique sequence[b]	Comments	References
E. 1. 67 kDa 'low K_m form'[c]	67	no	none	Not yet clear if CaM sensitivity directly or indirectly mediated	7
F. 1. 75 kDa brain form[c]	75	no	none	Appears to have greater specificity for cGMP than other CaM-PDEs	8
II. cGMP-stimulated PDE family				Also called cGS-PDE or type II PDE in other studies. No selective inhibitors available	9
A. 105 kDa adrenal/heart subfamily	105	yes, part	RRQPAAS RDLFAQEPV	Expressed at very high concentrations in adrenal glomerulosa cells	10
B. 105 kDa membrane-bound type[c]	105	no	none	Gives slightly different peptide map than soluble cGMP-stimulated isoenzymes	9
C. 66 kDa liver form[c]	66	no	none	May be proteolytic product of one of larger forms or a form specific to liver	11
III. cGMP-inhibited PDE family				Also called cGI-PDE and variously type III PDE or type IV PDE in other studies. All selectively inhibited by cGMP and cardiotonic drugs like milrinone	12
A. 110 kDa heart/platelet type(s)[c]	110	no	none	Regulated by cAMP-dependent phosphorylation	13
B. S49-K30a type[c]	105	no	none	Probably a slightly mutated form of 110 kDa type	14

(continued)

Table 1.1 *(Cont.)*

PDE families	Molecular mass (kDa)	Primary sequence[a]	A unique sequence[b]	Comments	References
III. C. dense vesicle form[c]	63	no	none	May be proteolytic product of 110 kDa cGI-PDE or smaller form specific to liver	11
D. adipose tissue type[c]	135	no	none	Regulated by insulin-dependent phosphorylation	12a
IV. **cAMP-specific PDE family**				All are selectively inhibited by Ro 20-1724 and have a high preference for cAMP as substrate. Most probably inhibited by rolipram	15, 16
A. cAMP-specific subfamily A					
1.		yes	PLVDFFCET CSKPWLVG	Called RD1 and ratPDE2 in original descriptions. One of three or more alternatively spliced products of cAMP-specific PDE IV$_A$ gene	16
2.		yes	NA	Originally called RD2. Apparently formed by alternative splicing of the cAMP-specific PDE IV$_A$ gene to give an N-terminal truncated version of the PDE. Not yet confirmed by Northern blot or PCR analysis	16

Table 1.1 (*Cont.*)

PDE families	Molecular mass (kDa)	Primary sequence[a]	A unique sequence[b]	Comments	References
IV. A. 3.		yes	NA	Originally called RD3. Apparently formed by alternative splicing of the cAMP-specific PDE IV$_A$ gene to give an N-terminal truncated version of the PDE. Not yet confirmed by Northern blot or PCR analysis	16
B. cAMP-specific subfamily B					
1.		yes	SLRIVRNNF TLLTNLHG	Called DPD in the original description; one of two or more alternatively spliced products of cAMP-specific PDE IV$_B$ gene	17
2.		yes, part	NA	Called ratPDE4 in the original descriptions; one of two or more alternatively spliced products of cAMP-specific PDE IV$_B$ gene	15
C. cAMP-specific subfamily C					
1.		yes, part	PMAQITGLR KSCHTSLT	Called ratPDE1 in original description. Homologous but not identical to cAMP-specific PDE IV$_A$ and IV$_B$ above. Expressed in germ cells and kidney	15

(continued)

Table 1.1 *(Cont.)*

PDE families	Molecular mass (kDa)	Primary sequence[a]	A unique sequence[b]	Comments	References
IV. D. cAMP-specific subfamily D					
1.		yes, part	NA	Called ratPDE3.1 in original description. Homologous but not identical to the cAMP-specific IV$_{A'}$ IV$_B$ and IV$_C$ PDEs above. Stimulated by FSH in Sertoli cells	15
2.		yes, part	NA	Called ratPDE3.2 in original description. Apparently an alternative splice product of the cAMP-specific PDE IV$_{D1}$ gene	15
V. cGMP-specific PDE family				All selectively inhibited by dipyridamole or M&B 22-948 (Zaprinast) and bind cGMP with high affinity	18
A. Lung/platelet subfamily[c]	~93	no	none	Often co-elutes from DEAE with CaM-dependent PDEs in tissues where both are expressed	19
B. Rod photoreceptor subfamily	~99	yes		Activated by light–transducin–GTP cascade. Mediates visual transduction. Large number of different γ subunits	18

Table 1.1 (*Cont.*)

PDE families	Molecular mass (kDa)	Primary sequence[a]	A unique sequence[b]	Comments	References
V. B. 1. rod membrane-associated PDE(s)	99	yes	α..GEVTAEE VEKFLDSNV β..SPSEGQ VHRFLDQNP	Made up of α, β and γ subunits. Not yet clear if both α and β are catalytically active. γ subunit inhibits PDE activity	18
		yes, part			
2. rod soluble PDE	~99	no	none	Same apparent subunit structure as rod membrane-associated PDE but with additional δ subunit	18
C. Cone photoreceptor subfamily	99	yes	α'..GISQE TVEKYLEAN	Made up of α, β γ and δ subunits; γ subunit may be larger than in rod PDEs	18, 20

[a] Indicates whether primary protein or cDNA sequence has been reported.

[b] These amino acid sequences are currently thought to be unique to the isoenzyme in question. In most cases they are arbitrarily taken from the N-terminal region where that region has been reported. As more individual subfamily members are identified, other unique sequences may need to be determined. *None* indicates that no sequence is yet reported; *NA* indicates that data are as yet not available because appropriate difference regions were not reported in the original publications. It is expected that these data will be reported shortly.

[c] It is not yet clear if these PDEs should be classified as a separate subfamily or as an alternative splice product of an established subfamily. They are somewhat arbitrarily assigned as separate subfamilies here based largely on kinetic and inhibitor data. Firm placement will require sequence analysis.

I, II, III, etc. = family
A, B, C, etc. = subfamily
1, 2, 3, etc. = splice product (type or form)
α, β, γ, etc. = individual subunit

subfamily members. Indeed, the readers interested in developing such probes are referred to the rapidly proliferating number of papers reporting primary protein and cDNA sequences for the various PDEs. Several of the chapters in this book provide relevant information in this regard (Chapters 9, 10 and 11).

It should also be pointed out that a rather large number of PDE activities have been characterized that do not fit well into the classification scheme shown in Table 1.1. Additionally, there are also PDEs that do not preferentially hydrolyze purine 3′,5′ nucleoside monophosphates. For the most part these other PDEs have not yet been purified to homogeneity nor identified as cDNA clones. Therefore, information about them is limited and no sequence information is available. Some may turn out to be closely related to one or more of the families or subfamilies described above. Others may well turn out to be different and require that additional PDE families be added to the growing list. For example, the number of distinct PDEs in liver tissue has been particularly difficult to determine, due in part to the extreme degree of purification needed to reach homogeneity in this tissue. The fact that liver also contains many different cell types and is well known to contain large numbers of proteases also has contributed to difficulties in this regard. Readers are referred to Chapters 3, 4 and 8 for more complete descriptions of the PDE activities in liver and to Chapter 6 for a description of pyrimidine 3′,5′ nucleoside monophosphate PDE(s).

Finally, the reader should be aware that at least one other family of PDEs that will hydrolyze cAMP has been described in yeast and in the slime mold, *Dictyostelium discoideum*. cDNA clones for these isoenzymes do not have regions homologous to the catalytic domain conserved in all other purine 3′, 5′ nucleoside monophosphate-specific PDEs for which data are now available. It is as yet unknown whether PDEs homologous to these yeast and *Dictyostelium* enzymes are present in mammalian tissues. As the one in *Dictyostelium* is intimately involved in signal transduction, it seems quite possible that one or more mammalian homologs will be found.

1.3 FUNCTIONAL SIGNIFICANCE OF MULTIPLE PDE ISOENZYME FAMILIES

For the most part, our understanding of the functional reasons for the existence of so many different isoenzymes of cyclic nucleotide PDE is still more a matter of speculation than fact. However, some reasons are already clear and it seems likely that others will soon follow. For example, we already know that many of the individual isoenzymic forms are differentially expressed in selected cell types and that this is required for the function of the cell (see, for example, Chapters 8 and 10). The best-studied examples are for the photoreceptor isoenzymes, where different subfamilies are present in rods and cones. These photoreceptor PDEs are not expressed even in adjacent neuronal layers of the retina. Studies in the author's laboratory indicate that a similar situation is found in the zona glomerulosa layer of the adrenal cortex where the cGMP-stimulated PDE is highly concentrated. It seems likely that high levels of expression for other PDE isoenzymes will be found in other cell types as more information about localization becomes available.

Most investigators in the field feel that another reason for multiple isoenzymes relates to differential regulation of individual PDEs by a variety of intracellular signals. For example, having one enzyme with a particular set of kinetic properties that can respond directly to intracellular Ca^{2+} levels provides a mechanism for rapidly connecting cAMP and cGMP metabolism to all of those agents that alter Ca^{2+} homeostasis in the cell. The various isoenzymes of Ca^{2+}-CaM-dependent PDEs nicely fill this role. Since many different PDEs now appear to be regulated by cyclic nucleotide-dependent and/or Ca^{2+}-dependent phosphorylation, another way of helping to integrate cellular responses to a variety of signals is provided. These multiple types of regulation also offer possible molecular mechanisms for obtaining different time-scales of response to a signal. Therefore, it seems likely that the cyclic nucleotide PDEs provide one of the major mechanisms for integrating the control of multiple signals that regulate intracellular metabolism and function. They surely provide an important avenue allowing multiple signals to converge on cAMP- and cGMP-mediated processes. A large number of different PDEs obviously increases the opportunity for precise, cell-type-specific regulation of cyclic nucleotide flux in any given cell.

Finally, it should be mentioned that cyclic nucleotide PDEs may have roles in addition to controlling the steady-state levels of cAMP and cGMP in the cell. For example, it is already known that many of the isoenzymes exhibit changes in conformation upon binding of cyclic nucleotide. It is possible that these changes in shape are important to other functions of the PDEs and that the hydrolytic reaction is a way of reversing the shape change. A similar situation is known for the role of GTP binding to the 'G-proteins' involved in many signal transduction pathways. Similarly, it is now known from the work of Walseth, Goldberg and colleagues and of Barber and colleagues that the pools of cAMP and cGMP turn over very rapidly in most cell types [21–23]. It is also known that the enthalpy of hydrolysis for cAMP and cGMP is very large (10–14 kcal/mol) [24]. At least part of that energy is potentially available to do work. The analogy can be made that just as many different isoenzymes of ATPase have evolved to couple the high energy of hydrolysis of ATP to many different processes, so also may many isoenzymes of PDE have evolved for similar purposes. This is not to say that the PDEs do not help to control steady-state levels of cyclic nucleotides, since they obviously do. They may, however, have additional functions.

1.4 PHARMACOLOGIC IMPLICATIONS OF MULTIPLE PDEs

Just as it is now felt that having a large number of different isoenzymes of PDE allows the whole animal to provide a finely tuned, highly cell-type-selective control of cyclic nucleotide flux in a tissue, it is also felt that this same diversity may provide an avenue allowing finely tuned, highly cell-type-specific manipulation of cyclic nucleotide levels by pharmacologic agents. It should be noted that studies of this type are in their infancy, since we are just beginning to appreciate how many different isoenzymes exist. Not until we know exactly what isoenzymes are expressed in which cell types and how they are controlled can rational screening and intervention procedures really be designed. As might be expected from the fact that we have only known about multiple subfamilies of PDEs and multiple alternative splice products within subfamilies for a very short time, there are very few pharmacologic agents available that are known to distinguish between subfamilies and none that distinguish between members of a subfamily. Two different chapters in this book address some of the current applications of PDE inhibitors to physiological problems and readers are directed to these chapters for current information in this area. However, it is clear that the PDE system offers much promise for the development of therapeutically important agents, but also that much more work is needed to exploit the full potential of multiple PDE isoenzymes as targets for pharmacologic intervention.

1.5 CONCLUSIONS

It is hoped that the information provided in this and the following chapters will allow the reader to gain a more detailed understanding of the roles, regulation and functions of the various isoenzymes of cyclic nucleotide phosphodiesterase. This is a rapidly expanding field and no book can expect to be current for very long. However, a substantial effort has been made to obtain chapters from authors who are doing work that should lay the groundwork for many studies in the coming years. The author is particularly grateful for sequence information received before primary publication from many of the authors, as it has allowed many of the sequence comparisons and nomenclature designations to be put on a firmer experimental footing.

ACKNOWLEDGEMENTS

Work from the author's laboratory described in this chapter was supported by grants from the National Institute of Health, DK 21723 and EY 08197.

REFERENCES

1. Charbonneau, H. (1990) In *Cyclic Nucleotide Phosphodiesterases: Structure, Regulation and Drug Action* (Beavo, J. and Houslay, M. D., eds), John Wiley & Sons, Chichester, pp. 267–295.

2. Novack, J. P., Charbonneau, H., Walsh, K. A., and Beavo, J. A. (1990) *Proc. Natl Acad. Sci. USA* (submitted).

3. Charbonneau, H., Novack, J. P., Hunt, D., Beavo, J. A., and Walsh, K. A. (1990) *Biochemistry* (submitted).
4. Wang, J. H., Sharma, R. K., and Mooibroek, M. J. (1990) In *Cyclic Nucleotide Phosphodiesterases: Structure, Regulation and Drug Action* (Beavo, J. and Houslay, M. D., eds), John Wiley & Sons, Chichester, pp. 19–59.
5. Sharma, R. K., and Wang, J. H. (1986) *J. Biol. Chem.*, **261**, 14160–14166.
6. Grewal, J., Ahn, H. S., and Mutus, B. (1990) *J. Pharmacol. Exp. Ther.* (submitted).
7. Geremia, R., Rossi, P., Moconi, D., Pezzotti, R., and Conti, M. (1984) *Biochem. J.*, **217**, 693–700.
8. Shenoliker, S., Thompson, W. J., and Strada, S. J. (1985) *Biochemistry*, **255**, 5916–5923.
9. Manganiello, V. C., Tanaka, T., and Murashima, S. (1990) In *Cyclic Nucleotide Phosphodiesterases: Structure, Regulation and Drug Action* (Beavo, J. and Houslay, M. D., eds), John Wiley & Sons, Chichester, pp. 61–85.
10. Trong, H., Sonnenburg, W., Beier, N., Charbonneau, H., Beavo, J. A., and Walsh, K. (1990) *Biochemistry* (submitted).
11. Houslay, M. D., and Kilgour, E. (1990) In *Cyclic Nucleotide Phosphodiesterases: Structure, Regulation and Drug Action* (Beavo, J. and Houslay, M. D., eds), John Wiley & Sons, Chichester, pp. 185–224.
12. Harrison, S. A., Reifsnyder, D. H., Gallis, B., Cadd, G. G., and Beavo, J. A. (1986) *Mol. Pharmacol.*, **25**, 506–514.
13. Macphee, C. H., Reifsnyder, D. H., Moore, T. A., Lerea, K. M., and Beavo, J. A. (1988) *J. Biol. Chem.*, **263**, 10353–10358.
14. Bourne, H. R., Brothers, V. M., Kaslow, H. R., Groppi, V., Walker, N., and Steinberg F. (1984) *Adv. Cyclic Nucleotide Protein Phosphorylation Res.*, **16**, 185–194.
15. Conti, M., and Swinnen, J. V. (1990) In *Cyclic Nucleotide Phosphodiesterases: Structure, Regulation and Drug Action* (Beavo, J. and Houslay, M. D., eds), John Wiley & Sons, Chichester, pp. 243–266.
16. Davis, R. L. (1990) In *Cyclic Nucleotide Phosphodiesterases: Structure, Regulation and Drug Action* (Beavo, J. and Houslay, M. D., eds), John Wiley & Sons, Chichester, pp. 227–241.
17. Colicelli, J., Birchmeier, T., O'Neill, K., Riggs, M., and Wiggler, M. (1989) *Proc. Natl Acad. Sci. USA*, **86**, 3599–3603.
18. Gillespie, P. G. (1990) In *Cyclic Nucleotide Phosphodiesterases: Structure, Regulation and Drug Action* (Beavo, J. and Houslay, M. D., eds), John Wiley & Sons, Chichester, pp. 163–184.
19. Francis, S. H., Thomas, M. K., and Corbin, J. D. (1990) In *Cyclic Nucleotide Phosphodiesterases: Structure, Regulation and Drug Action* (Beavo, J. and Houslay, M. D., eds), John Wiley & Sons, Chichester, pp. 117–140.
20. Li, T., Volpp, K., and Applebury, M. L. (1990) *Proc. Natl Acad. Sci. USA* (submitted).
21. Goldberg, N. D., Ames, A., Gander, J. E., and Walseth, T. F. (1983) *J. Biol. Chem.*, **258**, 9213–9219.
22. Goldberg, N. D., Walseth, T. F., Eide, S. J., Krick, T. P., Kuehn, B. L., and Gander, J. E. (1984) *Adv. Cyclic Nucleotide Protein Phosphorylation Res.*, **16**, 363–379.
23. Barber, R., Ray, K. W., and Butcher, R. W. (1980) *Biochemistry*, **19**, 2560–2567.
24. Rudolph, S. A., Johnson, E. M., and Greengard, P. (1971) *J. Biol. Chem.*, **246**, 1271–1273.

SPECIFIC FORMS OF CYCLIC NUCLEOTIDE PHOSPHODIESTERASES

PART B

SPECIFIC FORMS OF
CYCLIC NUCLEOTIDE
PHOSPHODIESTERASES

2

CALMODULIN-STIMULATED CYCLIC NUCLEOTIDE PHOSPHODIESTERASES

Jerry H. Wang, Rajendra K. Sharma and Marilyn J. Mooibroek

Department of Medical Biochemistry, University of Calgary, 3330 Hospital Drive NW, Calgary, Alberta T2N 4N1, Canada

2.1   INTRODUCTION

Calmodulin-stimulated cyclic nucleotide phosphodiesterase (PDE) was discovered in close parallel with calmodulin (CaM) itself during the late 1960s and early 1970s [1]. In a series of publications, Cheung [2–4] unequivocally established the existence, in bovine brain, of a protein activator which increased the activity of a partially purified cAMP-dependent PDE by several-fold. Independently, Kakiuchi and colleagues [5,6] demonstrated, in rat brain, a Ca^{2+}-stimulated cyclic nucleotide PDE and a protein-activating factor which enhanced the Ca^{2+} sensitivity of the enzyme. Later, Teo et al. [7] obtained a PDE preparation from bovine heart which was dependent on the presence of both Ca^{2+} and the protein activator for activity and demonstrated that the protein activator was a Ca^{2+} binding protein [8]. Thus, the protein activator and the activating factor were shown to be identical proteins, for which Cheung et al. coined the word 'calmodulin' [9].

Calmodulin-stimulated cyclic nucleotide PDE is one of the most intensively studied and best characterized of the multiple PDEs. As PDE activation was the first function shown for CaM, this enzyme was widely used for the elucidation of the general mechanisms of CaM action ([10–13] for review). The enzyme was among the first cyclic nucleotide PDEs, as well as CaM-dependent proteins, purified to homogeneity and characterized in terms of molecular properties. As a result of improved purification and the availability of specific monoclonal antibodies, it has become clear during the last few years that the CaM-stimulated PDE exists as tissue-specific and immunologically distinct isoenzymes [14–17]. Studies of the well-defined and highly purified isoenzymes have revealed subtle differences in kinetic and regulatory properties among them and shed new light

Cyclic Nucleotide Phosphodiesterases: Structure, Regulation and Drug Action
Edited by J. Beavo and M. D. Houslay   © 1990 John Wiley and Sons Ltd

on their function. Rapid progress is now being made in the elucidation of their primary structures [18,19]. It is hoped that the molecular basis of the isoenzymes will soon be elucidated. Some earlier reviews dealing with this family of PDEs include those of Lin and Cheung [20], Sharma et al. [21] and Cheung and Storm [22].

2.2 OCCURRENCE AND DISTRIBUTION

The CaM-dependent PDE is an enzyme of relatively low abundance. Mammalian brain is the richest source of the enzyme, at a level of about 10 mg/kg tissue (bovine brain) [23]. In brain, the enzyme is enriched in certain populations of neurons, particularly the large pyramidal cells of the cerebral cortex (layer III), the pyramidal cells of the hippocampus and olfactory nucleus and in Purkinje cells of the cerebellum [24]. Mammalian (bovine) heart, the second most abundant source of the enzyme, contains only about 10–20% of that found in brain [25]. In addition to mammalian brain and heart, CaM-stimulated cyclic nucleotide PDE activity has been detected in many other tissues and cell types. Almost all mammalian tissues examined, with a few exceptions, e.g. rat splenic lymphocytes [26], human peripheral blood lymphocytes [27] and monocytes [28], contain CaM-stimulated PDE activity [29–32]. Also, it has been demonstrated in a number of other sources, including chicken gizzard [33], quail oviduct [34], frog ovary [35], *Drosophila* head [36–38], mollusc nervous system [39] and *Neurospora crassa* [40]. The detection of this enzyme in diverse sources ranging from human to *N. crassa* indicates that it has a widespread, if not ubiquitous, distribution in eukaryotes. Though it is demonstrable, the low abundance of the enzyme in most tissues may account for the slow progress in the purification of this enzyme from tissues other than brain and heart.

The simplest procedure used to demonstrate the existence of this enzyme was to show that the tissue extract contained EGTA and/or CaM-antagonist-inhibited PDE activity and that the inhibition could be reversed by Ca^{2+} and/or CaM respectively. However, the amount of CaM-stimulated PDE activity relative to the independent PDE activity varies greatly among tissues [30–32,41]. For example, at high cAMP concentrations, the major PDE activity of crude bovine brain extract is Ca^{2+}-dependent, whereas only 40% of the PDE activity of bovine heart extract exhibits Ca^{2+} dependence [8]. The CaM-stimulated PDE activity may evade detection in tissues containing relatively high levels of Ca^{2+}-independent PDE activity. A number of tissues, such as pancreas [42], lung [16], sperm [43] and neutrophil [44,45] were initially determined not to contain CaM-stimulated PDE but later shown, upon closer examination, to contain this form of the enzyme. A more unequivocal demonstration of the CaM-stimulated PDE therefore, requires the separation of the multiple forms of the enzyme by column chromatography procedures. Anion exchange chromatography is a commonly used procedure as, in most tissues, the CaM-stimulated form elutes at a lower ionic strength than the Ca^{2+}-independent PDEs [21,31,32,41]. Good separation requires the inclusion of EGTA in the chromatography buffers to prevent the association of the enzyme with CaM [46]. It is also important to control the activity of endogenous proteinases, since CaM-stimulated PDE is highly susceptible to proteolysis, resulting in a CaM-independent form of the enzyme.

Calmodulin-stimulated PDE is usually considered as a cytosolic enzyme, since it has been prepared largely from the soluble fraction of tissue extracts. This is in agreement with the demonstration of a homogeneous distribution of the enzyme in somatic cytoplasm of neurons by immunocytochemical techniques [24]. However, careful subcellular distribution studies have indicated that small amounts of CaM-stimulated PDE activity (10–20%) in brain, heart, liver and kidney may be found associated with particulate fractions [47,48]. Indeed, in rat sperm CaM-stimulated PDE appears to exist exclusively in a membrane-associated state [43,49]. The exact localization of the particulate enzyme, while still poorly established, appears to vary from tissue to tissue. For example, CaM-stimulated PDE activity has been shown to be enriched in synaptic densities of rat brain [48] and nuclear fractions from rat heart [50]. It is not clear whether the membrane-associated and the soluble CaM-stimulated PDEs represent different isoenzymes. No attempts have been made to purify and characterize the membrane-associated CaM-stimulated PDE from these tissues.

There have been numerous studies on the ontogenetic development of cyclic nucleotide PDE activity, but only a few of these have attempted to discriminate the multiple forms of the enzyme. It was reported that cAMP-PDE activity changed from Ca^{2+}-dependent to Ca^{2+}-independent states during the embryonic development of chicks [51]. On the other hand, the appearance of CaM-stimulated PDE activity emerges during late stages of embryonic development of fetal rabbit lung and liver, while the multiple PDE pattern is maintained during the embryonic development of fetal rabbit brain, heart, and kidney [52]. In a recent study, Epstein et al. [53] showed that CaM-stimulated PDE represented the main form of the enzyme in chick ventricular myocardium throughout embryonic development. The specific activity of the cAMP-PDE was found to increase gradually during embryogenesis, reaching highest values just before hatching, and then declining by about 30% within one week. In contrast, adenylate cyclase and CaM did not change in specific activity during embryogenesis. This observation suggests that steady-state levels of cAMP in chicken heart is controlled mainly by cAMP-PDE activity [53].

The cellular content of CaM-stimulated PDE may be altered by environmental factors. Kincaid et al. [54] showed that injection of 3-acetylpyridine into rats to destroy climbing fibers resulted in a marked decrease in CaM-stimulated PDE activity in the cerebellum without significantly affecting the amount of total CaM binding proteins or CaM-stimulated phosphatase (calcineurin). Changes in the levels of this enzyme have also been observed in certain pathological conditions. An increase in the amount of CaM-stimulated PDE activity has been reported in thyroids of individuals with Graves' disease [55]. Also, human lymphocytes do not appear to contain CaM-stimulated PDE but a lymphoblastoid cell line (RPMI 8392) has been shown to contain this enzyme as the main form of PDE [56].

2.3 ISOENZYMES

Several lines of evidence in the early literature suggested the existence of CaM-stimulated cyclic nucleotide PDE isoenzymes. Firstly, a number of studies demonstrated the separation of multiple CaM-stimulated PDE activity peaks from mammalian tissue extracts by one of the protein separation procedures. Secondly, while homogeneous preparations of bovine brain and heart CaM-stimulated cyclic nucleotide PDEs were obtained during the early 1980s in several laboratories, the reported subunit molecular masses for these preparations varied significantly. Lastly, the kinetic properties of CaM-stimulated PDE described by different investigators, often on enzyme preparations from different sources, were markedly different, with K_m values sometimes differing by orders of magnitude. However, few systematic studies were undertaken until recently to address the question of CaM-stimulated PDE isoenzymes.

2.3.1 Demonstration of isoenzymes by monoclonal antibodies

Sharma et al. [15] were the first to show definitively the existence of immunologically distinct PDE isoenzymes. A bovine brain CaM-stimulated PDE preparation purified by a procedure involving DEAE–cellulose, Affigel blue, CaM affinity and G-200 Sephadex column chromatographies [23] was demonstrated by alkaline SDS-PAGE analysis to contain two polypeptides of molecular mass 63 kDa and 60 kDa. The two polypeptides were established to be subunits of different PDE isoenzymes by a number of experiments [15]. First, a panel of monoclonal antibodies tested against the two polypeptides were found to fall into two classes, one class reacting with both polypeptides and the other being specific for the 60 kDa polypeptide. Second, micropeptide mapping by the Cleveland procedure [57] indicated that the two polypeptides were distinct molecular species and could not have been derived from proteolytic artifacts during enzyme purification. Third, sucrose density centrifugation of immunocomplexes of the enzyme and a 60 kDa polypeptide-specific antibody, mAb A2 or C1, resulted in three PDE activity peaks: peaks I, II and III in order of decreasing sedimentation rate. Analysis of the peak fractions by SDS-PAGE revealed that the three peaks contained 63 kDa, 63 kDa plus 60 kDa and 60 kDa polypeptides respectively.

Fourth, the PDE preparation was separated on a mAb C1 immunoaffinity column into two fractions, the flow-through fraction containing the 63 kDa peptide only and the column-bound fraction greatly enriched with 60 kDa peptide. Lastly, the two fractions of PDE showed different relative substrate specificities with respect to cAMP and cGMP.

The brain 60-kDa specific monoclonal antibodies A2 and C1 crossreacted with bovine heart [58] and bovine lung [16] PDEs. A survey of bovine tissues using the mAb C1 immunoabsorbent showed that antibody-reactive PDE isoenzymes existed in all the tissues examined, including liver, kidney, spleen and uterus [58]. A bovine heart PDE monoclonal antibody, ACAP-1, produced by Hanson and Beavo [59], also crossreacted with the 60 kDa but not the 63 kDa isoenzyme from bovine brain. Thus, on the basis of immunological properties, bovine CaM-stimulated PDE isoenzymes fall into two categories: a category of isoenzymes reactive to the 60-kDa-isoenzyme-specific monoclonal antibodies and a category of non-reactive ones. This classification is probably not restricted to bovine tissues, since rat brain PDE has also been fractionated into mAb-ACAP-1-reactive and non-reactive fractions, but, interestingly, the two isoenzymes of rat brain have similar molecular masses of about 63 kDa [59].

Clearly, the use of monoclonal antibodies as analytic tools for isoenzyme detection and differentiation has serious limitations. For example, bovine heart, lung and the brain 60 kDa PDEs all react with the brain 60-kDa-isoenzyme-specific antibodies, but they have been shown by other criteria to be different isoenzymes [21]. It is also likely that CaM-stimulated PDE isoenzymes which do not react with the brain 60-kDa-specific antibodies can be further subdivided. Thus, the immunologic classification of CaM-stimulated PDE isoenzymes is limited by the availability of discriminatory antibodies. In addition, although isoenzymes in the same group may be structurally more similar than those in different groups, this is by no means proven. Even less clear is the relationship between immunologic reactivity and the functional properties of these isoenzymes.

2.3.2 Purification

The existence of isoenzymes may be demonstrated more directly by protein structure comparisons among different enzyme preparations. Such an approach requires extensive purification of the individual isoenzymes. In the late 1970s and early 1980s several investigators [23,25,60–62] purified CaM-stimulated PDE from bovine brain and heart to close to homogeneity. Most of the purification procedures involved CaM affinity chromatography in combination with several conventional column chromatographies. The purification procedures were often tedious, protracted and with low enzyme yields. The purified samples displayed a high specific cAMP hydrolysis activity, with values in the order of several hundred micromoles per min per mg protein. Most preparations showed a single predominant peptide on SDS-PAGE with molecular weights in the range of 57 000 to 63 000. The low efficiency of the purification procedures and the knowledge that the enzyme was susceptible to limited proteolysis obscured the possible significance of the observed variation in molecular weights.

The availability of monoclonal antibodies has greatly improved the purification procedure, thereby facilitating the identification of CaM-stimulated PDE isoenzymes. By using a rapid purification procedure involving an immobilized monoclonal antibody of bovine heart PDE, mAb ACAP-1, Hansen and Beavo [14] purified CaM-stimulated PDE from both bovine brain and bovine heart to close to homogeneity. The rapid purification minimized the possibility of post-homogenization proteolysis. Direct comparison of the enzyme preparations from the two tissues showed that the heart PDE had a lower subunit molecular mass than the brain enzyme, thus suggesting that brain and heart contain distinct CaM-stimulated PDE isoenzymes. An immobilized monoclonal antibody, mAb C1, specific for bovine brain 60 kDa PDE, has been used to demonstrate the existence of a mammalian lung-specific CaM-stimulated PDE [16]. In contrast to the enzymes from other tissue extracts, the CaM-stimulated PDE immunoprecipitated from the bovine lung extract contained tightly associated CaM, which could not be removed by extensive washing of the immunoprecipitates in the presence of EGTA. Bovine lung contains very low amounts of CaM-stimulated PDE, so purification of the lung isoenzyme to close to homogeneity required a procedure consisting of both CaM affinity and monoclonal antibody immuno-affinity chromatographies as well as several conventional column chromatographies [16]. A number of procedures, including those involving the use of monoclonal antibody immunoaffinity chromatography, for the purification of CaM-stimulated PDE from various tissues, have been recently described in detail [63–66].

2.3.3 Structural characterization

The physical properties of the homogeneous preparations of CaM-stimulated cyclic nucleotide PDE as well as several other highly purified preparations are summarized in Table 2.1. Since the isoenzymic forms and/or isoenzyme compositions of the earlier preparations of the brain enzyme are not known, these enzyme preparations are listed individually.

The majority of the homogeneous preparations of CaM-stimulated PDE share certain structural properties. They are homodimeric proteins composed of subunits of molecular mass about 60 kDa and are capable of binding 2 mol CaM per mol PDE isozyme in the presence of Ca^{2+} [67,68]. However, the individual isoenzymes show subtle variations in this common subunit structure and CaM binding property. The subunit molecular masses of the two bovine brain isoenzymes and the bovine heart and bovine lung isoenzymes are slightly

Table 2.1 Molecular mass and subunit structure of CaM-stimulated PDEs

Source of enzyme	Purity	Molecular mass (kDa)		Subunit molecular mass (kDa)	References
		$-$CaM	$+$CaM		
Bovine brain[a]	Pure[b]	126	ND	59	61
		120	159	60	23
		120	ND	59	62
	Partially pure	150		74	69
Bovine heart	Pure	ND	155	57, 59	25, 14
Rat testes	Partially pure	70	180	68	17
Rat pancreas	Partially pure	175			
		116			42
Bovine brain 63 kDa isoenzyme				63	15, 59
Bovine brain 60 kDa isoenzyme				60, 61	14, 15
Bovine lung isoenzyme	Pure	120[c]	160	58	16

[a] Isoenzymic form is not known.
[b] Major protein band on SDE-PAGE gel is identified unequivocally as representing the PDE and specific cAMP hydrolysis activity is in the order of hundreds of micromoles per min per mg. ND = not determined.
[c] Subunit CaM is dissociated by using chaotropic agents

different, with values, as determined from the SDS-PAGE mobilities, of 63 kDa, 60 kDa, 58 kDa and 59 kDa, respectively. As stated before, in contrast to other isoenzymes, the lung isoenzyme contains CaM as a subunit. Among the preparations which have not been purified to near-homogeneous states and possess low specific enzyme activity relative to those of the essentially pure preparations is one from bovine brain which has an unusually high molecular mass of 150 kDa in its nondenatured state and a subunit molecular mass determined by SDS-PAGE of 75 kDa [69]. Another variation is seen with a PDE recently purified extensively from bovine testes which has been reported to exist as a monomeric protein with a molecular mass of 70 kDa but which dimerizes in the presence of Ca^{2+} and CaM to a molecular species of molecular mass 180 kDa, presumably with a subunit composition of $CaM_2.PDE_2$ [17].

The amino acid sequences of bovine brain 60 kDa and bovine heart PDE isoenzymes have been determined to be almost identical except that the 60 kDa brain isoenzyme contains an insertion sequence proximal to the N-terminus [19,70]. This result substantiates that bovine heart and the 60 kDa bovine brain PDEs are isoenzymes, and indicates that they arise from differential mRNA splicing. A comparison of the primary structure of CaM-stimulated PDE with that of bovine heart cGMP-stimulated cAMP-PDE, as well as those deduced from the cDNA sequences of *Drosophila dnc*[+] gene [71] and yeast *phosphodiesterase 2* gene [72], has revealed a homologous segment of 200–270 amino acids in each of the four proteins. These homologous regions are suggested to comprise the catalytic domains of the PDEs. Regions of the enzymes showing no structural relatedness probably represent functional domains unique to the individual enzymes. The homologous region in CaM-stimulated PDE is in the C-terminal half of the protein [19].

2.3.4   Identification by protein fractionation procedures

Earlier characterization of multiple cyclic nucleotide PDEs often employed protein fractionation procedures to separate the PDE activity in a tissue extract into multiple activity peaks. In most cases, only one of the multiple PDEs was shown to be stimulated by Ca^{2+}–CaM, but a few studies demonstrated the existence of two such PDE fractions [73–76]. While this is suggestive of CaM-stimulated PDE isoenzymes, such a suggestion based on a single protein fractionation procedure should be accepted with caution for several reasons. First, the two peaks of activity may represent a single enzyme species existing in its free and CaM-bound states [46]. Second, cross-contamination of different forms of PDE could result in the appearance of CaM-stimulated activity in more than one enzyme fraction [73,77]. Third, an enzyme species could associate with other proteins or cellular components or undergo self-association to appear as multiple forms [78]. Fourth, multiple peaks of activity could be an artifact of limited proteolysis during tissue extraction [79]. Nevertheless, the possibility that certain tissues contain more than one CaM-stimulated PDE isoenzyme, which can be separated by conventional chromatographic procedures, warrants careful re-examination.

Rat pancreas [42] and testes extracts [80] were shown to contain two and three CaM-stimulated PDE activity peaks, respectively, upon anion exchange chromatography. The pancreas PDEs were purified a few thousand-fold but, as judged by their respective specific enzyme activities, were still not pure [42]. Molecular masses of the highly purified preparations, determined by gradient disc gel electrophoresis, were 116 kDa and 175 kDa respectively (Table 2.1). At least one of the testes isoenzymes was purified almost 1000-fold [17]. Molecular characterization of this isoenzyme has shown it to be clearly different from those of the other well-characterized isoenzymes listed in Table 2.1 (see Section 2.3.3).

2.3.5   Kinetic characteristics

The kinetic properties of the CaM-stimulated PDEs from a wide variety of tissues and animal species have been examined in many different laboratories. Some of the early results compiled by Lin and Cheung [20] showed that the reported kinetic parameters differed markedly between different studies. For example, the K_m values for both cAMP and cGMP varied over two orders of magnitude among the early reports. In addition, the reported effects of Ca^{2+}–CaM on the kinetic properties of CaM-stimulated PDE also showed considerable discrepancies, ranging from a decrease of K_m, an increase of V_{max} or an effect on both kinetic parameters [20]. The reason for the variations in the reported kinetic properties is not known. The enzyme preparations used in these early studies were of different purity, and therefore might have been contaminated with different amounts of other PDEs. Varying assay conditions might also have had an effect on the kinetic properties. For example, imidazole was used in some assays and this compound has been shown to increase both the K_m and V_{max} of bovine brain CaM-stimulated PDE activity [62]. Undoubtedly, some of the reported variation in kinetic properties was also due to different isoenzymes being characterized.

The more recent kinetic studies were mostly carried out on highly purified preparations of CaM-stimulated PDE. Table 2.2 presents kinetic data obtained from studies of homogeneous preparations and a few selected highly purified preparations of CaM-stimulated PDE. In a couple of cases where the kinetic studies were carried out both in the presence and in the absence of imidazole, imidazole was shown to increase the K_m value 5–15-fold.

CaM-stimulated PDE isoenzymes may be classified into two general groups on the basis of kinetic criteria: a group showing high affinity for cGMP, but low affinity for cAMP, and a group showing high affinities for both cAMP and cGMP. All the well-characterized bovine brain and heart isoenzymes belong to the first group and, as described above (Sections 2.3.1 and 2.3.2), they represent several different isoenzymic forms. Also included in this group are the purified enzymes from bovine lung and coronary artery and the partially purified enzymes from bovine carotid artery, porcine coronary artery, rat adrenal gland and one form of the rat testis CaM-stimulated PDE (Table 2.2). In the second group of isoenzymes is the peak I CaM-stimulated rat testis PDE, which has been extensively purified, characterized and shown to be a distinct molecular entity [17], and several of the other partially purified enzymes.

Within each of these two general groups is the potential for subgroups. For example, the four well-defined bovine isoenzymes, all belonging to the first group, appear to be separable into two subgroups on the basis of their affinities for cAMP and cGMP. The 60 kDa brain, heart and lung isoenzymes show almost identical kinetic properties, while the 63 kDa brain isoenzyme shows about three times higher affinity towards both cGMP and cAMP and has a higher V_{max} for cGMP than for cAMP. Similarly, it is likely that more members of the second group remain to be identified.

Table 2.2 Kinetic properties of CaM-stimulated cyclic nucleotide PDE

Source of enzyme	K_m^a (μM)		V_{max} (ratio)	Reference
	cAMP	cGMP	(cAMP/cGMP)	
Bovine heart	215.0[b], 39.0[c]	9.0[b]	3.3[b]	46
Bovine brain 63 kDa isoenzyme	11.0	1.2	0.3	220
Bovine brain 60 kDa isoenzyme	32.0	2.7	1.8	16
Bovine lung	42.0	2.75	1.7	16
Rat brain low molecular mass isoenzyme			1.2	59
Rat brain high molecular mass isoenzyme			0.68	59
Bovine brain[d]	2.9	2.7	0.07	69
Pig coronary arteries	70.0	3.0	0.1	226
Rat testes				
Peak I	1.0	1.0	—	
Peak II	30.0	3.0	—	
Peak III	—	1.5	—	80
Rat pancreas				
Peak I	0.35	0.16	1.4	
Peak II	9.0	1.7	1.1	42
Bovine epididymal spermatozoa	7.5, 95.0	—	—	43
Human cardiac ventricle	0.75	1.0	1.0	226
Guinea pig left ventricle	0.8	0.9, 53.0	1.1	227
Bovine coronary arteries	2.5, 140.0	3.0, 17.0	0.18	228

[a] K_m values are given for CaM-achieved enzyme reaction.
[b] Enzyme assay was carried out in the presence of imidazole.
[c] Wu et al., unpublished observation.
[d] The enzyme appears to be significantly different from other bovine brain CaM-stimulated PDE isoenzymes, in having a higher molecular mass, 150 kDa and high relative V_{max} for cGMP.

2.4 ACTIVATORS AND INHIBITORS

The catalytic properties of CaM-stimulated PDE have been well characterized using the enzyme preparations from either the mammalian brain or heart. These two isoenzymes share many catalytic properties, such as metal ion requirements, activators and inhibitors, and there are no compelling reasons to suggest that these common properties are not shared by other CaM-stimulated PDEs. Nonetheless, such a possibility cannot be ruled out.

2.4.1   Activation by calmodulin

The observed CaM-stimulated cyclic nucleotide PDE reaction is composed of a basal activity which is independent of CaM and a Ca^{2+} and CaM-stimulated activity. Both the basal and the CaM-stimulated activities are dependent on millimolar concentrations of Mg^{2+}, while the CaM-stimulated reaction requires the presence of Ca^{2+} in addition to Mg^{2+} [20–22]. The Mg^{2+} requirement of the enzyme reaction and the Ca^{2+} requirement of the CaM stimulation can both be replaced by Mn^{2+}. In addition, low concentrations of a number of other metal ions, including certain heavy metals such as Hg^{2+}, Pb^{2+} and Cd^{2+}, may substitute for Ca^{2+} in supporting the CaM activation of the enzyme [81–83]. In view of the second messenger role of Ca^{2+} in biological systems, Ca^{2+} is considered the physiological metal activator; however, the possibility that heavy metal toxicity is mediated by CaM has been suggested [82,83].

For most of the CaM-stimulated PDE preparations from mammalian tissues, the stimulated PDE activity is 6–20-fold greater than that of the basal activity. Yet enzyme preparations with stimulated activity only two-fold [20] or as high as over 50-fold [61] that of the basal activity have also been described. The exact reason for the wide variation in enzyme stimulation by CaM is not known, but a number of factors may contribute to this variation. Most tissues contain both CaM-stimulated and CaM-independent PDEs. An incomplete removal of the independent PDE from the CaM-dependent enzyme preparation will increase the apparent basal activity with a consequent decrease in the observed CaM stimulation. Also, as indicated before, CaM-stimulated PDE is highly susceptible to proteolysis, resulting in an activated, CaM-independent form. Therefore, contamination of the preparation by protease-activated forms of the PDE will result in a lower apparent CaM stimulation. The extent of CaM stimulation is also dependent on assay conditions. For example, CaM-stimulated activity can be enhanced about 100% by imidazole, whereas the basal activity of the enzyme is only slightly affected [46]. Thus, inclusion of imidazole in the assay will increase the observed CaM stimulation. The extent of CaM stimulation may depend on the specific nucleotide substrate and the substrate concentration used in the assay. Since CaM may decrease the K_m in addition to increasing the V_{max} of the cAMP hydrolysis reaction [20], a lower extent of CaM activation is expected when the enzyme reaction is carried out at a saturating versus a non-saturating substrate concentration.

Calmodulin and Ca^{2+} show synergistic interactions in the activation of PDE, so an increase in CaM or Ca^{2+} concentration results in an increase in the apparent enzyme affinity towards Ca^{2+} or CaM respectively [84–86]. Huang et al. [87] have shown that CaM activation of bovine brain PDE displays a hyperbolic dose-dependence curve, suggesting that the two subunits do not undergo co-operative interaction with regard to CaM binding. On the other hand, Cox et al. [88] have observed a positive homotropic interaction in CaM stimulation of PDE with a Hill coefficient of about 2. The reason for this discrepancy is not known. The activation of the enzyme by Ca^{2+} has been invariably shown to display a high degree of homotropic positive co-operativity [84,85,87,89]. The molecular and kinetic bases of the enzyme activation by Ca^{2+} and CaM will be discussed in Section 2.6.

2.4.2 Other activators

Many compounds of different structures have been shown to activate CaM-stimulated PDE in place of CaM in a Ca^{2+}-independent manner. These activators may be classified into four structural classes: proteins and peptides, lipids, detergents and other synthetic organic compounds. The Ca^{2+}-dependent activation of PDE is mediated specifically by CaM. High concentrations of other proteins of the 'EF hand' Ca^{2+}-binding protein family, such as parvalbumin and troponin C, do not substitute for CaM in the enzyme activation [90,91]. One 'EF hand' protein, oncomodulin, has been shown to stimulate PDE in a Ca^{2+}-dependent manner, but it is almost 1000-fold less potent than CaM [92]. Several lines of evidence suggest that the apparent enzyme activation is not due to contamination of the oncomodulin preparation by trace amounts of CaM [92,93]. Among these, the most convincing was the observation that incubation of oncomodulin under mild oxidation conditions resulted in the dimerization of the protein due to the oxidation of the single thiol group in the protein to a disulfide bond. Protein dimerization was accompanied by a dramatic increase in PDE-activating potency [93]. This characteristic cannot be attributed to contaminating CaM, since mammalian CaM does not contain a thiol group and its activity is not affected by mild oxidation conditions. Oncomodulin is an oncodevelopmental protein existing in placenta and certain tumor cells, but not in normal adult tissues [94]. It is not clear whether its CaM-like activity is relevant to the physiological function of this protein.

A 19 kDa protein from *S. aureus* conditioned media has been shown to activate CaM-stimulated PDE in place of CaM in a Ca^{2+}-independent manner. This protein is believed to be a protease derivative of the exotoxin of the bacteria [95]. Gustin, a 37 kDa Zn^{2+} binding protein from human parotid saliva, is another protein shown to stimulate CaM-stimulated PDE in a Ca^{2+}-independent manner [96]. This protein appears to be involved in the function of taste buds. The physiological significance of the PDE stimulation by the 19 kDa bacterial protein and gustin is, however, not clear. Poly-(L)-aspartate but not poly-(L)-glutamic acids can also activate CaM-stimulated cyclic nucleotide PDE in a Ca^{2+}-independent manner [97].

A number of acidic phospholipids, including phosphatidylinositol, phosphatidylserine, various lysophospholipids, several unsaturated fatty acids, gangliosides, and an *E. coli* glycolipid, lipid X (N_2, O3-diacylglucosamine 1-phosphate) [78,86,98–102], have been shown to activate the PDE in a Ca^{2+}-independent manner. Effective concentrations of these lipids for enzyme activation are above their critical micellar concentrations [100].

The activation of PDE by various amphiphilic substances suggests that the activators bind to the enzyme, at least in part, by hydrophobic interactions. A number of detergents, including Zwittergent and SDS, also stimulate the PDE at low concentrations [103]. This observation has raised the possibility that compounds such as lysophospholipids and unsaturated fatty acids may activate the PDE, partly due to their detergent-like properties.

A group of synthetic compounds, the quinazoline sulfonamide derivatives [104], and a vitamin E derivative, α-tocopherylphosphate [105], have also been shown to activate the PDE independently of Ca^{2+}. Although most of the known enzyme activators do not appear to be physiologically significant, some of these may be valuable tools for probing the structure and function of the PDE.

2.4.3 Inhibitors

A large number of compounds of diverse structure are capable of inhibiting CaM-stimulated cyclic nucleotide PDE. On the basis of their effects on the enzyme reaction, the inhibitors may be divided into two general categories: those counteracting CaM stimulation of the enzyme and those inhibiting the enzyme reaction directly. Inhibitors in the first category may be further classified into three groups according to their mode of action. Since the stimulatory activity of CaM is dependent on the binding of Ca^{2+}, agents interfering with Ca^{2+} binding form the first group of inhibitors. The second group of inhibitors abolish CaM-stimulated PDE activity by binding to CaM. The third group exert their actions by binding to the PDE to block the interaction between CaM and the enzyme.

Those inhibitors which counteract CaM activation of the enzyme by binding to CaM are by far the most extensively studied. They are represented by a group of highly heterogeneous compounds in terms of both biological functions and molecular structures. Among them are many hormones, neurotransmitters, toxins, metabolites, detergents and pharmacologic agents of varied therapeutic actions. In terms of molecular classification, the group contains proteins, peptides, amines, lipids, and drugs of various chemical classes. Several representative inhibitors of this group are listed in Table 2.3 to illustrate their diversity in structure and function.

Weiss et al. [77] originally demonstrated that the antipsychotic drug trifluoperazine inhibited CaM-stimulated PDE activity preferentially over the basal enzyme activity. Subsequently, it was shown that the inhibitor underwent Ca^{2+}-dependent high-affinity binding to CaM [106]. Later studies in numerous laboratories have led to the discovery of a multitude of other pharmacologic and toxicological agents capable of blocking CaM activation of PDE. Comprehensive listing of these CaM antagonists can be found in several review articles

Table 2.3 CaM-binding inhibitors

Representative inhibitors	Function classification	Structural classification	References[a]
Trifluoperazine	Antipsychotic agent	Phenothiazine	103, 107, 109, 110
W7	Muscle relaxant	Napthalenesulfonamide	103, 107, 109, 110
Prenylamine	Antianginal agent	Diphenylpropylamine	110
Vinblastin	Antitumor agent	Vinca alkaloid	107, 110
Phenoxybenzamine	α-Adrenergic antagonist	Benzyl-β-chloroethylamine	109
Butaclamol	Antipsychotic agent	Benzocycloheptapyridoisoquinoline	109
Pimozide	Antipsychotic agent	Diphenylbutamine	109
Felodipine	Ca^{2+} channel antagonist	Dihydropyridine	107, 109, 110
Imipramine	Antidiarrheal agent	Diphenylalkylpiperidine	109
DDT	Insecticide	Chlorinate ethane	109
R-24571 (calmidazolium)		Miconazole analogue	103, 107, 109
Triton X-100	Detergent	Polyoxyethylene	103, 107, 109
Mellitin	Insect toxin	Peptide	109
Mastporan	Toxin	Peptide	113
β-Endorphin	Neurotrasmitter	Peptide	107, 109
Glucagon	Hormone	Peptide	108, 111
Cyclosporin	Immunosuppressant	Peptide	114
Spermine	Metabolite	Polyamine	230
Histone	Nuclear protein	Protein	109
Calcineurin	CaM-dependent enzyme	Protein	109
Sphingosine	Metabolite of sphingolipid	Amino alcohol	231

[a] References given are review articles when possible.

[103,107–110], and together the lists contain close to a hundred compounds. Van Belle [103] stated in an article in 1984 that a screening of 5650 compounds for CaM antagonist activity, carried out at Janssen Pharmaceutical, had uncovered 1200 compounds with a PDE inhibitory potency comparable to that of trifluoperazine. However, the identities of these CaM antagonists were not given. The widely disparate pharmacologic and biological activities of the known CaM antagonists suggests that the major functional activities of these compounds do not involve CaM antagonist action.

Among the large number of peptide CaM antagonists are some hormones, neurotransmitters [108,111], insect toxins [112,113] and an immunosuppressant [114] (Table 2.3). The biological activity of these peptides, however, is not thought to depend on their CaM antagonist activities. These peptide antagonists inhibit CaM-stimulated PDE with IC_{50} values ranging from nanomolar to micromolar. The most potent are the toxic peptides of bee and wasp, melittin and masporan respectively, which bind CaM with affinities similar to those of CaM-regulated enzymes.

Most CaM-regulated proteins associate with CaM in a mutally exclusive fashion. Consequently, the activation of CaM-stimulated PDE can be blocked by other CaM-regulated proteins under conditions of limiting CaM. Indeed, CaM-stimulated phosphatase was originally discovered as an inhibitor of PDE [115,116]. The ability to inhibit PDE is still used as one of the criteria of CaM-regulated proteins [117,118].

The most common inhibitors which interfere with the binding of Ca^{2+} to CaM are the Ca^{2+}-specific chelators. Thus, EGTA is routinely used to demonstrate the existence of Ca^{2+}-stimulated PDE in tissue extracts and to arrest CaM-stimulated PDE activity. Other inhibitors in this group include metal ions which bind at the Ca^{2+} binding sites of CaM but do not support PDE activation. A novel CaM antagonist 3-(2-benzothiazolyl-4-5-dimethoxy-N-[3-14-phenylpiperidinyl)propyl] benzenesulfonamide (HT-74) which inhibits CaM-stimulated PDE by affecting the binding of Ca^{2+} to CaM has been described by Tanaka et al. [119].

Inhibitors which block CaM stimulation of PDE by directly interacting with the enzyme include several derivatives of CaM. For example, a tryptic peptide, peptide 77–148, which has lost the ability to activate the PDE, can block CaM stimulation of the enzyme, suggesting that it retains the ability to bind the enzyme [120]. This suggestion is supported by the observation that immobilized CaM peptide 77–148 is capable of binding the brain PDE in a Ca^{2+}-dependent manner [121]. A few CaM derivatives containing a covalently bound CaM antagonist have also been shown to block CaM activation of the enzyme [122,123].

Inhibitors of the second general category, those affecting PDE activity directly, include the common PDE inhibitors caffeine, papaverine and theophylline. These inhibitors show little selectivity among the multiple PDEs. Progress has been made in developing inhibitors showing good selectivity for different forms of the enzyme. Kramers et al. [124] have shown that alkyl substitution of isobutylmethyl-xanthine (IBMX) at position 7 or 8 greatly enhances the selectivity of the inhibitor towards CaM-stimulated PDE. For example, 8-methoxymethyl-IBMX, a commonly used specific inhibitor of CaM-stimulated PDE, has an inhibitory potency for CaM-stimulated PDE 30–50 times higher than that for CaM-independent PDEs [125]. Other relatively specific inhibitors of CaM-stimulated PDE include the IBMX derivative 2-O-propoxyphenyl-8-azapurin-6-one M & B 22948 [126,127] and two inhibitors, Vopocetine and HA 558, described by Hidaka and co-workers [128,129].

The classification of CaM-stimulated PDE inhibitors into CaM-stimulation-directed and enzyme-activity-directed categories is not clearcut. Many CaM antagonists at somewhat elevated concentrations inhibit basal PDE activity as well as the enzyme activity stimulated by mechanisms other than CaM, such as phospholipid activation or partial proteolysis [130]. A few CaM antagonists, including some of the dihydropyridine Ca^{2+} channel blockers, display good specificity against CaM stimulation in other CaM-stimulated enzyme reactions, e.g. that of myosin light chain kinase, but inhibit both the basal and the CaM-stimulated activities of most CaM-stimulated PDEs with equal potency [131,132].

A few studies have been carried out to examine the possible differential susceptibilities of CaM-stimulated PDE isoenzymes to inhibitors. Matsushima et al. [133] showed that while dihydropyridines, such as nicardipine, inhibited the basal and CaM-stimulated activities of the CaM-stimulated PDE from rabbit brain with equal potency, they showed much higher inhibitory potency towards the activated rabbit heart PDE than the basal enzyme activity. Bovine lung isoenzyme, in contrast to other CaM-stimulated PDEs, was found not to be affected by CaM antagonists [16], due to the presence of CaM as a subunit of the isoenzyme.

2.5 DEMONSTRATION OF CaM—STIMULATED PDE ACTIVITY IN VIVO

Since the predominant isoenzymes of CaM-stimulated PDE show higher affinities for cGMP in vitro than for cAMP, it has been suggested that they are primarily cGMP PDEs [134]. This appears to be true in bovine artery. The potent and relatively selective inhibitor for CaM-stimulated PDE, methyl-3-isobutyl-8-methylxanthine, caused a marked increase in intracellular cGMP concentration in bovine coronary artery but did not significantly change cAMP levels [135]. The inhibitor also failed to significantly potentiate the elevation of cAMP concentration brought about by isoproterenol. These results suggest that, in this tissue, the CaM-stimulated PDE does not contribute significantly towards cAMP hydrolysis even when the cAMP concentration is increased as a result of β-adrenergic receptor activation.

However, in several other tissues CaM-stimulated cAMP-PDE activity has been suggested to occur under different cell activation conditions. As reviewed by Erneux et al. [136] the thyrotropin-induced accumulation of cAMP in dog thyroid and the β-adrenergic receptor-mediated accumulation of cAMP in 1321 N_1 cells, a human astrocytoma cell line, were both inhibited by muscarinic agonists by a mechanism which was suggested to involve the activation of CaM-stimulated PDE rather than the inhibition of adenylate cyclase as had been previously shown in a variety of tissues [137–139]. This suggestion was based on a number of criteria. First, the muscarinic-agonist-dependent inhibition of cAMP accumulation was eliminated by various PDE inhibitors with potencies parallel to those exhibited for the inhibition of CaM-stimulated PDE activity in vitro, [140–144]. Second, the inhibition of cAMP accumulation was greatly reduced in the absence of Ca^{2+} or in the presence of Co^{2+} or Mn^{2+} to block the influx of Ca^{2+} [140,141,145]. Third, the Ca^{2+} ionophore A23187 could mimic muscarinic agonists in bringing about the reduction of cAMP accumulation. Fourth, adenylate cyclase activity of the isolated cell membranes was not inhibited by muscarinic agonists, and the agonist effect on intracellular cAMP levels was not influenced by pertussis toxin [146–149]. Finally, the rate of cAMP degradation in cells which had been stimulated by an agonist to raise intracellular cAMP concentrations and then treated with an antagonist to block further stimulation was shown to be increased markedly by a muscarinic agonist [140,150]. Although not as well documented, the activation of CaM-stimulated PDE has also been suggested to account for the muscarinic-receptor-mediated inhibition of prostaglandin E_1-induced cAMP accumulation in WI-38 human diploid fibroblasts [151,152] and the inhibition of β-adrenergic-stimulated cAMP accumulation in rat prostate tissue [153].

The muscarinic-receptor-mediated modulation of cellular cyclic nucleotide metabolism appears to be more complex in human thyroid cells in primary culture than in dog thyroid, in that the muscarinic agonist carbamylcholine, in addition to inhibiting the thyrotropin-induced cAMP accumulation, can increase both cAMP and cGMP concentrations in the thyroid cells by itself [154]. Both the inhibition of cyclic nucleotide accumulation and its stimulation are believed to be mediated by the muscarinic receptor, since they are blocked by the specific antagonist atropine. Similar to dog thyroid, the inhibition of cAMP accumulation in human thyroid cells is suggested to be mediated by CaM-stimulated PDE; this is based on a number of criteria, including the effects of Ca^{2+} and PDE inhibitors on the intracellular cAMP concentration.

An increased PDE activity has also been suggested to be coupled to other receptors. Chronic β-adrenergic stimulation of rat erythrocytes induces a triphasic change in cAMP concentration: a transient rise in cAMP concentration, subsiding in about 20 min, followed by a second, more gradual increase in cAMP concentration which reaches a stable plateau after about 60 min of hormonal treatment [155]. Upon treating the cells to deplete cell Ca^{2+}, this complex pattern of cAMP flux changed into a simple pattern of a rapid increase in cAMP concentration reaching a stable plateau in about 15–20 min. The elevated cAMP concentration at the plateau could be rapidly reduced to basal concentrations if Ca^{2+} and A23781 were added to the cell medium [155]. These observations suggest that CaM-stimulated PDE also has important roles in the control of β-adrenergic-receptor-mediated cAMP fluxes.

Buxton and Brunton [156] showed that the norepinephrine-stimulated increase in cAMP concentrations in rat cardiomyocytes was enhanced by an α_1-adrenergic antagonist, prazosin, and that the effect of prazosin could be abolished by PDE inhibitors. Thyrotropin or forskolin-induced cAMP accumulation in cultured rat thyroid cells [157] has been shown to be inhibited by α_1-adrenergic agonists in a PDE-dependent manner. These observations suggest that CaM-stimulated PDE activity can be coupled to the activation of α_1-adrenergic receptors. Luteinizing hormone-stimulated cAMP accumulation in cultured rat granulosa cells has been reported to be inhibited by gonadotropin-releasing factor in a Ca^{2+}-dependent and IBMX-antagonized manner, suggesting that CaM-stimulated PDE is coupled to the activation of the gonadotropin-releasing factor receptor [158].

Several in vivo studies have suggested CaM-stimulated cAMP-PDE activity to be involved in the regulation of ion channel activities. The electrical activity of the bursting pacemaker neuron R15 of *Aplasia* can be stimulated by various cAMP-mediated neuromodulators such as serotonin and an egg-laying hormone [159]. Although the cell stimuli are coupled to the activation of adenylate cyclase, a subsequent increase in cell Ca^{2+} concentration also occurs. Among the various electrical activities of the neuron is an inward K^+ current (I_R) which is augmented by cAMP [160] but depressed by Ca^{2+} [161]. The mechanism of this Ca^{2+}-dependent inactivation of the K^+ current has been suggested to involve the activation of CaM-stimulated PDE, since it can be abolished by the PDE inhibitor methyl-3-isobutyl-8-methylxanthine [162]. Similarly, CaM-stimulated PDE has been implicated as the underlying mechanism for the feedback inhibition of cAMP-stimulated ion current in neurons of *Pleurobrachaea* [163]. In ventral cells of the buccal ganglion, cAMP stimulates a slow Na^+ channel [164]. The amplitude and duration of the current response can be enhanced by the CaM antagonists trifluoperazine and W7, N-(6-aminohexyl)-5-chloro-1-naphthalene-sulfonamide or by intracellular injection of EGTA, but reduced by periods of depolarization [163]. Thus, in spite of the lack of direct measurements of cAMP concentration, the suggestion that CaM-stimulated PDE participates in the regulation of the ion channel activities is reasonable. The cAMP-stimulated channel activities may be suggested to be regulated by a feedback regulatory loop which consists of a sequence of reactions including the stimulation of a Ca^{2+} channel by cAMP to increase Ca^{2+} concentrations in the cell and the activation of the cAMP PDE activity to catalyze the hydrolysis of cAMP to return the neuron to its resting state.

Saitoh et al. [165] have developed a procedure which attempts to demonstrate and to assay for the in vivo CaM stimulation of the PDE directly. The procedure involves a rapid tissue extraction in the presence of trifluoperazine followed by an immediate and brief assay of the enzyme activity at a low temperature. The presence of trifluoperazine prevents the in vitro formation of activated PDE, and the use of low temperatures and a short enzyme assay time minimize the dissociation of the CaM–PDE complex formed in vivo. The procedure was applied to the study of the activation state of PDE in porcine artery strips under various conditions. It was found that conditions expected to increase cytosolic Ca^{2+} concentration resulted in an increased level of CaM-stimulated PDE activity [165]. This result suggests that the in vivo activated state of the enzyme can be preserved and measured by using this procedure.

2.6   MECHANISM OF CaM STIMULATION OF PDEs

The first postulated mechanism of Ca^{2+}-dependent stimulation of PDE by CaM described the reaction as occurring in two steps: an initial step of Ca^{2+} binding to CaM to induce a change in CaM conformation and a subsequent step involving the association of Ca^{2+}-bound CaM with the PDE to bring about enzyme activation [166]. Over the years, tremendous advances have been made in elucidating the various mechanistic aspects of enzyme activation. These advances fall into three general areas: those dealing with the Ca^{2+} binding properties and Ca^{2+}-induced conformational change of CaM, those dealing with the structural basis of the interaction of CaM with CaM-regulated enzymes, and those concerning the kinetics and energetics of the interactions between Ca^{2+}, CaM and the enzyme.

2.6.1 Ca^{2+} binding and Ca^{2+}-induced conformational change of calmodulin

Although studies in this area contribute directly to the elucidation of the mechanism of PDE activation, a comprehensive review of the vast body of literature on this subject is beyond the scope of this chapter. Thus, only a summary of the current state of the area will be given. There are many excellent reviews on this and related subjects [167–171].

Calmodulin contains four Ca^{2+} binding sites which are characteristic of the 'EF hand' structure. These sites are designated as Ca^{2+} binding sites I, II, III and IV in order of their location from the N-terminus. The crystal structure of a mammalian CaM in its Ca^{2+}-bound state has been elucidated [172]. The crystalline CaM assumes a dumbbell shape, with each lobe of the dumbbell containing a pair of 'EF hand' Ca^{2+} binding sites. The two lobes are connected by a long central helix and do not show any other interactions. The two Ca^{2+} binding sites within each lobe appear to interact through hydrogen bonds. Equilibrium Ca^{2+} binding curves of CaM suggest that the four Ca^{2+} binding sites are non-interacting and equivalent in Ca^{2+} affinity. On the other hand, kinetic studies of Ca^{2+} dissociation using stopped flow measurements [173–176] or proton or $^{43}Ca^{2+}$ NMR techniques [177,178] have revealed two classes of CaM-bound Ca^{2+} with dissociation rates of $10–40 \text{ s}^{-1}$ and $500–1000 \text{ s}^{-1}$ respectively, suggesting the existence of Ca^{2+} binding sites of distinct affinities. The discrepancy in conclusions derived from the kinetic and equilibrium binding studies is more apparent than substantial. By assuming that CaM contains two pairs of Ca^{2+} binding sites with affinities differing by approximately an order of magnitude and that the two sites of each pair display unique positive co-operativity in Ca^{2+} binding, a theoretical curve can be generated to fit a binding curve of four apparently equivalent and non-interaction sites [179]. In addition to the four high-affinity Ca^{2+} binding sites, CaM possesses two to four low-affinity cation binding sites which are not specific for Ca^{2+} [180]. The function of these low-affinity binding sites is not known.

The Ca^{2+}-induced conformational change of CaM has been studied using a wide range of techniques and approaches. In particular, various NMR techniques have been used to probe structural changes in specific regions of the molecule ([167, 169] for reviews). In general, the structural change of CaM induced by Ca^{2+} can be divided into two discrete stages depending on the level of Ca^{2+} saturation. The first stage is completed at an average Ca^{2+} binding stoichiometry of 2 mol Ca^{2+} per mol CaM, while the second stage is completed upon Ca^{2+} saturation. This observation supports the suggestion of two classes of Ca^{2+} binding sites in CaM. The difference in affinities of the two classes of sites is sufficiently large to allow almost total occupancy of the high-affinity sites (completing the first-stage structural change) before signficant Ca^{2+} binding to the low-affinity sites occurs. Since the change at the first stage appears to involve largely the C-terminal half of the molecule, it is suggested that sites III and IV are the high-affinity sites and sites I and II are of lower affinity.

Calmodulin fragments have been used extensively to provide important insights into the Ca^{2+} binding properties and the Ca^{2+}-induced conformational change of CaM. Controlled trypsin hydrolysis of CaM in the presence of Ca^{2+} produces two fragments, CaM 1–77 and CaM 78–148, with high yields [120,181]. Each of the fragments represents about half of the molecule and contains a pair of Ca^{2+}-binding sites. Both fragments retain Ca^{2+} binding and drug binding abilities. The fragments have been characterized by UV, CD and various NMR spectroscopies. Spectra of intact CaM are almost indistinguishable from those obtained from a 1:1 mixture of the two fragments or those synthesized from spectra of the two separate fragments, either in the presence or absence of Ca^{2+}. The results indicate that both the conformation and the Ca^{2+}-induced conformational change of the intact CaM are preserved in the fragments, and therefore suggest strongly that the two halves of the molecule do not interact during Ca^{2+} binding ([167] for review). On the basis of these and other studies summarized above, the properties of Ca^{2+} binding to CaM may be described as follows. CaM possesses four high-affinity Ca^{2+} binding sites which are organized into two pairs. Sites I and II at the N-terminal lobe and sites III and IV at the C-terminal lobe are low- and high-affinity pairs, respectively. While the two lobes of the molecule do not show significant interactions in Ca^{2+} binding, sites within each pair undergo positive co-operativity. One of the most important consequences of the Ca^{2+}-induced conformational change of CaM is that it enables the protein to associate with the regulated proteins. Important changes in CaM relevant to this function will be discussed in the following sections.

2.6.2 Calmodulin and phosphodiesterase interaction: structural considerations

Calmodulin-stimulated PDE is but one of a large number of enzymes and proteins which show Ca^{2+}-dependent association with CaM. The majority of these proteins appear to bind at common or overlapping sites on CaM. Calmodulin antagonist drugs have been used to probe the nature of these CaM binding sites. The structural heterogeneity of these drugs (Table 2.3) indicates the broad stereospecificity of the drug binding sites. Most of the drugs, however, are similar in that they contain a bulky hydrophobic moiety. Quantitative correlation between the PDE inhibition potency and the hydrophobicity of different series of antagonists has been demonstrated, suggesting that hydrophobic interactions play important roles in the CaM–PDE interaction [109,182]. Most of these drugs also contain an amino group and therefore carry a positive charge at neutral pH. The binding of trifluoperazine is pH-dependent and occurs only at pH's above the isoelectric point of the drug [183]. These observations suggest that ionic interactions also play important roles in the association of the protein with the drug molecules.

The crystal structure of CaM shows two hydrophobic clefts, one on each half of the molecule [170]. This agrees with the observation that CaM shows high-affinity binding of 2 mol phenothiazine per mol [106]. The binding of phenothiazine, as well as other drugs such as felodipine, to the two sites exhibits positive co-operativity [184,185]. Thus, while the binding of Ca^{2+} to the two lobes of CaM shows little interaction, the Ca^{2+}-induced drug binding to the two halves of the protein is positively co-operative. In addition, in the presence of CaM antagonist drugs, Ca^{2+} binding becomes highly co-operative with a Hill coefficient approaching 4 [184–186].

For obvious reasons, peptides which undergo Ca^{2+}-dependent associations with CaM have been used as models of CaM-dependent proteins. In particular, melittin and masporan, whose CaM affinities approach those of the high-affinity CaM-stimulated enzymes, including the PDE, have been suggested to mimic the CaM binding domains of these enzymes. This is supported by the observation that affinity purified anti-melittin antibodies crossreact with CaM-stimulated PDE and myosin light chain kinase and block CaM stimulation of these enzymes [187]. Like CaM antagonist drugs, the CaM binding peptides all contain net positive charges and a preponderance of hydrophobic residues. Melittin and masporan can assume an amphiphilic α-helical structure in aqueous solutions with the hydrophobic and basic residues on opposite sides of the helical wheel [188–190]. Peptide fragments representing CaM binding domains have been isolated from a number of CaM-dependent proteins, and also shown to be capable of forming amphiphilic α-helices [171,190,191].

On the basis of structural considerations, the binding of the model peptides is suggested to involve the central helix and to span over both domains of CaM [192,193]. This is in agreement with recent proton NMR studies on the interaction between melittin and deuterated CaM [193] and between CaM and the CaM binding domain peptide of skeletal muscle myosin light chain kinase [194,195]. Similarly, studies of differential lysine reactivities have shown that the rates of acylation of lysine residues 21, 75 and 148, among the seven lysines of CaM, are reduced upon binding of myosin light chain kinase or CaM-stimulated phosphatase to CaM, suggesting that the protein binding sites of CaM extend over both halves of the protein [196–198].

Although CaM binding proteins share many common characteristics, recent studies have revealed that there are also important differences in the interaction of CaM with the target proteins. First of all, the affinities of these proteins for CaM vary widely, with K_a values ranging from subnanomolar to micromolar. Secondly, CaM may undergo different conformational changes upon binding different proteins. For example, an anti-CaM monoclonal antibody shows much higher affinity for PDE-bound CaM than for free CaM or CaM bound to certain other proteins such as the CaM-stimulated phosphatase [59]. Thirdly, as determined from CaM modification studies, there appear to be subtle differences in the structural determinants of CaM required for the activation of different enzymes.

A large number of chemical and proteolytic derivatives of CaM, as well as genetically engineered CaM mutants, have been prepared and tested for their effects on various CaM-dependent enzymes. Several recent examples are presented in Table 2.4. A specific structural modification of CaM may affect the protein's ability to interact with a target protein in a number of different ways. First, the modification may abolish the ability of CaM to interact with the target proteins so that the enzyme activity is not affected by the modified protein under any condition. Second, the modified CaM may lose its ability to activate the enzyme but can still bind to the enzyme and block the enzyme activation by native CaM. Third, the CaM activity may be impaired by the structural modification resulting in a lower maximal enzyme activation, A_{max}, a higher activation constant, K_a, or both. The observation that the activation of the various CaM-dependent enzymes is affected differently by specific structural

Table 2.4 Effect of CaM derivatives on CaM-stimulated PDE

CaM derivatives	Effect on PDE	Effect on selected other enzymes				Reference
		Phosphorylase kinase	Myosin light chain kinase	Calcineurin	NAD$^+$ kinase	
CaM fragment 1–77	No effect	Activation	No effect	No effect	ND	120
CaM fragment 78–148	Inhibition	Activation	Inhibition	No effect	ND	120
CaPP$_1$-CaM[a]	Inhibition	Activation	Inhibition	Activation	ND	123
Methyl CCNU-CaM[a]	Activation with higher K_a	ND	ND	ND	ND	
POS-TP$_2$-CaM[a]	Activation with higher K_a	ND	ND	ND	No effect	229
CaML$_{16}$ or CaML$_{19}$[b]	Activation with higher K_a	No effect	Activation with higher K_a	Activation with higher K_a and lower A_{max}	ND	200
UV8[b]	Activation	ND	Activation with higher K_a and lower A_{max}	ND	No effect	202
CaMPM[b]	Activation with higher K_a	ND	Activation with higher K_a	Activation	ND	201
CaMIM[b]	Activation with higher K_a	ND	Activation with higher K_a	Activation	ND	201

[a] The abbreviations used are: CAPP$_1$-CaM, norchlorpromazine isothiocyanate modified CaM, POS-TP$_2$-CaM, 10-(1-propronyloxysuccinamide)-2-trifluoromethyl-phenothiazine modified CaM with a stoichiometry of 2 mol/ml, methyl CCNU-CaM, 1-(2-chloroethyl)-3-(4-methylcyclohexyl)-1-nitrosourea modified CaM.
[b] Genetically engineered CaM mutants. Specific mutants are described in the text.

modifications of CaM (Table 2.4) suggests that structural requirements of CaM for the activation of each of the target enzymes are unique.

The structural determinants of CaM which are important for the activation of CaM-stimulated PDE have not been elucidated. Various proteolytic derivatives of CaM, including the ones representing the N-terminal half, 1–77, the C-terminal half, 78–148, and the N-terminal two-thirds, 1–107, have been shown to have little or no PDE-activating activity, suggesting that the whole CaM molecule is required for the enzyme activation [120] (Table 2.4). On the other hand, many amino acid residues of CaM can be modified or replaced without complete abolition of the PDE-activating activity. Walsh and Stevens [199] originally showed that chemical modifications of the sole histidine, the two tyrosines or four of the six arginine residues did not markedly reduce the ability of CaM to activate CaM-stimulated PDE, whereas carbodiimide modification of carboxylic groups and N-succinimide modification of four of the nine methionines inactivated CaM. Three of the four modified methionine residues, 71, 72 and 76 are present in the central helix region. The suggestion that the PDE-activating activity can tolerate modifications of various amino acid residues of CaM appears to be generally supported by recent studies. For example, CaM mutants with 16 and 19 amino acid substitutions, CaML16 and CaML19 respectively, activate CaM-stimulated PDE to the same extent as the wild-type CaM [200]. Although methionine residues on the central helix appear essential for the activation of the PDE [199], a number of CaM mutants containing specific structural modifications of the central helix activate the PDE readily. Thus, either a disruption of the α-helical structure by substituting proline at threonine 79 (CaMPM), or increasing the length of the central helix by inserting four additional residues (CaMIM), does not abolish the protein's ability to activate the PDE [201]. In addition, substitution of a cluster of glutamate residues 82, 83 and 84 by lysine residues (UV8) has little effect on the activity of CaM towards the PDE but completely inactivates the NAD^+-kinase-activating activity and greatly reduces the MLCK-activating activity [202].

2.6.3 The structure and function of phosphodiesterase

The structure of CaM-stimulated PDE has been elucidated only at the primary structure level [19,70]. However, a number of schematic representations have been proposed on the basis of the enzymes' catalytic characteristics to explain the molecular mechanisms of CaM activation. The common hypothesis of these schemes is that the enzyme can exist in an activated or a non-activated state. Each state is a manifestation of different interactions between three structural domains of the enzyme: the catalytic, inhibitory and Ca^{2+} binding domains. Thus, the inhibitory and catalytic domains of the non-activated enzyme undergo strong interactions to result in a low enzyme activity, whereas in the various activated forms of the enzyme, this interaction is disrupted or weakened.

The stimulation of the enzyme by CaM may be conceived of as consisting of the binding of CaM at the CaM binding domain to induce a conformational change of the enzyme which results in the disruption of the interaction between the inhibitory and catalytic domains. The relationship between the CaM binding domain and the inhibitory domain in the PDE is not clear. The simplest mechanism is for CaM to compete with the catalytic domain for the binding to the inhibitory domain. In this case, the CaM binding and the inhibitory domains are the same or closely proximal.

The domain structure of CaM-stimulated PDE has been investigated by chemical modification studies. Modification of bovine brain CaM-stimulated PDE by an arginine-specific reagent, p-hydroxyphenylglyoxal, rendered the enzyme refractory to CaM stimulation, whereas the basal enzyme activity was not affected. The observation suggests that one or more arginines in the CaM binding domain are essential for the interaction between the enzyme and CaM [203]. Porcine brain CaM-stimulated PDE could be irreversibly inactivated by UV irradiation in the presence of 4-azido-7-phenylpyrazolo-[1-5a]-1,3,5-triazine. The enzyme inactivation was suggested to result from photoaffinity labeling of the active site, since the substrate cGMP, or a competitive inhibitor, IBMX or papaverine, could attenuate the enzyme inactivation whereas CaM had no effect [204]. A more detailed description of the domain structure of CaM-stimulated PDE is given in Chapter 1.

As discussed earlier, CaM-stimulated PDE can be activated and rendered CaM-independent by limited proteolysis using a variety of proteases, including the highly non-specific pronases [79,205,206]. The resulting active fragments, of molecular mass 35–45 kDa depending on the protease used, are highly resistant to further digestion and incapable of binding CaM. These fully active fragments may be suggested to contain the catalytic domain free of both the CaM binding and inhibitory domains.

The activation of CaM-stimulated PDE by compounds of diverse structure may also be understood by the proposed domain structure of the enzyme. The association of the inhibitory and catalytic domains of the enzyme may involve multiple interactions. Different activators such as polyaspartate, phospholipids, unsaturated fatty acids etc. may disrupt the association at different interactive points. The stimulation of the enzyme by low concentrations of detergents may be attributed to the general destabilizing effect of the reagent on noncovalent interactions of the protein molecule.

As discussed above (Section 3.3), almost all the well-characterized CaM-stimulated PDE isoenzymes are dimeric proteins except for the peak I rat testis isoenzyme, which is a monomer but dimerizes upon binding CaM. The relationship between the dimeric structure and the activity or CaM activation of the enzyme is not clear. The various, proteolytically derived active forms of CaM-stimulated PDE are monomeric, suggesting that the dimeric structure is not essential for the enzyme activation. Radiation inactivation of a purified sample of bovine brain CaM-stimulated PDE showed that the inactivation of the basal and CaM-stimulated activities of the enzyme followed the decay curves of 60 kDa and 105 kDa protein species respectively [207]. This observation suggested that the brain PDE existed as an equilibrium mixture of monomers and dimers and, that while the monomer was active without CaM, the dimer depended on CaM for activity [207]. Further studies are clearly needed to verify such a suggestion.

2.6.4 Kinetics and energetics of Ca^{2+}-dependent calmodulin stimulation

The two-step mechanism postulated initially for the stimulation of PDE by Ca^{2+} and CaM has been useful in focusing investigations on specific aspects of the enzyme-activation process, such as the Ca^{2+} binding properties of CaM. However, the mechanism does not take into consideration all the potential reactions in the activation process, and therefore cannot serve as a framework for quantitative descriptions of the relationship between Ca^{2+}, CaM and PDE. A more general model which considers the sequential binding of Ca^{2+} to CaM and the interaction of PDE with the various $Ca^{2+}-CaM$ complexes has been proposed [67] (Figure 2.1). Dissociation constants for some of the reactions in Figure 2.1 are given, including those for the Ca^{2+} binding to free CaM, and the interaction between PDE and fully liganded CaM. In the absence of Ca^{2+}, the affinity of CaM for PDE is too low to determine. The dissociation constant of this reaction has been estimated to be greater than 10^{-4} M [67,208,209].

Although other constants in Figure 2.1 have not been determined, all equilibrium constants in the scheme are interdependent. The interaction of PDE and Ca^{2+} with CaM are linked ligand binding processes and therefore undergo free energy coupling. For example, since the binding of Ca^{2+} to CaM increases the affinity of CaM for the PDE, i.e. $K_a \ll K_e$, the affinity of CaM for Ca^{2+} is expected to increase upon CaM–enzyme association: i.e. $K'_1K'_2K'_3K'_4 \ll K_1K_2K_3K_4$. Such an effect of the target protein on the Ca^{2+} binding

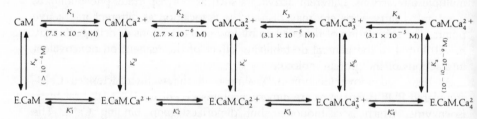

Figure 2.1 Energy coupling between Ca^{2+}-binding of CaM and the association of CaM with PDE K_1-K_4, dissociation constants of Ca^{2+} and CaM; $K'_1-K'_4$, dissociation constants of protein complexes and Ca^{2+}; and K_a-K_e, dissociation constants of PDE and liganded CaM

properties of CaM has been demonstrated by many investigators using various CaM targets [30,209–212], including bovine heart PDE [209]. A thermodynamic scheme, similar to that shown in Figure 2.1, has been proposed to describe the free energy coupling of the interaction of Ca^{2+} PDE with CaM [213].

An analysis of the mechanism of free energy coupling of Figure 2.1 has revealed certain unique properties of the enzyme-activation reaction. The very large increase (10^5–10^6-fold) in the affinity of CaM for PDE, upon Ca^{2+} binding by CaM (Figure 2.1), indicates that an equivalent enhancement in CaM affinity towards Ca^{2+} occurs when the enzyme is bound to CaM, i.e. $K_e/K_a = K_1K_2K_3K_4/K_1'K_2'K_3'K_4'$. An implicit postulate is that the binding of Ca^{2+} to each of the four sites contributes fractionally to the overall increase in the affinity of CaM for PDE, and, consequently, the PDE-induced increase in the Ca^{2+} affinities of the four binding sites of CaM can be kept below two orders of magnitude. This is significant, since during a Ca^{2+}-mediated cell activation, the cytosolic Ca^{2+} concentration is raised by about two orders of magnitude.

The free energy coupling between the interaction of CaM with PDE and Ca^{2+} also contributes to the strong positive co-operativity of Ca^{2+} in the stimulation of PDE. Upon Ca^{2+} binding to the first site, the interaction between CaM and the enzyme is enhanced, which in turn enhances Ca^{2+} binding to the second site. Thus, a stepwise enhancement in the affinity of CaM for the PDE brings about a stepwise increase in the CaM affinity for Ca^{2+}. The observation that the equilibrium Ca^{2+} binding curve of CaM does not show significant co-operativity but becomes strongly co-operative in the presence of a CaM antagonist indicates that the two lobes of CaM communicate with each other upon binding of the target protein.

Huang et al. [87] carried out a steady-state kinetic study of the activation of the PDE by Ca^{2+} and CaM to determine which of the various CaM–PDE complexes possesses the activated enzyme activity. Initial velocities of the PDE reaction determined over a wide range of CaM and Ca^{2+} concentrations were found to conform to an equation which was derived on the basis of Figure 2.1 with the assumption that the fully liganded CaM–PDE complex, CaM–Ca_4^{2+}–PDE, was the predominant activated enzyme species. The result suggested that the activation of the PDE required the binding of all four Ca^{2+} to CaM. A more detailed kinetic analysis of the PDE activation, using flow calorimetry to monitor the reaction when the concentrations of Ca^{2+} and CaM were continuously changed, arrived at the same conclusion [208]. Cox et al. [88], on the other hand, carried out a steady-state kinetic study of CaM activation of the PDE to address a somewhat different question: which of the four Ca^{2+}–CaM complexes was responsible for PDE activation? Initial velocity measurements of the PDE reaction were carried out over a wide range of CaM concentrations at several constant Ca^{2+} concentrations. The relative concentrations of the various Ca^{2+}–CaM species were then calculated on the basis of the Ca^{2+} binding curve of CaM for the reactions at 50% maximal activation where the activating species would be expected to be present at a constant concentration regardless of the total CaM and Ca^{2+} concentrations. It was found that the concentrations of two species, $CaM-Ca_3^{2+}$ and $CaM-Ca_4^{2+}$, had to be summed to achieve a constant concentration, suggesting that both complexes are capable of activating PDE. The apparent discrepancy in conclusions derived from the two studies may be due to the different questions posed. Due to the strong positive co-operativity in the Ca^{2+} activation, the complex $CaM-Ca_3^{2+}-PDE$ converts readily to $CaM-Ca_4^{2+}-PDE$ upon association with the enzyme.

Chau et al. [176] carried out stopped flow measurement using a Ca^{2+} indicator to determine the kinetics of Ca^{2+} release from the fully liganded CaM–PDE complex. The oscillogram revealed a time-course of multiple exponential functions indicative of multiple rates of Ca^{2+} release. The complex curve was not resolved for the computation of individual kinetic constants, but the rate of the most rapidly dissociating Ca^{2+} was estimated to have a half-life of 150–200 ms, corresponding to a rate constant of 3.5–4.6 s^{-1}. When the kinetics of inactivation of a fully activated PDE by EGTA was determined by using a chemical quenching stopped flow instrument, the rate constant of the inactivation reaction was found to be 4.5 s^{-1}. This observation supported the suggestion that $CaM-Ca_4^{2+}-PDE$ is the predominant activated enzyme species.

The requirement for binding of all four Ca^{2+} to CaM for the enzyme activation suggests that an association of the PDE with CaM does not by itself result in enzyme activation. The fact that the bovine lung isoenzyme, which contains CaM as a subunit, still requires Ca^{2+} for activation [16] strongly supports this suggestion. However, the brain PDE has been shown to become activated upon cross-linking to CaM, irrespective of the presence of Ca^{2+} [214]. This observation does not necessarily contradict the above suggestion, since the cross-linking reaction might have frozen the enzyme in an activated conformation, in addition to linking it to CaM.

2.7 FINE TUNING OF CaM REGULATION

Although the regulatory activity of CaM appears to be governed by a general common mechanism, recent studies have revealed differential interactions of CaM with its target proteins (see Section 2.5.2). Characterization of the regulatory properties of a few CaM-stimulated PDE isoenzymes has revealed the existence of well-developed mechanisms for the fine tuning of CaM activities towards individual isoenzymes.

2.7.1 Differential CaM affinities

The CaM-stimulated PDE isoenzymes purified from bovine tissues, brain 60 kDa, brain 63 kDa, heart and lung PDEs, show different affinities towards CaM. The two brain isoenzymes have K_as of CaM activation of approximately 1 nM, which is 10–20 times higher than that of the heart isoenzyme [59, 92]. The lung isoenzyme has the highest affinity towards CaM, since it contains CaM as a subunit [16]. Similarly, the porcine brain CaM-stimulated PDE has been shown to have a 10–20-fold lower CaM affinity than the isoenzyme from porcine artery [215].

Since Ca^{2+} and CaM exhibit positive heterotropic co-operativity in the activation of PDE, isoenzymes showing differential CaM affinities are expected to display differential Ca^{2+} sensitivity and/or different modes of co-operative interaction between CaM and Ca^{2+}. At an identical CaM concentration, bovine heart PDE is stimulated by much lower Ca^{2+} concentrations than the bovine brain 60 kDa isoenzyme. In contrast to the isoenzymes from brain and heart, the activation of the bovine lung isoenzyme by Ca^{2+} is independent of CaM concentration [216]. The physiological significance of the differential CaM affinities remains unclear. As mammalian brain contains more than ten times higher concentrations of CaM than mammalian heart and smooth muscle, it has been suggested that the differential CaM affinity of the tissue-specific isoenzymes is a mechanism by which the CaM regulatory reactions are adapted to the respective tissues [21,59]. While the suggestion is reasonable, it requires further scrutiny. If it is true, some of the other CaM-stimulated enzymes should also exist as brain and heart-specific isoenzymes with similar differential CaM affinities. The existence of CaM as a subunit in the lung PDE has been suggested to provide the enzyme with a competitive advantage of CaM. Unlike other PDE isoenzymes, the lung isoenzyme is not inhibited by CaM antagonists such as trifluoperazine, nor by other CaM-dependent enzymes such as the CaM-stimulated phosphatase [16]. It is not at all clear why such a competitive advantage is specifically required by the lung PDE. In any case, the existence of tissue-specific isoenzymes of distinct CaM-activation properties which are conserved in different mammalian species suggests strongly that such differential regulation represents a fine-tuning mechanism for CaM action.

2.7.2 Regulation of phosphodiesterase isoenzymes by phosphorylation

Protein phosphorylation, as well as, perhaps, other post-translational modifications such as carboxymethylation [217], appears to be a common mechanism for the fine tuning of CaM regulation. The phosphorylation of mammalian brain CaM-stimulated cyclic nucleotide PDE by both cAMP-dependent protein kinase [23] and CaM-dependent protein kinase-II [218] had been described before the discovery of the 60 kDa and 63 kDa bovine brain PDE isoenzymes. More recent studies using purified isoenzyme preparations revealed that the 60 kDa and the 63 kDa isoenzymes were phosphorylated by cAMP-dependent protein kinase [219] and CaM-dependent protein kinase-II respectively [220–222]. The phosphorylation reactions are highly specific; a number of other protein kinases tested, including phosphorylase kinase, myosin light chain kinase, casein kinase-I and -II, protein kinase-C and a spleen protein tyrosine kinase, were unable to phosphorylate either of the isoenzymes [58]. Salient features of the regulation of the two PDE isoenzymes by phosphorylation mechanisms are summarized in Table 2.5. Although the two isoenzymes are phosphorylated by different protein kinases, both can be dephosphorylated by the CaM-stimulated phosphatase [219,221]. Thus both the phosphorylation and the dephosphorylation reactions of the two isoenzymes are under the control of second messenger molecules. In addition, the phosphorylation of the 60 kDa PDE by cAMP-dependent protein kinease can be blocked by Ca^{2+} and CaM, which exert this effect by binding to the 60 kDa isoenzymes [219].

In both cases, the phosphorylation of the brain PDE isoenzymes is accompanied by a decrease in CaM affinity of the enzyme, with the effect being much greater on the 60 kDa isoenzyme than on the 63 kDa isoenzyme [219,220]. The phosphorylation-induced change in the CaM activation may also be manifested as a change in Ca^{2+} sensitivity of the isoenzyme. At a saturating concentration of CaM (in the micromolar range), the activation of the phosphorylated isoenzyme requires significantly higher concentrations of Ca^{2+} than the corresponding nonphosphorylated isoenzyme. Other catalytic properties of the isoenzymes are not affected by the phosphorylation reactions [219,220].

Table 2.5 Regulation of PDE by Ca^{2+} and cAMP

Isoenzyme	Regulation
60 kDa	1. Activation by Ca^{2+} and CaM 2. Phosphorylation by cAMP-dependent protein kinase to result in an increase in the Ca^{2+} concentration required for PDE activation 3. Blockage of PDE phosphorylation by Ca^{2+} and CaM 4. Reversal of the phosphorylation by CaM-dependent phosphatase
63 kDa	1. Activation by Ca^{2+} and CaM 2. Phosphorylation by CaM-dependent protein kinase to result in an increase in Ca^{2+} concentration required for PDE activation 3. Reversal of the phosphorylation by CaM-dependent PDE

2.7.3 Regulatory significance of phosphodiesterase phosphorylation

In vivo phosphorylation has not been demonstrated for either of the bovine brain PDE isoenzymes; however, the observation that the phosphorylation of the PDE isoenzymes is protein-kinase-specific and the effects are specifically on CaM activations suggests that the PDE phosphorylations are physiologically significant. Although the multiple Ca^{2+}- and cAMP-dependent regulatory activities towards the isoenzyme phosphorylation reactions (Table 2.5) appear bewilderingly complex and often paradoxial at first glance, they can be postulated as a set of uniquely organized interacting reactions for each of the isoenzymes which endows the cells with clear regulatory advantages in controlling intracellular cAMP concentrations. The working hypotheses describing the organization of the various regulatory activities are schematically presented in Figure 2.2. These working hypotheses are based on the assumption that CaM-stimulated PDEs act as cAMP-hydrolyzing enzymes during cell activation conditions where the concentrations of both Ca^{2+} and AMP are increasing. A number of studies reviewed above (Section 2.5) strongly support this suggestion.

The key feature of the working hypotheses is that the dynamic interaction between Ca^{2+} and cAMP signals is taken into consideration. Since the increase in Ca^{2+} and cAMP concentrations during cell activation are transitory, reactions regulated by these second messengers are continuously adjusted in accordance with the change in messenger concentrations. One consequence of the dynamic interaction between Ca^{2+} and cAMP is that the multiple regulatory activities towards the PDE phosphorylations may be temporally separated during cell activation. Such temporal separation of the regulatory activities is an important consideration in the working hypotheses for both isoenzymes.

For the regulation of the 60 kDa isoenzyme (Figure 2.2A), an increase in cAMP concentration at the onset of cell stimulation results in the activation of cAMP-dependent protein kinase and the phosphorylation of the PDE isoenzyme. Phosphorylation partially suppresses CaM activation of the PDE, so that the rise in cAMP concentration is unhindered. At a later stage of cell activation, when the cell Ca^{2+} concentration is increased further, the phosphatase reaction is activated to reverse the PDE phosphorylation. The dephosphorylated PDE is then activated by CaM. Since the Ca^{2+}–CaM blocks cAMP-dependent protein-kinase-catalyzed phosphorylation of the PDE isoenzyme, the dephosphorylated isoenzyme will not be rephosphorylated even if the cAMP concentration in the cell is still high. The concerted action of these three regulatory mechanisms by Ca^{2+}–CaM, i.e. stimulation of phosphatase, stimulation of PDE and inhibition of PDE phosphorylation, brings about a rapid decline in intracellular cAMP concentration.

The importance of temporal separation of the regulatory reactions is especially apparent in considering the organization of the multiple regulatory activities towards the 63 kDa PDE isoenzyme. The three distinct CaM-dependent mechanisms that regulate the 63 kDa PDE (Table 2.5) can produce opposing effects. The direct action of CaM is opposed by the kinase-mediated CaM action, and CaM-stimulated kinase and phosphatase reactions are oppositely directed. Superimposed on such antagonistic CaM actions is the stimulation of adenylate cyclase by CaM [223], which counteracts the PDE reaction. Figure 2.2B shows that regulatory advantageous interactions between these antagonistic reactions can be postulated if the CaM activation of adenylate cyclase and the protein kinase

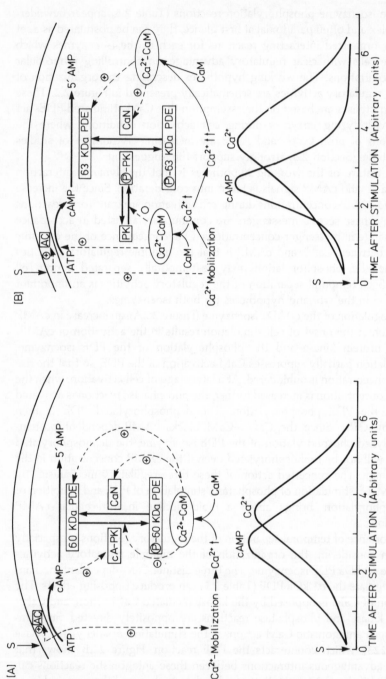

Figure 2.2 Hypotheses of the temporally separated regulations of 60 kDa (A) and 63 kDa (B) phosphodiesterase isoenzymes by Ca^{2+} and cAMP. Symbols: AC, adenylate cyclase; PDE, phosphodiesterase; CaN, CaM-stimulated phosphatase; CA-PK, cyclic AMP-dependent protein kinase; PK, protein kinase; CaM, CaM-dependent protein kinase II; P-phosphorylated; +, activation; −, inhibition. Upper diagrams, organization of regulatory reactions; lower diagrams, simulated Ca^{2+} and cAMP fluxes

occur prior to those of the phosphatase and the PDE isoenzyme. It is postulated that the initial increase in cell Ca^{2+} concentration results in the activation of CaM-dependent protein kinase-II, leading to the phosphorylation of the PDE isoenzyme. The CaM activation of the PDE isoenzyme is therefore blocked. This PDE inhibition, in concert with the activation of adenylate cyclase by CaM, brings about a rapid increase in cAMP concentration. As the cell Ca^{2+} concentration increases further, the phosphatase is activated to dephosphorylate the PDE isoenzyme, allowing isoenzyme activation by CaM, which leads to a decline in cAMP concentration. Although it is not known how such temporal separation of the CaM-regulated reactions may be achieved in the cell, one possible mechanism is by differential Ca^{2+} sensitivity. For example, brain adenylate cyclase has been shown to be activated by much lower concentrations of Ca^{2+} than brain PDE [224] and therefore can be activated prior to the PDE during a transient increase in cell Ca^{2+}. Also, recent studies from a number of laboratories have shown that CaM-dependent protein kinase-II can be auto-phosphorylated and converted into an active, Ca^{2+}-independent form within seconds ([225] for review). It has been suggested that temporal separation of the protein kinase and phosphatase reactions can be achieved by utilizing the autophosphorylation reaction of the protein kinase [221].

REFERENCES

1. Appleman, M. M., Rall, T. W., and Dedman, J. R. (1985) *J. Cyclic Nucleotide Protein Phosphorylation Res.*, **10**, 417–421.
2. Cheung, W. Y. (1967) *Biochem. Biophys. Res. Commun.*, **29**, 478–482.
3. Cheung, W. Y. (1970) *Biochem. Biophys. Res. Commun.*, **38**, 533–538.
4. Cheung, W. Y. (1971) *J. Biol. Chem.*, **246**, 2859–2869.
5. Kakiuchi, S., and Yamazaki, R. (1970) *BBRC*, **41**, 1104–1110.
6. Kakiuchi, S., Yamazaki, R., and Nakajima, H. (1970) *Proc. Jap. Acad.*, **46**, 587–592.
7. Teo, T. S., Wang, T. H., and Wang, J. H. (1973) *J. Biol. Chem.*, **248**, 588–595.
8. Teo, T. S., and Wang, J. H. (1973) *J. Biol. Chem.*, **248**, 5950–5955.
9. Cheung, W. Y., Lynch, T. J., and Wallace, R. W. (1978) *Adv. Cyclic Nucleotide Res.*, **9**, 233–251.
10. Wang, J. H., and Waisman, D. M. (1979) *Curr. Top. Cell. Regul.*, **15**, 47–107.
11. Wolff, D. J., and Brostrom, C. O. (1979) *Adv. Cyclic Nucleotide Res.*, **11**, 27–88.
12. Cheung, W. Y. (1980) *Science*, **207**, 19–27.
13. Klee, C. B., and Vanaman, T. C. (1982) *Adv. Protein Chem.*, **35**, 213–321.
14. Hansen, R. S., and Beavo, J. A. (1982) *Proc. Natl Acad. Sci. USA*, **79**, 2788–2792.
15. Sharma, R. K., Adachi, A. -M., Adachi K., and Wang, J. H. (1984) *J. Biol. Chem.*, **259**, 9248–9254.
16. Sharma, R. K., and Wang, J. H. (1986) *J. Biol. Chem.*, **261**, 14160–14166.
17. Rossi, P., Giorgi, M., Geremia, R., and Kincaid, R. L. (1988) *J. Biol. Chem.*, **263**, 15521–15527.

18. Kubo, K., Matsuda, Y., Kase, H., and Yamada, K. (1984) *Biochem. Biophys. Res. Commun.*, **124**, 315–321.
19. Charbonneau, H., Novack, J. P., MacFarland, R. T., Walsh, K. A., and Beavo, J. A. (1987) In *5th International Symposium on Ca²⁺ Binding Proteins* (Norman, A. W., Vanaman, T. C. and Means, A. R., eds), New York, pp. 505–517.
20. Lin, Y. M., and Cheung, W. Y. (1980) In *Calcium and Cell Function* (Cheung, W. Y., ed.), Vol. 1, Academic Press, New York, pp. 79–100.
21. Sharma, R. K., Mooibroek, M., and Wang, J. H. (1988) In *Molecular Aspects of Cellular Regulation* (Cohen, P. and Klee, C. B., eds), Vol. 5, Elsevier, Amsterdam, New York, Oxford, pp. 265–295.
22. Cheung W. Y., and Storm, D. R. (1982) In *Hand Book of Experimental Pharmacology* (Nathanson J. A. and Kebabian, J. W., eds), Vol. 58/I, Springer-Verlag, Berlin, Heidelberg, New York, pp. 301–323.
23. Sharma, R. K., Wang, T. H., Wirch, E., and Wang, J. H. (1980) *J. Biol. Chem.*, **255**, 5916–5923.
24. Kincaid, R. L., Balaban, C. D., and Billingsley, M. L. (1987) *Proc. Natl Acad. Sci. USA.*, **84**, 1118–1122.
25. La Porte, D. C., Toscano, W. A., and Storm, D. R. (1979) *Biochemistry*, **18**, 2820–2825.
26. Hait, W. N., and Weiss, B. (1977) *Biochim. Biophys. Acta.*, **497**, 86–100.
27. Thompson, W. J., Ross, C. P., Pledger, W. J., Strada, S. J., Banner, R. L., and Hersh, E. M. (1976) *J. Biol. Chem.*, **251**, 4922–4929.
28. Thompson, W. J., Ross, C. P., Strada, S. J., Hersh, E. M., and Lavis, V. R. (1980) *Cancer Res.*, **40**, 1955–1960.
29. Kakiuchi, S., Yamazaki, R., Teshima, Y., Uenishi, K., and Miyamoto, E. (1975) *Biochem, J.*, **146**, 109–120.
30. Wells, J. N., and Hardman, J. G. (1977) *Adv. Cyclic Nucleotide Res.*, **8**, 119–143.
31. Appleman, M. M., Ariano, M. A., Takemoto, D. J., and Whitson, R. H. (1982) In *Handbook of Experimental Pharmacology* (Nathanson, J. A. and Kebabian, J. W., eds), Vol. 58/I, Springer-Verlag, Berlin, Heidelberg, New York, pp. 261–299.
32. Beavo, J. A., Hanson, R. S., Harrison, S. A., Hurwitz, R. L., Martins, T. J., and Mumby, M. C. (1982) *Mol. Cell. Endocrinol.*, **28**, 387–410.
33. Birnbaum, R. J., and Head, J. E. (1983) *Biochem. J.*, **215**, 627–636.
34. Dumas, M.-Y., Fanidi, A., Pageaux, J.-F., Courion, C., Nemoz, G., Prigent, A.-F., Pacheco, H., and Laugier, C. (1988) *Endocrinology*, **122**, 165–172.
35. Jedlicki, E., Orellane, O., Allende, C. C., and Allende, J. E. (1985) *Arch. Biochem. Biophys.*, **241**, 215–224.
36. Solti, M., Davey, P., Kiss, I., Londesborough, J., and Friedrich, P. (1983) *Biochem. Biophys. Res. Commun.*, **111**, 652–658.
37. Yamanaka, M. K., and Kelly, L. E. (1981) *Biochim. Biophys. Acta*, **674**, 277–286.
38. Walter, M. F., and Kiger, J. A. (1984) *J. Neurosci.*, **4**, 495–501.
39. Calhoon, R. D., and Gillette, R. (1983) *Brain Res.*, **271**, 371–374.
40. Perez, R. O., Tuinen, D. V., Marme, D., and Turian, G. (1983) *Biochim. Biophys. Acta*, **758**, 84–87.
41. Beavo, J. A. (1988) In *Advances in Second Messenger and Phosphoprotein Research* (Greengard, P. and Robinson, G. A., eds), Vol. 22, Raven Press, 1–39.
42. Vandermeers, A., Vandermeers-Piret, M.-C., Rathe, J., and Christophe, J. (1983) *Biochem. J.*, **211**, 341–347.

43. Wasco, W. M., and Orr, G. A. (1984) *Biochem. Biophys. Res. Commun.*, **118**, 636–642.
44. Engerson, T., Legendre, J. L., and Johnes, H. P. (1986) *Inflammation*, **10**, 31–35.
45. Grady, P. G., and Thomas, L. L. (1986) *Biochim. Biophys. Acta*, **885**, 282–293.
46. Ho, H. C., Teo, T. S, Desai, R., and Wang, J. H. (1976) *Biochim. Biophys. Acta*, **429**, 461–473.
47. Kakiuchi, S., Yamazaki, R., Teshima, Y., Uenishi, K., Yasuda, S., Kashiba, A., Sobue, K., Ohshima, M., and Nakajima, T. (1978) *Adv. Cyclic Nucleotide Res.*, **9**, 253–264.
48. Grab, D. J., Carlin, R. K., and Siekevitz, P. (1981) *J. Cell Biol.*, **89**, 433–439.
49. Chaudhry, P. S., and Casillas, E. R. (1988) *Arch. Biochem. Biophys.*, **262**, 439–444.
50. Ahluwalia, G., Rhoads, A. R., and Lulla, M. (1984) *Int. J. Biochem.*, **16**, 483–488.
51. Tanigawa, Y., and Simoyama, M. (1976) *Biochem. Biophys. Res. Commun.*, **73**, 19–24.
52. Singer, A. L., Dunn, A., and Appleman, M. M. (1978) *Arch. Biochem. Biophys.*, **187**, 406–413.
53. Epstein, P. M., Andrenayak, D. M., Smith, C. J., and Pappano, A. J. (1987) *Biochem. J.*, **243**, 525–531.
54. Kincaid, R. L., Balaban, C. D., and Billingsley, M. L. (1987) *J. Cyclic Nucleotide Protein Phosphorylation Res.*, **12**, 473–486.
55. Yagura, T., Nagata, I., Kuma, K., and Uchino, H. (1985) *J. Clin. Endocrinol. Metab.*, **60**, 1180–1186.
56. Epstein, P. M., Moraski, S. Jr, and Hachiso, R. (1987) *Biochem. J.*, **243**, 533–539.
57. Cleveland, D. W., Fischer, S. G., Krischner, M. W., and Laemmli, U. K. (1977) *J. Biol. Chem.*, **252**, 1102–1106.
58. Sharma, R. K., and Wang, J. H. (1989) Unpublished observations.
59. Hansen, R. S., and Beavo, J. A. (1986) *J. Biol. Chem.*, **261**, 14636–14645.
60. Ho, H. C., Wirch, E., Stevens, F. C., and Wang, J. H. (1977) *J. Biol. Chem.*, **252**, 43–50.
61. Morrill, M. E., Thompson, S. T., and Stellwagen, E. (1979) *J. Biol. Chem.*, **254**, 4371–4374.
62. Klee, C. B., Crouch, T. H., and Krinks, M. H. (1979) *Biochemistry*, **18**, 722–729.
63. Hansen, R. S., Charbonneau, H., and Beavo, J. A. (1988) *Methods Enzymol.*, **159**, 543–557.
64. Kincaid, R. L., and Vaughan, M. (1988) *Methods Enzymol.*, **159**, 557–573.
65. Draetta, G., and Klee, C. B. (1988) *Methods Enzymol.*, **159**, 573–581.
66. Sharma, R. K., and Wang, J. H. (1988) *Methods Enzymol.*, **159**, 582–594.
67. Wang, J. H., Sharma, R. K., Huang, C. Y., Chau, V., and Chock, P. B. (1980) *Ann. NY Acad. Sci.*, **356**, 190–204.
68. Richman, P. G., and Klee, C. B. (1978) *J. Biol. Chem.*, **253**, 6323–6326.
69. Schenolikar, S., Thompson, W. J., and Strada, S. J. (1985) *Biochemistry*, **24**, 672–678.
70. Charbonneau, H., Beier, N., Walsh, K. A., and Beavo, J. A. (1986) *Proc. Natl Acad. Sci. USA*, **83**, 9308–9312.
71. Chen, C.-N., Denome, S., and Davis, R. L. (1986) *Proc. Natl Acad. Sci. USA*, **83**, 9313–9317.
72. Sass, P., Field, J., Nikawa, J., Toda, T., and Wigler, M. (1986) *Proc. Natl Acad. Sci. USA*, **83**, 9303–9307.
73. Uzunov, P., and Weiss, B. (1972) *Biochim. Biophys. Acta*, **284**, 220–226.
74. Hidaka, H., Yamaki, T., and Yambe, H. (1978) *Arch. Biochim. Biophys.*, **187**, 315–321.

75. Hidaka, H., Yamaki, T., Ochiai, Y., Asano, T., and Yambe, H. (1977) *Biochim. Biophys. Acta*, **484**, 398–407.
76. Donnelly, T. E. Jr (1977) *Biochim. Biophys. Acta*, **480**, 193–203.
77. Weiss, B., Fertal, R., Figlin, R., and Uzunov, P. (1974) *Mol. Pharmacol.*, **10**, 612–615.
78. Pichard, A. L., and Cheung, W. Y. (1976) *J. Biol. Chem.*, **251**, 5726–5731.
79. Tucker, M. M., Robinson, J. B., and Stellwagen, E. (1981) *J. Biol. Chem.*, **256**, 9051–9058.
80. Purvis, K., Olsen, A., and Hansson, V. (1981) *J. Biol. Chem.*, **256**, 11434–11441.
81. Cox, J. L., and Harrison, S. D. Jr (1983) *Biochem. Biophys. Res. Commun.*, **115**, 106–111.
82. Habermann, E., Crowell, K., and Janicki, P. (1983) *Arch. Toxicol.*, **54**, 61–70.
83. Chao, S. H., Suzuki, Y., Zysk, J. R., and Cheung, W. Y. (1984). *Mol. Pharmacol.*, **26**, 75–82.
84. Brostrom, C. D., and Wolff, D. J. (1974) *Arch. Biochem. Biophys.*, **165**, 714–727.
85. Dedman, J. R., Potter, J. D., Jackson, R. L., Johnson, J. D., and Means, A. R. (1977) *J. Biol. Chem.*, **252**, 8415–8422.
86. Walters, J. D., and Jirsa, R. C. (1988) *FEBS Lett.*, **236**, 312–314.
87. Huang, C. Y., Chau, V., Chock, P. B., Wang, J. H., and Sharma, R. K. (1981) *Proc. Natl Acad. Sci. USA*, **78**, 871–874.
88. Cox, J. A., Malnoe, A., and Stein, E. A. (1981) *J. Biol. Chem.*, **256**, 3218–3222.
89. Crouch, T. H., and Klee, C. B. (1980) *Biochemistry*, **19**, 3692–3698.
90. Comte, M., Maulet, Y., and Cox, J. A. (1983) *Biochem. J.*, **209**, 269–272.
91. LeDonne, N. C., and Coffee, C. J. (1979) *J. Biol. Chem.*, **254**, 4317–4320.
92. Mutus, B., Karuppiah, N., Sharma, R. K., and MacManus, J. P. (1985) *Biochem. Biophys. Res. Commun.*, **131**, 500–506.
93. Mutus, B., Palmer, E. J., and MacManus, J. P. (1988) *Biochemistry*, **27**, 5615–5622.
94. MacManus, J. P., Whitfield, J. F., Boynton, A. L., Durkin, J. P., and Swierenga, S. H. H. (1982) *Oncodev. Biol. Med.*, **3**, 79–90.
95. Alakhov, V. Yu, Emelyanendo, E. I., Shakhparonov, M. I., and Dudkin, S. M. (1985) *Biochem. Biophys. Res. Commun.*, **132**, 591–597.
96. Law, J. S., Nelson, N., Watanabe, K., and Henkin, R. I. (1987) *Proc. Natl Acad. Sci. USA*, **84**, 1674–1678.
97. Tanaka, T., Ito, M., Ohmura, R., and Hidaka, H. (1985) *Biochemistry*, **24**, 5281–5284.
98. Wolff, D. J., and Brostrom, C. O. (1976) *Arch. Biochem. Biophys.*, **173**, 720–731.
99. Niggli, V., Adunyah, E. S., and Carafoli, E. (1981) *J. Biol. Chem.*, **256**, 8588–8592.
100. Gietzen, K., Xu, Y.-H., Galla, H.-J., and Bader, H. (1982) *Biochem. J.*, **207**, 637–640.
101. Hidaka, H., Yamaki, T., Ochiai, Y., Asano, T., and Yamabe, H. (1977) *Biochim. Biophys. Acta*, **484**, 398–407.
102. Davis, C. W., and Daly, J. W. (1980) *Mol. Pharmacol.*, **17**, 206–211.
103. Van Belle, H. (1984) *Adv. Cyclic Nucleotide Protein Phosphorylation Res.*, **17**, 557–567.
104. Tanaka, T., Yamada, E., Sone, T., and Hidaka, H. (1983) *Biochemistry*, **22**, 1030–1034.
105. Sakai, T., Yamanaka, H., Tanaka, R., Makino, H., and Kasai, H. (1977) *Biochim. Biophys. Acta*, **483**, 121–134.
106. Levin, R. M., and Weiss, B. (1977) *Mol. Pharmacol.*, **13**, 690–697.
107. Roufogalis, B. D. (1982) In *Calcium and Cell Function* (Cheung, W. Y., ed.), Vol. 3, Academic Press, New York, pp. 129–159.

108. Weiss, B., Prozialeck, W., Cimino, M., Barnette, M. S., and Wallace, T. L. (1980) *Ann. NY Acad. Sci.,* **356**, 319–345.
109. Prozialeck, W. C. (1983) *Annu. Rep. Med. Chem.,* **18**, 203–212.
110. Asano, M., and Hidaka, H. (1984) In *Calcium and Cell Function* (Cheung, W. Y., ed.), Vol. 5, Academic Press, New York, pp. 123–164.
111. Malencik, D. A., and Anderson, S. K. (1983) *Biochemistry,* **22**, 1995–2001.
112. Comte, M., Maulet, Y., and Cox, J. A. (1983) *Biochem. J.,* **209**, 269–274.
113. Malencik, D. A., and Anderson, S. R. (1983) *Biochem. Biophys. Res. Commun.,* **114**, 50–56.
114. Colombani, P. M., Robb, A., and Hess, A. D. (1985) *Science,* **228**, 337–339.
115. Wang, J. H., and Desai, R. (1976) *Biochem. Biophys. Res. Commun.,* **72**, 926–932.
116. Wang, J. H., and Desai, R. (1977) *J. Biol. Chem.,* **252**, 4175–4184.
117. Jedlicki, E., Orellana, O., Allende, C., and Allende, J. E. (1985) *Arch. Biochem. Biophys.,* **241**, 215–222.
118. Sharma, R. K. (1990) *J. Biol. Chem.,* **265**, 152–156.
119. Tanaka, T., Ohmura, T., Yamakade, T., and Hidaka, H. (1982) *Mol. Pharmacol.,* **22**, 408.
120. Newton, D. L., Oldewurtel, M. D., Krinks, M. H., Shiloach, J., and Klee, C. B. (1984) *J. Biol. Chem.,* **259**, 4419–4426.
121. Ni, W.-C., and Klee, C. B. (1985) *J. Biol. Chem.,* **260**, 6974–6981.
122. Newton, D., and Klee, C. B. (1984) *FEBS Lett.,* **165**, 269–272.
123. Prozialeck, W. C., Wallace, J. C., and Weiss, B. (1987) *J. Pharmacol. Exp. Ther.,* **243**, 171–179.
124. Kramer, G. L., Garst, J. E., Mitchel, E. S., and Wells, J. N. (1977) *Biochemistry,* **16**, 3316–3321.
125. Wells, J. N., and Miller, J. R. (1988) *Methods Enzymol.,* **159**, 489–496.
126. Ruckstuhl, M., and Landry, Y. (1981) *Biochem. Pharmacol,* **30**, 697–702.
127. Broughton, B. J., Chaplen, P., Knowles, P., Lunt, E., Pain, D. L., and Wooldridge, K. R. H. (1974) *Nature,* **251**, 650–652.
128. Hidaka, H., and Endo, T. (1984) *Adv. Cylcic Nucleotide Protein Phosphorylation Res.,* **16**, 245–259.
129. Hidaka, H., Tanaka, T., and Itoh, H. (1984) *Trends Pharmacol. Sci.,* **5**, 237–239.
130. Itoh, H., and Hidaka, H. (1984) *J. Biochem.,* **96**, 1721–1726.
131. Epstein, P. M., Fiss, K., Hachisu, R., and Andreyak, D. M. (1982) *Biochem. Biophys. Res. Commun.,* **105**, 1142–1148.
132. Umekawa, H., Yamakawa, K., Numoki, K., Tairu, N., and Hidaka, H. (1988) *Biochem. Pharmacol.,* **37**, 3377–3381.
133. Matsushima, S., Tanaka, T., Saitoh, M., Watanabe, M., and Hidaka, H. (1987) *Biochem. Biophys. Res. Commun.,* **148**, 1468–1474.
134. Kakiuchi, S., Yamazaki R., Teshima, Y., and Uenishi, K. (1973) *Proc. Natl Acad. Sci. USA,* **70**, 3526–3530.
135. Lorenz, K. L., and Wells, J. N. (1983) *Mol. Pharmacol.,* **23**, 424–430.
136. Erneux, C., Van Sande, J., Miot, F., Cochaux, P., Decoster, C., and Dumont, J. E. (1985) *Mol. Cell. Endocrinol.,* **43**, 123–134.
137. Watanabe, A. M., McConnaughey, M. M., Strawbridge, R. A., Fleming, J. W., Jones, L. R., and Besch, H. R. (1978) *J. Biol. Chem.,* **253**, 4833–4836.
138. Sabol, S. L., and Nirenberg, M. (1979) *J. Biol. Chem.,* **254**, 1913–1920.

139. Hebdon, G. M., LeVine, H., Sahyoun, N. E., Schmitges, C. J., and Cuatrecasas, P. (1981) *Proc. Natl Acad. Sci. USA,* **78**, 120–123.

140. Meekar, R. B., and Harden, T. K. (1982) *Mol. Pharmacol.,* **23**, 384–392.

141. Van Sande, J., Decoster, C., and Dumont, J. E. (1975) *Biochem. Biophys. Res. Commun.,* **62**, 168–175.

142. Miot, F., Dumont, J. E., and Erneux, C. (1983) *FEBS Lett.,* **151**, 273–276.

143. Van Sande, J., Decoster, C., and Dumont, J. E. (1979) *Mol. Cell. Endocrinol.,* **14**, 45–57.

144. Tanner, L. I., Harden, T. K., Wells, J. N., and Martin, M. W. (1987) *Mol. Pharmacol.,* **29**, 455–463.

145. Decoster, C., Mockel, J., Van Sande, J., Unger, J., and Dumont, J. E. (1980) *Eur. J. Biochem.,* **104**, 199–208.

146. Cochaux, P., Van Sande, J., and Dumont, J. E. (1982) *Biochim. Biophys. Acta,* **721**, 39–46.

147. Cochaux, P., Van Sande, J., and Dumont, J. E. (1985) *FEBS Lett.,* **179**, 303–306.

148. Evans, T., Smith, M. M., Tanner, L. I., and Harden, T. K. (1984) *Mol. Pharmacol.,* **26**, 395–404.

149. Hughes, A. R., Martin, M. W., and Harden, T. K. (1984) *Proc. Natl Acad. Sci. USA,* **81**, 5680–5684.

150. Van Sande, J., Erneux, C., and Dumont, J. E. (1977) *J. Cyclic Nucleotide Res.,* **3**, 335–345.

151. Nemeck, G. M., and Honeyman, T. W. (1989) *J. Cyclic Nucleotide Protein Phosphorylation Res.,* **15**, 421–428.

152. Barber, R., Ray, K. P., and Butcher, R. W. (1980) *Biochemistry,* **19**, 2560–2567.

153. Shima, S., Komoriyama, K., Hirai, M., and Kouyama, H. (1983) *Biochem. Pharmacol.,* **32**, 529–533.

154. Brandi, M. L., Rotella, C. M., Tanini, A., Toccafondi, R., and Aloj, S. M. (1987) *J. Endocrinol. Invest.,* **10**, 451–458.

155. Clayberger, C. A., Goodman, D. B. P., and Rasmussen, H. (1981) *J. Membrane Biol.,* **58**, 191–201.

156. Buxton, I. O., and Brunton, L. L. (1985) *J. Biol. Chem.,* **26**, 6733–6737.

157. Berman, M. I., Jardack, B., Thomas, C. G. Jr, and Nayfeh, S. N. (1987) *Arch. Biochem. Biophys.,* **253**, 249–256.

158. Ranta, T., Knecht, M., Darbon, J. M., Baukal, A. J., and Catt, K. J. (1983) *Endocrinology,* **113**, 427–429.

159. Levitan, E. S., Kramer, R. H., and Levitan, I. B. (1987) *Proc. Natl Acad. Sci. USA,* **84**, 6307–6311.

160. Benson, J. A., and Levitan, I. B. (1983) *Proc. Natl Acad. Sci. USA,* **80**, 3522–3525.

161. Kramer, R. H., and Levitan, I. B. (1988) *J. Neurosci.,* **8**, 1796–1803.

162. Kramer, R. H., Levitan, E. S., Wilson, M. O., and Levitan, I. B. (1988) *J. Neurosci.,* **8**, 1804–1813.

163. Green, D. J., and Gillette, R. (1988) *J. Neurophysiol.,* **59**, 248–258.

164. Rasmussen, H., and Barrett, P. Q. (1984) *Physiol. Rev.,* **6**, 938–984.

165. Saitoh, Y., Hardman, J. G., and Wells, J. N. (1985) *Biochemistry,* **24**, 1613–1620.

166. Wang, J. H., Teo, T. S., Ho, H. C., and Stevens, F. C. (1975) *Adv. Cyclic Nucleotide Res.,* **5**, 179–194.

167. Forsen, S., Vogel, H. J., and Drakenberg, T. (1986) In *Calcium and Cell Function* (Cheung, W. Y., ed.), Vol. 8, Academic Press, New York, pp. 113–157.

168. Klee, C. B. (1988) In *Molecular Aspects of Cellular Regulation* (Cohen, P. and Klee, C. B., eds), Vol. 5, Elsevier, Amsterdam, New York, Oxford, pp. 35–56.

169. Evans, J. S., Levine, B. A., Williams, R. J. P., and Worwald M. R. (1988) In *Molecular Aspects of Cellular Regulation* (Cohen, P. and Klee, C. B., eds), Vol. 5, Elsevier, Amsterdam, New York, Oxford, pp. 57–82.

170. Babu, Y. S., Bugg, C. E., and Cook, W. J. (1988) In *Molecular Aspects of Cellular Regulation* (Cohen, P. and Klee, C. B., eds), Vol. 5, Elsevier, Amsterdam, New York, Oxford, pp. 83–89.

171. Blumenthal, D. K., and Krebs, E. G. (1988) In *Molecular Aspects of Cellular Regulation* (Cohen, P. and Klee, C. B., eds), Vol. 5, Elsevier, Amsterdam, New York, Oxford, pp. 341–356.

172. Babu, Y. S., Sack, J. S., Greenbough, T. J., Bugg, C. E., Means, A. R., and Cook, W. J. (1985) *Nature*, **315**, 37–40.

173. Bailey, P., Ahlstrom, S., Martin, S. R., and Forsen, S. (1984) *Biochem. Biophys. Res. Commun.*, **120**, 185–191.

174. Suko, J., Wyskovsky, W., Pidlick, J., Hauptner, R., Plank, B., and Hellmanh, G. (1986) *Eur. J. Biochem.*, **152**, 425–434.

175. Malenick, D. A., Anderson, S. R., Shalitin, Y., and Schimerlik, M. I. (1981) *Biochem. Biophys. Res. Commun.*, **101**, 390–395.

176. Chau, V., Huang, C. Y., Chock, P. B., Wang, J. H., and Sharma, R. K. (1982) *Calmodulin and Intracellular Ca^{2+} Receptors* (Kakiuchi, S., Hidaka, H. and Means, A. R., eds), Plenum Press, New York, London, pp. 199–218.

177. Seaman, K. (1980) *Biochemistry*, **19**, 207–215.

178. Anderson, T., Krakenberg, T., Forsen, S., and Thulin, E. (1982) *Eur. J. Biochem.*, **126**, 501–505.

179. Wang, C. L. A. (1985) *Biochem. Biophys. Res. Commun.*, **130**, 426–430.

180. Milos, M., Schaer, J. J., Comte, J. J., and Cox, J. A. (1986) *Biochemistry*, **25**, 6279–6287.

181. Drabikowski, W., Kuzniki, J., and Grabarek, Z. (1977) *Biochim. Biophys. Acta*, **485**, 124–133.

182. Norman, J. A., Drummond, A. H., and Moser, P. (1982) *Mol. Pharmacol.*, **16**, 1089–1096.

183. Weiss, B., Projialeck, W., Cimino, M., Barnette, M. S., and Wallace, T. L. (1980) *Ann. NY Acad. Sci.*, **356**, 319–345.

184. Johnson, J. D. (1983) *Biochem. Biophys. Res. Commun.*, **112**, 787–793.

185. Mills, J. S., Bailey, B. L., and Johnson, J. D. (1985) *Biochemistry*, **24**, 4897–4902.

186. Newton, D. L., Burke, T. K., Rice, K. C., and Klee, C. B. (1983) *Biochemistry*, **22**, 5472–5476.

187. Kaetzel, M. A., and Dedman, J. R. (1987) *J. Biol. Chem.*, **262**, 3726–3729.

188. DeGrado, W. F., Kezdy, F. J., and Kaiser, E. T. (1981) *J. Am. Chem. Soc.*, **103**, 679.

189. Cox, J. A., Comte, M., Fitton, J. E., and DeGrado, W. F. (1985) *J. Biol. Chem.*, **260**, 2527–2534.

190. Erickson-Viitanen, S., and DeGrado, W. F. (1987) *Methods Enzymol.*, **139**, 455–478.

191. Kincaid, R. L., Nightingale, M. S., and Martin, B. M. (1988) *Proc. Natl Acad. Sci. USA*, **85**, 8983–8987.

192. O'Neill, K. T., and DeGrado, W. F. (1985) *Proc. Natl Acad. Sci. USA*, **82**, 4954–4958.

193. Sceholzer, S. H., Cohen, M., Putkey, J. R., Means, A. R., and Crespi, H. L. (1986) *Proc. Natl Acad. Sci. USA*, **83**, 3634–3638.

194. Klevit, R. E., Blumenthal, D. K., Wemmer, D. E., and Krebs, E. G. (1985) *Biochemistry*, **24**, 8152–8157.

195. DeGrado, W. F., Vitanen-Erickson, S., Wolfe, H. Jr, and O'Neill, K. T. (1987) *Proteins*, **2**, 20–34.

196. Wei, Q., Jackson, A. E., Pervaiz, S., Carraway, K. L. III, Lee, E. Y. C., Puett, D., and Brew, K. (1988) *J. Biol. Chem.*, **263**, 19541–19544.

197. Manalan, A. S., and Klee, C. B. (1987) *Biochemistry*, **26**, 1382–1390.

198. Jackson, A. E., Carraway, K. L., Puet, D., and Brew, K. (1986) *J. Biol. Chem.*, **261**, 12226–12232.

199. Walsh, M. P., and Stevens, F. C. (1977) *Biochemistry*, **16**, 2742–2749.

200. Putkey, J. A., Draetta, G. F., Slaughter, G. R., Klee, C. B., Cohen, P., Stull, J. T., and Means, A. R. (1986) *J. Biol. Chem.*, **261**, 9896–9903.

201. Putkey, J. A., Ono T., VanBerkum, M. F. A., and Means, A. R. (1988) *J. Biol. Chem.*, **263**, 11242–11249.

202. Craig, T. A., Watterson, M. D., Prendergast, F. G., Haiech, J., and Roberts, D. M. (1987) *J. Biol. Chem.*, **262**, 3278–3284.

203. Morrill, M. E., Thompson, S. T., and Stellwagen, E. (1979) *J. Biol. Chem.*, **254**, 4371–4374.

204. Sullivan, T. A., Duemler, B. H., Kuttesch, N. J., Keravis, T. M., and Wells, J. N. (1987) *J. Cyclic Nucleotide Res.*, **11**, 355–364.

205. Cheung, W. Y. (1969) *Biochim. Biophys. Acta*, **191**, 303–315.

206. Kincaid, R. L., Stith-Coleman, I. E., and Vaughan, M. (1985) *J. Biol. Chem.*, **260**, 9009–9015.

207. Kincaid, R. L., Kempner, E., Manganiello, V. C., Osborne, J. C., and Vaughan, M. (1981) *J. Biol. Chem.*, **256**, 11351–11355.

208. Gregori, L., Gillever, P. M., Doan, P., and Chau, V. (1985) *Curr. Topics Cell. Regul.*, **27**, 447–454.

209. Olwin, B. B., and Storm, D. R. (1985) *Biochemistry*, **24**, 8081–8086.

210. Yazawa, M., Ikura, M., Hikichi, K., Ying, L., and Yagi, K. (1987) *J. Biol. Chem.*, **262**, 10951–10954.

211. Olwin, B. B., Edelman, A. M., Krebs, E. G., and Storm, D. R. (1984) *J. Biol. Chem.*, **259**, 10949–10955.

212. Maulet, T., and Cox, J. A. (1983) *Biochemistry*, **22**, 5680–5686.

213. Kellar, C. H., Olwin, B. B., Hiedeman, W., and Storm, D. R. (1982) In *Calcium and Cell Function* (Cheung, W. Y., ed.), Vol. 3, Academic Press, New York, pp. 103–127.

214. Kincaid, R. L. (1984) *Biochemistry*, **23**, 1143–1147.

215. Keravis, T. M., Duemler, B. H., and Wells, J. M. (1987) *J. Cyclic Nucleotide Protein Phosphorylation Res.*, **11**, 365–372.

216. Wang, J. H., Sharma, R. K., Yokoyama, N., and Mooibroek, M. M. (1989) In *Information Transduction and Processing Systems from Cells to Whole Body* (Hatase, O. and Wang, J. H., eds), Elsevier, Amsterdam, pp. 3–10.

217. Wolff, D. J., and Brostrom, C. O. (1974) *Arch. Biochem. Biophys.*, **163**, 349–358.

218. Fukunaga, K., Yamamoto, H., Tanaka, E., Iwasa, T., and Miyamoto, E., (1983) *Life Sci.*, **35**, 493–499.

219. Sharma, R. K., and Wang, J. H. (1985) *Proc. Natl Acad. Sci. USA*, **82**, 2603–2607.

220. Sharma, R. K., and Wang, J. H. (1986) *J. Biol. Chem.*, **261**, 1322–1328.
221. Zhang, G. Y., Sharma, R. K., and Wang, J. H. (1990) *J. Biol. Chem.* (in press).
222. Hashimoto, Y., Sharma, R. K., and Soderling, T. R. (1989) *J. Biol. Chem.*, **264**, 10884–10887.
223. Brostrom, C. O., Huang, Y., Breckenridge, B. M., and Wolff, D. J. (1975) *Proc. Natl Acad. Sci. USA*, **72**, 64–68.
224. Piascik, M. T., Wisler, P. L., Johnson, C. L., and Potter, J. D. (1980) *J. Biol. Chem.*, **255**, 4176–4181.
225. Nairn, A. C., Hemming, H. C., and Greengard, P. (1985) *Annu. Rev. Biochem.*, **54**, 931–976.
226. Keravis, T. M., Duemler, B. H., and Wells, J. M. (1987) *J. Cyclic Nucleotide Protein Phosphorylation Res.*, **11**, 365–372.
227. Reeves, M. L., Leigh, B. K., and England, P. J. (1987) *Biochem. J.*, **241**, 535–541.
228. Weishaar, R. E., Burrows, S. D., Kobylarz, D. C., Quade, M. M., and Evans, D. B. (1986) *Biochem Pharmacol.*, **35**, 787–800.
229. Mann, D. M., Vanaman, T. C. (1988) *J. Biol. Chem.*, **263**, 11284–11290.
230. Walter, J. D., and Johnson, J. D. (1988) *Biochim. Biophys. Acta*, **957**, 138–142.
231. Jefferson, A. B., and Shulman, H. (1988) *J. Biol. Chem.*, **263**, 15241–15244.

3

CYCLIC GMP-STIMULATED CYCLIC NUCLEOTIDE PHOSPHODIESTERASES

Vincent C. Manganiello,* Takyuki Tanaka,† and Seiko Murashima‡

*Section of Biochemical Physiology, Laboratory of Cellular Metabolism, NHLBI, NIH, Bethesda, MD 20892, USA
†Department of Neurosurgery, University of Nagoya School of Medicine, Funo-cho, Chikusa-ku, Nagoya, Japan
‡Second Department of Internal Medicine, Mie University School of Medicine, 1515 Kamihama-cho, Tsu-Shi, Mie 514, Japan

3.1 INTRODUCTION

Cyclic nucleotide phosphodiesterases (PDEs) constitute a complex group of enzymes which are found in differing proportions and varying amounts in different mammalian cells and tissues [1–6]. At least six or seven distinct types or major classes of PDEs, several of which have been purified to apparent homogeneity, can be distinguished on the basis of their regulatory mechanisms and biochemical, pharmacologic, immunological, and physical properties [7]. Individual types also differ in their cellular distribution and subcellular localization [1–7]. Information is beginning to accumulate as to the evolutionary and structural relationships, in terms of amino acid sequences and gene regulation, between the different PDE classes [7,8]. Each PDE class may, however, represent a family of isoenzymes; in fact three or four putative isoenzymes of the Ca^{2+}- and calmodulin (CaM)-sensitive PDE [7,9–12] and at least two putative isoenzymes of the cGMP-stimulated PDE [13] families have been described.

One type of PDE, the so-called cGMP-stimulated PDE, was initially described in rat liver supernatant [14] and in crude particulate fractions from several rat tissues, especially brain [15], and in extracts from thymocytes [16], L cell fibroblasts [17], adipose tissue [18], and human platelets [19]. The enzyme was then partially purified from rat heart and bovine adrenal supernatants and rat liver supernatant and particulate fractions [20–23], and its kinetic properties described.

Cyclic Nucleotide Phosphodiesterases: Structure, Regulation and Drug Action
Edited by J. Beavo and M. D. Houslay   © 1990 John Wiley and Sons Ltd

Allosteric regulation of cAMP and cGMP hydrolysis is the hallmark of this PDE class, which hydrolyzes both cAMP and cGMP with positively co-operative kinetics. By virtue of the positively co-operative kinetics, at subsaturating cyclic nucleotide concentrations the hydrolysis of one cyclic nucleotide can be stimulated by the other. As first noted by Appleman and co-workers [20,21], cGMP is the preferred substrate for this PDE, with apparent K_m cGMP < cAMP. With partially purified preparations from rat liver, cGMP was demonstrated to be preferred both as substrate (apparent K_m cGMP < cIMP < cAMP) and effector (apparent K_{act} cGMP < cIMP < cAMP < cXMP); at appropriate concentrations of tritiated substrates and unlabeled effectors, cGMP, cIMP, and cAMP each stimulated the hydrolysis of the other two cyclic nucleotides [23]. This PDE has been designated as the 'cGMP-stimulated cyclic nucleotide' or 'cGMP-stimulated cAMP' PDE, because, in early studies which utilized physiological cyclic nucleotide concentrations, stimulation of cAMP hydrolysis by the preferred effector cGMP was more apparent than was stimulation of cGMP hydrolysis by cAMP.

The unique kinetic characteristics of the cGMP-stimulated PDE suggest an important intracellular role in regulation of cyclic nucleotide metabolism and cyclic-nucleotide-mediated processes. Because of the positively co-operative kinetic behavior of the cGMP-stimulated PDE, low hydrolytic rates at low substrate concentrations accelerate in a sigmoidal fashion with increasing substrate, perhaps allowing for rapid and reversible responses to, and regulation of, increases in intracellular cyclic nucleotide concentrations. Furthermore, at estimated physiological concentrations of cAMP and cGMP [24–29], this PDE is more likely to function as a cGMP-stimulated cAMP-PDE, suggesting that under physiological conditions, elevations in tissue cGMP content could increase cAMP degradation by the cGMP-stimulated PDE, reduce cell cAMP content and inhibit cAMP-mediated processes.

In addition to the cGMP-stimulated PDE, at least one other PDE type might function as a convergence point in the mutual regulation of cAMP and cGMP action. Activation of the cGMP-stimulated PDE by increases in cellular cGMP which resulted in increased cAMP hydrolysis could provide a molecular basis for the opposing actions of cAMP and cGMP in some biological systems. On the other hand, another PDE type with a high affinity for cAMP and cGMP has also been described in a number of tissues [7]; cAMP hydrolysis by this PDE is sensitive to inhibition by physiological concentrations of cGMP and hence it has been designated as the 'cGMP-inhibited' cAMP-PDE [7]. Hormones and other effectors which bring about increases in intracellular cGMP might increase cAMP by inhibition of this 'cGMP-inhibited' cAMP-PDE and thus enhance cAMP-mediated processes. One can thus speculate that in different cell types, depending on the characteristics of guanylate cyclase, complement of cGMP-sensitive PDEs, cellular compartments or subcellular fractions, etc., alterations in cGMP content might oppose or enhance the action of cAMP.

3.2 PURIFICATION

Cyclic GMP-stimulated PDEs have been purified to apparent homogeneity from supernatants of extracts from bovine heart [30] and adrenal tissue [30,31], and calf liver [32], as well as from particulate fractions from rat liver [33] and rabbit [34] and bovine cerebral cortex [13]. It has been difficult to estimate the relative contribution of the cGMP-stimulated PDE to total hydrolysis of cAMP in tissue extracts because in most crude preparations the presence of the cGMP-stimulated PDE can be distinguished from other PDEs only by carefully selecting assay conditions, and in partially purified preparations it is difficult to dismiss the contributions of other PDEs, especially the cGMP-inhibited PDE. Using mono-clonal antibodies raised against the cGMP-stimulated PDE, Beavo and associates have attempted to estimate its contribution to total PDE activity in extracts from several bovine tissues [35,36]. These monoclonal antibodies immunoprecipitated but did not inhibit cGMP-stimulated PDE activity [35,36]. Cyclic cGMP-stimulated PDE was specifically and selectively immunoprecipitated from tissue extracts, and PDE activity (at saturating, i.e. 0.5 mM [^{3}H]cAMP, substrate concentrations) was assessed in the immunoprecipitates and supernatants. The cGMP-stimulated PDE accounted for a substantial portion of total cAMP hydrolysis in several bovine tissues, $\sim 80\%$ in adrenal tissue and liver, $\sim 60\%$ in spleen, $\sim 40\%$ in heart and lung, and 20–30% in testis and brain [35]. These values may underestimate the actual contribution of the cGMP-stimulated PDE in some of these tissues, because this approach may not have accounted for particulate cGMP-stimulated PDE activity. Recent findings indicate, for example, that $> 75\%$ of total cGMP-stimulated PDE activity in bovine [13] and rabbit [34] cerebral cortex is associated with particulate fractions.

The critical steps in the purification procedures developed for cGMP-stimulated PDE [13,30–34] have involved cyclic nucleotide affinity chromato-graphy. Martins et al. [30] found that cGMP-stimulated PDE activity from bovine cardiac or adrenal tissues was not retained at neutral pH by N^6-H$_2$N-(CH$_2$)$_2$-cAMP–Sepharose or C^8-H$_2$N(CH$_2$)$_2$-NH-cAMP–Sepharose (as were other cyclic nucleotide binding proteins), but was specifically adsorbed by cGMP-epoxy activated Sepharose and eluted with cGMP or cAMP. Particulate cGMP-stimulated PDEs solubilized from rabbit brain fractions with trypsin [34] or from bovine brain with detergents [13] were also purified by this series of cyclic nucleotide affinity resins. Successful utilization of these affinity columns has been based more on empirical selection than theoretical principles or understand-ing of cyclic nucleotide binding domains of the PDE. As pointed out by Martins et al. [30], the identification of the site of coupling between cGMP and epoxy activated Sepharose has not been determined and the capacity of the affinity resin for the cGMP-stimulated PDE was but a small fraction of the measured coupling density [30]. Since derivatives of cAMP modified at the N^6 but not the C^8 position were capable of activating the PDE and stimulating hydrolysis of cIMP

[23], and since C^8 derivatives activated cAMP-dependent protein kinase, it had been anticipated that C^8-NH_2-$(CH_2)NH$-cAMP–Sepharose would selectively remove cAMP-binding proteins and the cGMP-stimulated PDE would bind to the N^6-H_2N-(CH_2)-cAMP–Sepharose. At neutral pH, only the former prediction was sustained. Yamamoto et al. [32], however, did find that although the calf liver cGMP-stimulated PDE was not bound to N^6-$H_2N(CH_2)_2$-cAMP–Sepharose or C^8-$H_2N(CH_2)_2$-cAMP–Sepharose at neutral pH, it was retained by N^6-$H_2N(CH_2)_2$-cAMP–Sepharose and eluted with cGMP at pH 6. The mechanisms involved in this pH-dependent selective adsorption and elution were not investigated [32].

3.3 CHARACTERIZATION OF PURIFIED ENZYMES

3.3.1 Physical properties of enzymes from different sources

A number of the physical and kinetic characteristics of cGMP-stimulated PDEs purified from bovine cardiac and adrenal tissue [5,7,30], calf liver [32] and rabbit cerebral cortex [34] are presented in Table 3.1. During SDS-PAGE, cGMP-stimulated PDE purified from bovine cardiac tissue [30], bovine adrenal tissue [31]

Table 3.1 Properties of purified cGMP-stimulated PDE

Property	Bovine cardiac	Calf liver	Rabbit brain
Reference	5, 7, 30	32	34
Molecular mass			
Subunit	102–105 kDa	105 kDa	105 kDa
Native	204–230 kDa	200 kDa	400 kDa
Sedimentation coefficient (S)	7.4	6.9	
Stokes radius (Å)	64	67	63
Frictional ratio	1.6	1.7	
Apparent K_m^a (μM)			
cAMP	36	33	28
cGMP	11	15	16
V_{max} ((μmol/min)/mg)			
cAMP	120	170	160
cGMP	120	200	160
Hill coefficient (napp)			
cAMP	1.9	1.6–1.8	
cGMP	1.3	1.2–1.6	
K_a for cGMP (μM)			
(0.5–5.0 μM [^{3}H]cAMP)		0.5	0.35

and calf liver [32] supernatant fractions and from rabbit cerebral cortical particulate fractions (after release/solubilization by treatment with trypsin) [34] exhibited a single protein-staining band (~ 102–105 kDa). Nondenaturing gradient PAGE indicated a molecular mass of a ~ 240 kDa [30]. From hydrodynamic measurements, i.e. Stokes' radius (64–67 Å) and sedimentation coefficient ($S_{20, w}$ 6.9–7.4), a molecular mass in the range of ~ 200–204 kDa and a frictional ratio of 1.6–1.7 were calculated [5,7,30,32]. These findings suggest that native cGMP-PDEs may exist as non-spherical elongated dimers of similar, if not identical, subunits. A comparison of amino acid sequence information from several PDEs, including the Ca^{2+} and CaM-sensitive PDEs from bovine brain and the bovine cardiac cGMP-stimulated PDE, indicates the presence of a 200–270 amino acid residue segment that is similar but not identical [8]. Charbonneau et al. suggested that these related segments comprised PDE catalytic domains [8].

The purified rabbit brain cGMP-stimulated PDE exhibited an apparent molecular mass of ~ 380 kDa during gel filtration, suggesting that the rabbit brain enzyme is tetrameric or displays anomalous behavior during gel filtration due to aggregation or presence of carbohydrate or glycolipid moieties [34]. A cGMP-stimulated PDE with a much lower V_{max} than those listed in Table 3.1 and a molecular mass of 66 kDa has been purified from rat liver particulate fractions [33], but it is not certain if the smaller rat enzyme represents a native or proteolytically modified form of cGMP-stimulated PDE. Pyne and Houslay have recently reported that a crude preparation of 'insulin mediator' produced a 3–4-fold increase in the activities of purified particulate and soluble rat liver cGMP-stimulated PDEs without affecting other purified particulate cAMP-PDEs [37].

Recent evidence has indicated that a distinct cGMP-stimulated isoenzyme is present in bovine brain cerebral cortex [13]. In bovine brain homogenates $> 75\%$ of total cGMP-stimulated PDE activity was associated with washed particulate (23 000 g, 60 min) fractions, predominantly from cerebral cortical gray matter. After solubilization with the non-ionic detergent lubrol (1%), the particulate cGMP-stimulated PDE was purified by sequential chromatography on cAMP–agarose and cGMP–epoxy activated Sepharose columns [13]. The catalytic properties of the purified bovine brain particulate cGMP-stimulated PDE, i.e. V_{max} and K_m for cAMP and K_{act} for cGMP stimulation of cAMP hydrolysis, were similar to those of the liver soluble cGMP-stimulated PDE [13]. As evidenced by protein staining (Coomassie Blue) or Western immunoblots (polyclonal antibodies raised against purified calf liver soluble or bovine brain particulate cGMP-PDEs), the purified particulate cGMP-stimulated PDE exhibited a slightly greater molecular mass than supernatant forms isolated from bovine brain or calf liver. Exposure of the purified particulate cGMP-stimulated PDE to trypsin led to accumulation of immunoreactive peptides, some of which exhibited molecular mass values similar to those of the soluble PDEs [13]. Whether this has mechanistic implications as to the appearance or distribution of the soluble forms in intact cells is not known. In this regard it should be noted that the rabbit cerebral cortical PDE (with subunit molecular mass apparently identical to those of soluble bovine cardiac or calf liver cGMP-stimulated PDEs) was solubilized/ released from particulate fractions by limited proteolysis with trypsin [34].

Incubation of bovine brain particulate and calf liver soluble forms with Staph V8 protease generated four or five major peptides from both PDEs, with identical molecular mass values (Figure 3.1) [13]. Two peptides of ~ 25–30 kDa clearly differed in brain particulate and liver soluble proteolytic digests; this suggests at least some differences in the amino acid sequence of the brain particulate and liver soluble cGMP-stimulated PDEs [13]. As reported for the rabbit brain enzyme [34], the bovine brain and calf liver PDEs could be specifically photolabeled with [^{32}P]cGMP. Proteolytic digestion with V8 protease after photolabeling indicated that [^{32}P]cGMP was predominantly associated with two low molecular mass (~ 12 kDa) peptides and sometimes with fragments of 50–60 kDa [13]. The labeled peptides were shared by brain particulate and liver soluble PDEs. The [^{32}P]cGMP was not found in association with those peptides that differed in digests from brain particulate and liver soluble PDEs. Taken together, these

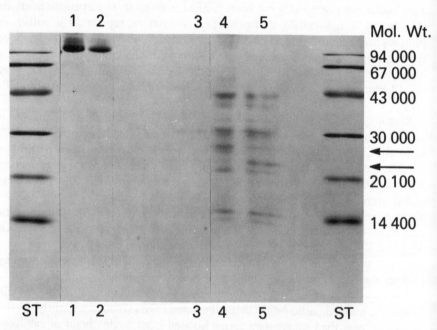

Figure 3.1 Purified PDEs (10 μg each) were digested according to the method of Cleveland et al. [57]. Lane 1, particulate cGMP-stimulated PDE without V8 protease; Lane 2, soluble cGMP-stimulated PDE without V8 protease; Lane 3, V8 protease (0.2 μg); Lane 4, particulate cGMP-stimulated PDE with V8 protease; Lane 5, soluble cGMP-stimulated PDE with V8 protease

findings suggest the existence of a cGMP-stimulated PDE isoenzyme family with conserved and different domains in brain particulate and liver soluble cGMP-stimulated PDEs. These results further imply that cGMP binding sites are located in the conserved regions [13].

3.3.2 Kinetic properties

Although detailed studies have not been carried out, cGMP does apparently bind to the purified cGMP-stimulated PDE with high affinity [30,31,38]. Martins et al. [30] reported that maximal binding (using a Millipore filtration technique) was achieved at ~ 4 μM cGMP with one mole of cGMP bound per mole of native enzyme. Whether this finding indicated that endogenously bound cGMP was not completely removed from at least one other site on the native enzyme (a dimer of molecular mass ~ 240 kDa) during purification or that a single cGMP binding site is generated in the dimeric holoenzyme by interactions of the monomer subunits is not known. K_d values for cGMP were not determined, perhaps because of some anomalies in the binding isotherms, i.e. a wide (i.e. 2.75) logarithmic interval of cGMP concentrations to achieve saturation of binding and curvilinear Scatchard transformations of the binding data [30]. According to Erneux and co-workers [31,38], binding of cGMP to the purified adrenal enzyme was stimulated by inclusion of IBMX (3-isobutyl-1-methylxanthine) in the binding assay and 70% of saturated $(NH_4)_2SO_4$ in the buffers used to terminate the assay and wash the filters. In these studies high-affinity cGMP binding (in the presence of 50 nM [^{3}H]cGMP) was displaced by unlabeled cGMP, cIMP and cAMP (50% displacement at 0.12, 1.0 and 5.7 μM, respectively [38]) and by other selected analogs of cGMP and cAMP, but not by 5'GMP or guanosine (up to 100 μM); K_d values, however, were not estimated in these studies [31,38].

In general, there is good agreement as to the kinetic properties of partially purified preparations [20–23], as well as of cGMP-stimulated PDEs purified from several sources (Table 3.1) [30–34]. The purified PDEs hydrolyze cGMP and cAMP with positively co-operative kinetics (Figure 3.2); cGMP is the preferred substrate and effector. cGMP-stimulated PDEs exhibit a higher affinity for cGMP than cAMP. In assays at 30°C, whereas at 0.5–1 μM substrate concentrations hydrolysis of cGMP was $\sim$ 10-fold greater than that of cAMP, the V_{max} for cGMP was only somewhat greater or about the same as for cAMP; N_{app} (Hill coefficient index for co-operative interactions) was 1.2–1.6 for cGMP and 1.6–1.9 for cAMP (Table 3.1).

The common observation of lower Hill coefficients for cGMP hydrolysis in most enzyme preparations [20–23,30–34] relates more to assay conditions than to the nature of the co-operative kinetic behavior for cGMP. Because of the higher affinity of the PDE for cGMP (apparent K_m cGMP < cIMP < cAMP [23], for example) and because cGMP is a more potent activator in stimulating cyclic nucleotide hydrolysis (apparent K_{act} cGMP < cIMP < cAMP < cXMP [23], for

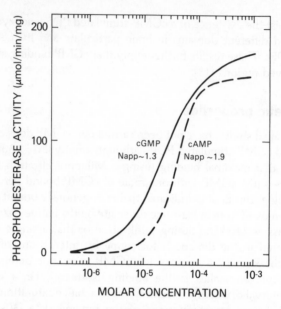

Figure 3.2 Schematic representation of hydrolysis of [³H]cGMP and [³H]cAMP by the purified cGMP-stimulated PDE from calf liver supernatant. N_{app} = Hill coefficient

example), in usual assays comparing hydrolysis of similar concentrations of [³H]cGMP and [³H]cAMP, cGMP activation of the PDE and effects on co-operative interactions occur at lower concentrations than required for activation of the PDE by cAMP. The actual similarities in co-operative behavior toward cAMP and cGMP were clearly observed when assay temperature was altered [39]. At 0–5°C, the apparent affinity for both cAMP and cGMP was increased and alteration in co-operative behavior resulted in a reduction of Hill coefficients to ~ 1–1.2 for both cGMP and cAMP (Figure 3.3) [39]. At higher temperature, i.e. at 45 °C, affinity for both cAMP and cGMP was reduced and Hill coefficients of ~ 1.9–2.0 for both cAMP and cGMP were observed (Figure 3.3) [39].

In assays at 30°C, with subsaturating [³H]cAMP (0.5 or 3.5 μM), hydrolysis was stimulated by cGMP with an apparent K_{act} of ~0.35–0.5 μM (Table 3.1). With the PDE purified from calf liver supernatant, maximal stimulation of cAMP hydrolysis (0.5 μM [³H]cAMP) by cGMP (2–5 μM) was dependent upon divalent cation in the assay, ~ 15–30-fold stimulation in the presence of Mg^{2+} (Figure 3.4), ~9-fold in the presence of Mn^{2+}, and ~6-fold with Ca^{2+} [32]. Stimulation of the hydrolysis of low cAMP concentrations by cGMP results primarily from a decrease in the K_m for cAMP, with little or no increase in V_{max} (Figure 3.5). Cyclic GMP induces a shift from positively co-operative (sigmoidal)

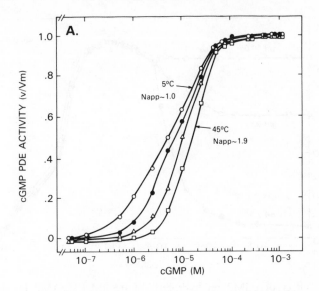

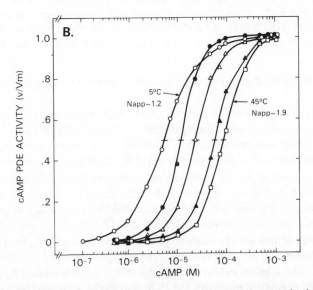

Figure 3.3   Effect of temperature on the kinetics of cAMP and cGMP hydrolysis by the purified cGMP-stimulated PDE from calf liver supernatant. V_{max} was determined at each assay temperature. v/V_{max} ratios were calculated at each substrate concentration and plotted versus log 5. ○, 5°C; ●, 20°C; △, 30°C; ▲, 40°C, □, 45°C. cGMP hydrolysis is presented in panel A; cAMP hydrolysis is presented in panel B. (From ref. 39, reproduced with permission.)

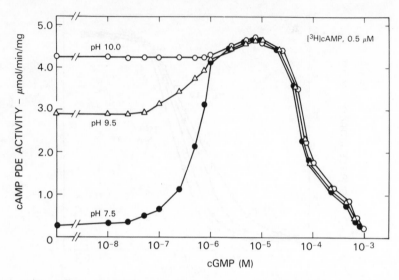

Figure 3.4 Effect of cGMP on hydrolysis of 0.5 μM [^{3}H]cAMP by the purified cGMP-stimulated PDE from calf liver supernatant. Control enzyme (●) or enzyme incubated at pH 9.5 (△) or pH 10 (○) for 5 min at 30°C in the presence of 5 mM MgCl$_2$ were assayed at pH 7.5 with 0.5 μM [^{3}H]cAMP in the absence or presence of the indicated concentrations of unlabeled cGMP. (From ref. 40, reproduced with permission.)

to linear Lineweaver-Burk plots with a decrease in the Hill coefficient for cAMP (Figure 3.5).

Maximum hydrolysis of 0.5 μM [^{3}H]cAMP was reduced at concentrations of cGMP above ~ 10–20 μM (the apparent K_m for cGMP), presumably due to unlabeled cGMP competing with [^{3}H]cAMP at catalytic sites (Figure 3.4). Incubation of the purified PDE at high pH (9.5–10) in the presence of MgCl$_2$ at 30°C prior to assay at pH 7.5 results in 'activation' of the cGMP-stimulated PDE [40]. As shown in Figure 3.4, after activation at pH 9.5 or 10, stimulation of cAMP hydrolysis by low concentrations of cGMP in the assay at pH 7.5 was either minimal or absent, respectively, but cAMP hydrolysis was inhibited by cGMP concentrations greater than 10 μM, near the apparent K_m for cGMP [40]. As seen in Figure 3.5, 30 μM cGMP stimulated hydrolysis of low concentrations of [^{3}H]cAMP (to a lesser extent than 1 μM cGMP), but inhibited hydrolysis of ~ 100 μM [^{3}H]cAMP. Thus, with the PDE maximally activated by low concentrations of cGMP (2–5 μM) (Figure 3.4), incubation at pH 10.0 (Figure 3.4), or by ~ 100 μM cAMP (Figure 3.5), cGMP at concentrations near its apparent K_m did not further stimulate but rather inhibited [^{3}H]cAMP hydrolysis. Inhibition by these higher cGMP concentrations was most likely related to competition at catalytic rather than allosteric regulatory sites and was reversed by higher concentrations of [^{3}H]cAMP; cGMP did not affect V_{max} for [^{3}H]cAMP (Figure 3.5).

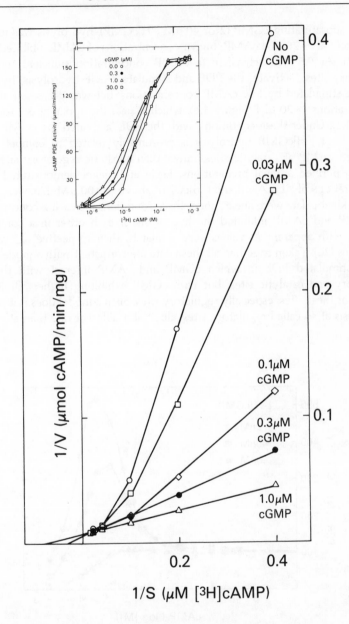

Figure 3.5 Effect of cGMP on the kinetics of hydrolysis of [³H]cAMP by the purified cGMP-stimulated PDE from calf liver supernatant. Lineweaver-Burk plot of data. Insert: Initial rates of hydrolysis of the indicated concentrations of [³H]cAMP in the absence (○) or presence of 0.3 μM (●), 1.0 μM (△) or 30.0 μM (□) cGMP (plotted as velocity versus log S)

With subsaturating cGMP (2 or 40 nM [³H]cGMP), hydrolysis was increased ~3–5-fold at ~20 μM cAMP; higher concentrations of cAMP inhibited (Figure 3.6) [32]. At 0.5 μM [³H]cGMP (a cGMP concentration sufficient to occupy regulatory sites, 'activate' the PDE and stimulate cAMP hydrolysis) hydrolysis was not stimulated by low cAMP concentrations, but was decreased at cAMP concentrations >20 μM (Figure 3.6), which is near the apparent K_m for cAMP (Table 3.1). Under these conditions, with the PDE 'activated' by cGMP, cAMP inhibition of [³H]cGMP hydrolysis is presumably related to competition by unlabeled cAMP at catalytic sites, rather than allosteric regulatory sites. With partially purified adrenal preparations, Egrie and Siegel demonstrated that at concentrations of [³H]cAMP and [³H]cGMP above ~100 μM, Lineweaver-Burk plots of kinetic data were linear; under these conditions increasing concentrations of cGMP and cAMP inhibited the hydrolysis of each other in a competitive manner, with apparent K_i values very similar to their respective K_m values as substrates [22]. Taken together, all these data are consistent with a model for the cGMP-stimulated PDE in which cGMP and cAMP interact with the same regulatory and catalytic sites, but with cGMP exhibiting higher affinity than cAMP for these sites, especially regulatory sites. Such a model does not preclude hydrolysis at so-called regulatory sites, albeit at a different rate than at catalytic sites.

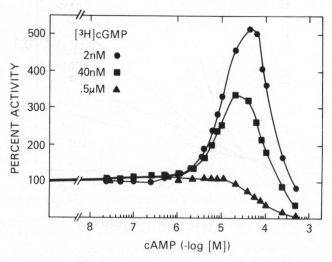

Figure 3.6 Effects of cAMP on hydrolysis of [³H]cGMP by the purified cGMP-stimulated PDE from calf liver supernatant. PDE was assayed with 2 nM (●), 40 nM (■), or 0.5 μM (▲) [³H]cGMP in the absence or presence of the indicated concentrations of cAMP. Activities are expressed relative to those with the same substrate in the absence of cAMP taken as 100%. (From ref. 32, reproduced with permission.)

Studies with both purified enzymes [30–34,38–43] and partially purified preparations [20–23,44–47] are thus consistent with the idea that cGMP, by binding to regulatory sites on the 'ligand-free' or 'low-affinity' state of the PDE, induces allosteric transition to the 'high-affinity' or 'activated' state which results in a reduction in K_m and Hill coefficient (N_{app}) for cAMP with little or no change in V_{max}, a shift from positively co-operative (sigmoidal) to normal Michaelis–Menten kinetics and increased hydrolysis of low concentrations of cAMP. The increased hydrolysis of low concentrations of cAMP is due primarily to increased affinity rather than changes in V_{max} for cAMP. Characteristics of the cGMP-stimulated or 'activated' state are summarized in Table 3.2. With a cGMP-stimulated PDE preparation partially purified from rat liver supernatant, Moss et al. [23] reported that cGMP, at concentrations which stimulated cAMP and cIMP hydrolysis, enhanced the susceptibility of the cGMP-stimulated PDE to chymotrypsin-induced inactivation. They concluded that binding of cyclic nucleotides (and appropriate analogs) induced conformational changes in the cGMP-stimulated PDE that were associated with not only increased catalytic activity (by virtue of decrease in K_m) but also enhanced susceptibility to proteolytic attack [23]. Cyclic AMP, which at the same concentrations was less effective than cGMP as an effector (activator) of the cGMP-stimulated PDE, did not accelerate chymotrypsin-induced inactivation [23]. In purifying the cGMP-stimulated PDE from rabbit brain, Whalin et al. [34], however, recently used trypsin to release the particulate enzyme with no apparent untoward effects on its allosteric properties. The implications of the different responses of these two different cGMP-stimulated PDEs to trypsin and chymotrypsin remain to be explored.

As might be expected for an enzyme subject to allosteric regulation, activity of the cGMP-stimulated PDE can be altered by a number of environmental conditions, especially temperature and pH [39,40], as well as potential therapeutic and physiological effectors. With purified preparations of the cGMP-stimulated PDE from calf liver, as the temperature of the PDE assay increased from 5°C to 45°C, V_{max} for both cAMP and cGMP hydrolysis increased (at 30°C, ΔH ($\Delta H = E_a + RT$, where E_a is energy of activation) was $\sim 11\,000$ Cal/mol for both) [39]. At subsaturating concentrations of cAMP and cGMP, however, the initial rates of hydrolysis increased as temperature decreased from 45°C to 5°C,

Table 3.2   Characteristics of the 'activated' PDE, i.e. cGMP-stimulated form

Increased hydrolysis of low substrate concentrations, below apparent K_m, i.e. $\sim 1\,\mu$M
[^{3}H]cAMP
Decreased apparent K_m for cAMP
Decreased Hill coefficient (N_{app}) for cAMP
Linear Lineweaver-Burk plots
No change in V_{max}
Increased susceptibility to chymotryptic inactivation

K_m for cAMP and cGMP decreased, and Hill coefficients decreased from ~ 2 to ~ 1 (Figure 3.3). At every assay temperature tested, cGMP stimulated hydrolysis of cAMP (0.5 μM [^{3}H]cAMP), with the apparent K_{act} for cGMP stimulation of cAMP hydrolysis lower at 5°C than at 45°C. PDE activity was more sensitive to inhibition by IBMX at 5°C than at 45°C, especially at low cAMP concentrations. These and other studies suggested that low temperature induced incomplete and readily reversible transitions to a state with increased affinity for substrates, effectors and inhibitors [39]. These results also support the notion that at physiological temperatures, the cGMP-stimulated PDE predominantly resides in a 'low-affinity' state, where at low substrate concentrations, sigmoidal co-operative responses in substrate—velocity relationships might provide a sensitive, 'on—off' regulation of cyclic nucleotide hydrolysis.

The allosteric and catalytic properties of the cGMP-stimulated PDE purified from calf liver supernatant have also been reported to be markedly influenced by alterations in pH [40]. In assays of different pH, with 0.5 μM [^{3}H]cAMP plus 1 μM cGMP or saturating substrate concentrations (250 μM) of cAMP or cGMP, hydrolysis was maximal at pH. 7.5–8.0. At these concentrations of substrate and effector, the cGMP-stimulated PDE would be 'activated'. However, at concentrations of cGMP and cAMP below the K_m and not sufficient to saturate regulatory sites and 'activate' the PDE, hydrolysis was maximal at pH 9.5. Because of this increase in basal hydrolysis at pH 9.5, cGMP stimulation of cAMP hydrolysis (0.5 μM [^{3}H]cAMP) was minimal at pH 9.5 and absent at pH 10. Hill coefficients for cAMP hydrolysis were ~ 1.8 at pH 7.5 and decreased as assay pH declined or increased. Thus assay pH (at 30°C) had complex effects on co-operative behavior (Hill coefficients, cGMP-stimulated cAMP hydrolysis) as well as on catalytic activity at both low substrate concentrations and in the 'activated' state (i.e. V_{max}, cGMP-stimulated cGMP-PDE activity) [40].

To minimize multiple effects of assay pH, the purified cGMP-stimulated PDE was first incubated at different pHs in the presence of $MgCl_2$ at 30°C, prior to assay at pH 7.5, since at this assay pH V_{max} and cGMP-stimulated activity were maximal [40]. Prior exposure to different pHs (6.5–10.0) did not alter V_{max} or cGMP-stimulated hydrolysis of cAMP (in assays at pH 7.5). After incubation at high pH (9.0–10.0), however, in assays at pH 7.5 there was a marked increase in hydrolysis of low concentrations of cAMP (but no increase in V_{max}), a decrease in K_m, and a decrease in the Hill coefficient (N_{app}) from ~ 1.7 to ~ 1.0. Incubation at pH 9.5 or 10 markedly increased hydrolysis of 0.5 μM [^{3}H]cAMP at pH 7.5, with reduction in relative stimulation by cGMP (Figure 3.4). After incubation at pH 10, hydrolysis of 0.5 μM [^{3}H]cAMP was maximally increased and was similar in the absence or presence of cGMP. Thus, maximal hydrolysis of low cAMP concentrations in assays at pH 7.5 was achieved by prior incubation of the PDE at pH 10 or by direct addition of cGMP to assays at pH 7.5. Activation induced by high pH was dependent on $MgCl_2$ and temperature, and could be prevented, but not readily reversed, by EDTA [40].

Direct addition of cGMP to assays at pH 7.5 increased hydrolysis of low cAMP concentrations, reduced K_m for cAMP, and caused a shift from positively co-operative to normal Michaelis–Menten kinetics. Thus after incubation at high pH in the presence of $MgCl_2$, the PDE expresses characteristics of the 'cGMP-stimulated' form, i.e. increased hydrolysis of low cAMP concentrations, little further response to added cGMP, a reduction in K_m for cAMP, a decrease in N_{app} and no change in V_{max}. These results support the view that allosteric activation of the cGMP-stimulated PDE is related more to changes in K_m than V_{max}. It is not known if incubation with $MgCl_2$ at high pH produces the same allosteric transitions as does addition of cGMP to assays at pH 7.5. As with changes in temperature, incubation at high pH seems to produce changes in allosteric and catalytic properties in the absence of substrate. These studies indicate that topographical features responsible for transitions of the cGMP-stimulated PDE to the activated state and for expression of catalytic activity can be regulated independently, and in the absence of substrates.

Stimulation of cAMP hydrolysis by cGMP has also been reported to be abolished or reduced by a number of agents, including some non-ionic detergents, sulfhydryl reagents, short-chain alcohols, and fatty acids [15,20–23,33,34,46,49]. With partially purified PDE preparations from bovine adrenal tissue, low concentrations of p-chloromercuribenzoate ($\sim 100\ \mu$M) were reported to stimulate cAMP hydrolysis to the same extent as cGMP [22]. Cyclic GMP stimulation of cAMP hydrolysis was abolished during solubilization of cGMP-stimulated PDE from rabbit brain cortex in the presence of triton X-100, octyl-β-D-glycopyranoside or deoxycholate [34], but not during solubilization from bovine cerebral cortex with lubrol [13]. Non-ionic detergents and short-chain alcohols were also reported to activate partially purified particulate and the soluble form from rat liver [21,23,46], and purified cGMP-stimulated PDE from calf liver [39].

Despite the information available describing the interesting kinetic properties of the cGMP-stimulated PDE, the structural domains responsible for the allosteric and catalytic properties of the cGMP-stimulated PDE have not been delineated, primarily because the routine yields of the protein from most tissues has not been sufficient for extensive structural analysis, given the technologies available until the recent past. Kinetic studies with cyclic nucleotide derivatives [23,31,38,43–47] and methylxanthine inhibitors [41,42,45] have defined some of the structural requirements for the effective interaction of substrates, effectors, and inhibitors, at allosteric and catalytic sites; these studies have thus allowed some inferences as to the topography of these domains.

3.3.3   Probes of the catalytic and allosteric domains

Using a series of cyclic nucleotide derivatives, Erneux and co-workers reported that cGMP was the most potent activator and preferred substrate, but that there was no obvious correlation between the rank order of effectiveness in the ability of various analogs to serve as substrates, effectors and inhibitors [44,46]. These findings supported the idea that regulatory (allosteric) and catalytic sites are distinct. For example, benzimadazole 3',5'-monophosphate (cBIMP), which was not hydrolyzed, was a more potent activator but a less potent inhibitor than purine riboside 3',5'-monophosphate (cPMP), which was a substrate for the cGMP-stimulated PDE [44]. These workers have also reported that the ability of various analogs to activate a purified bovine adrenal PDE and stimulate cAMP hydrolysis correlated with their effectiveness in displacing [³H]cGMP from high-affinity, presumably regulatory, binding sites, i.e. 3'-NH-cGMP > 3'-NH-cAMP ∼ 5'-NH-cGMP > cBIMP > cPMP [31,38]. There was no obvious correlation between their effectiveness in displacing [³H]cGMP and inhibition of cGMP hydrolysis. In these studies 5'-NH-cAMP was essentially inactive in stimulating PDE activity or displacing [³H]cGMP, but was the most effective inhibitor of cGMP hydrolysis. Whereas 3'-NH-cGMP, 3'-NH-cAMP, 5'-NH-cGMP and cBIMP were very effective in displacing [³H]cGMP from binding sites, only 5'-NH-cGMP was a potent inhibitor [31,38]. From their studies with a large number of derivatives, these workers have suggested that essential hydrogen bond interactions between amino acids at the allosteric regulatory site and specific groups of the purine moiety (the C-2 amino group as donor and N-7 as acceptor) may account for the relative effectiveness of cGMP at allosteric or regulatory sites [44,46,47]. The allosteric site is, however, not entirely specific for cGMP, since cAMP can activate the PDE and increase hydrolysis of cIMP and cGMP (Figure 3.6) [23,32], and since cAMP can apparently displace [³H]cGMP from presumed regulatory sites [38]. In addition, Erneux and co-workers have suggested that hydrolysis requires the presence of a negative charge at the cyclic phosphate and an oxygen atom in the equatorial position [43]. The relative specificity for cGMP at catalytic sites was attributed in large part to enhanced binding of cGMP due to greater polarization potential of the guanine base (dipole moment of cGMP > cAMP) towards polarizable amino acid side chains in the binding domain at the catalytic site [43,44].

Erneux and co-workers also successfully utilized a series of cyclic nucleotide analogs to compare the characteristics of the catalytic sites of partially purified preparations of the rat liver cGMP-stimulated and bovine brain CaM-sensitive PDEs [46]. In this study, K_i values of various analogs were compared for inhibition of cAMP hydrolysis by the 'activated' forms of the CaM-sensitive PDE (excess Ca^{2+} and CaM) or cGMP-stimulated PDE (in the presence of 3 μM cGMP or 10% ethanol). With the 'activated' PDEs, double reciprocal plots of kinetics of cAMP hydrolysis were linear in the absence and presence of analogs, suggesting that inhibition by the analogs was related to competitive inhibition at catalytic sites and allowing comparison as to structural determinants for inhibition on the basis of K_i values. To a certain extent, catalytic sites of the CaM-sensitive and cGMP-stimulated PDEs may contain some similar or conserved domains, since the potency for inhibition by several analogs was similar for both PDEs. On the other hand, modifications in the C-6 position in the purine ring suggested major differences in catalytic domains, with 6-chloropurine 3',5'-monophosphate and purine 3',5'-monophosphate ~6–10 times more potent in inhibiting the cGMP-stimulated PDE; 3'-amino-3'-deoxy-cGMP was ~30 times more potent in inhibiting the CaM-sensitive form, whereas 3'-amino-3'-deoxy-cAMP was more potent in inhibiting the cGMP-stimulated PDE [46]. From a comparison of amino acid sequences of several PDEs, Charbonneau et al. reached similar conclusions concerning catalytic domains [8]. They identified a segment of 200–270 amino acid residues containing conserved but non-identical sequences and suggested that these related segments comprised PDE catalytic domains [8].

Thus, different structural determinants of analogs seem to influence catalysis, allosteric activation, and inhibition of the cGMP-stimulated PDE, presumably reflecting different interactions of substrates and effectors (analogs) at allosteric regulatory and catalytic domains and distinct topographical features at these different domains. Allosteric and catalytic sites clearly recognize alterations in the pyrimidine and imidazole portions of the purine ring and ribose and cyclic phosphate moieties. Until the amino acid sequences and three-dimensional structures of allosteric and catalytic domains are determined, however, it may be unwise to emphasize differences between these sites, since, in many instances, except for a few compounds, structural modifications alter in parallel the effectiveness of an analog in serving as substrate and allosteric effector [23,43,44], and since cyclic nucleotide hydrolysis could take place at both catalytic and so-called regulatory sites.

Given the interaction of cyclic nucleotides at both regulatory and catalytic sites, inhibitors capable of perhaps interacting at both sites might also be expected to exhibit complex effects on enzyme activity, as reported by Erneux et al. [45] and Yamamoto et al. [41]. Several competitive inhibitors both stimulated and inhibited cGMP-stimulated PDE activity [41,45]. These effects were dependent on both drug and substrate concentrations. At appropriate inhibitor concentrations, IBMX, dipyridamole and papaverine stimulated cAMP hydrolysis between 0.5 and 2.5 μM [^{3}H]cAMP, with maximal stimulation being about two-fold. At higher cAMP concentrations, these three drugs were inhibitory, with maximum inhibition at 25–50 μM [^{3}H]cAMP, the apparent K_m for cAMP. Inhibition was reversed with higher cAMP concentrations, consistent with competitive-type inhibition [41]. Papaverine, IBMX and dipyridamole reduced the Hill coefficient for cAMP from 1.8 to 1.1–1.2 (without decreasing the apparent K_m for cAMP) and Lineweaver-Burk plots were nearly linear [41,45]. With 0.5 μM [^{3}H]cGMP or 0.5 μM [^{3}H]cAMP plus 1 μM cGMP, i.e. 'activated' enzyme, IBMX, papaverine and dipyridamole did not stimulate but inhibited hydrolytic activity. With increasing concentrations of [^{3}H]cGMP (>0.5 μM) as substrates, the drugs inhibited cGMP hydrolysis, with maximal inhibition at ~3–10 μM [^{3}H]cGMP (near the apparent K_m for cGMP). Inhibition was reversed at higher cGMP concentrations. The Hill coefficient for cGMP was reduced to 1.0 in the presence of the drugs, and Lineweaver-Burk plots for cGMP hydrolysis were linear [41].

With allosteric enzymes, a competitive inhibitor can actually increase catalysis of low concentrations of substrate if the inhibitor mimics substrate in effecting allosteric transitions [48]. In the case of cGMP-stimulated PDE, the stimulatory effect of the three competitive inhibitors was observed at low substrate concentrations and was associated with a decrease in co-operative interactions (reduction in Hill coefficient) [41,45]. At low substrate concentrations, interaction of a competitive inhibitor at catalytic sites is offset by its interactions at regulatory or catalytic sites which could bring about co-operative responses and enhance substrate binding and catalysis. Stimulation of catalysis by the inhibitor suggests that at low substrate concentrations, the inhibitor may be more effective than substrate in inducing allosteric transitions.

From the schematic representation in Figure 3.7, stimulatory effects of competitive inhibitors can be seen at low substrate concentrations, well below the apparent K_m. At these subsaturating concentrations, with regulatory sites partially filled, one cyclic nucleotide can stimulate hydrolysis of the other. Thus,

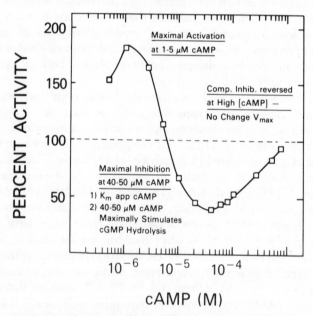

Figure 3.7 Schematic representation of the effects of some competitive-type inhibitors (IBMX, dipyridamole, papaverine) on hydrolysis of [³H]cAMP. Activity was assayed at the indicated concentrations of [³H]cAMP in the absence or presence of the inhibitors. PDE activity at each substrate concentration is presented as [(activity in the presence of inhibitor)/(activity in the absence of inhibitor)] × 100%.

under these conditions the inhibitors mimic substrate, induce co-operative interactions and enhance substrate binding and catalysis. Interaction of the competitive inhibitors at regulatory sites and/or catalytic sites could induce these allosteric transitions. Interaction at regulatory sites must be much less effective and weaker for IBMX than for cGMP, since IBMX (at concentrations as high as 200 μM [13] or 1 mM [31]) did not prevent binding of 40 nM [^{32}P]cGMP [13] or [^{3}H]cGMP [31] to presumed regulatory sites in the bovine brain [13] or bovine adrenal [31] cGMP-stimulated PDEs. At cAMP and cGMP concentrations sufficient to occupy regulatory sites and induce allosteric activation, i.e. 0.5 μM [^{3}H]cAMP + 1 μM cGMP, concentrations > 0.5–1 μM [^{3}H]cGMP or > 5–10 μM [^{3}H]cAMP, the competitive inhibitors were inhibitory [41]. Inhibition was maximal at ~ 5–10 μM cGMP and 20–40 μM cAMP, concentrations which were near the apparent K_m for cGMP and cAMP and which activated the PDE and produced maximal stimulation of 0.5 μM [^{3}H]cAMP and 2–40 nM [^{3}H]-cGMP respectively [41]. At these substrate concentrations, sufficient to 'activate' the PDE, inhibition by the drugs presumably reflected interactions and perturbations at catalytic sites. Further increases in substrate concentrations resulted in competition with the drugs at catalytic sites and in the reversal of their inhibitory effects [41]. Since complex effects of inhibitors can occur at low cyclic nucleotide concentrations, in tissues where the cGMP-stimulated PDE plays an important role in cyclic nucleotide metabolism, utilization of PDE inhibitors such as methylxanthines, papaverine, and dipyridamole could produce either no effect, or effects related to activation or inhibition of PDE activity.

As compared to theophylline (1, 3-dimethylxanthine), IBMX was much more effective in both stimulating and inhibiting hydrolysis of [^{3}H]cAMP [41]. This suggested that the isobutyl side chain, which could interact with a hydrophobic domain on the PDE, might be an important structural determinant in the effects of IBMX on PDE conformation and catalytic activity [41]. Kinetics of cAMP hydrolysis were therefore analyzed in the absence and presence of a number of methylxanthine derivatives according to a two-site competitive model for allosteric enzymes [42,48].

According to this model for allosteric enzymes:

(1)   The cGMP-stimulated PDE exists in either of two conformations which differ in affinity for substrate, i.e. the 'ligand-free' or 'low-affinity' state and the 'high-affinity' or 'activated' state.

(2)   Substrates and inhibitors compete for high- and low-affinity states.

(3)   Random binding of substrates, effectors or inhibitors to the low-affinity state induces allosteric transactions to the high-affinity states.

(4)   All forms of enzyme, enzyme–substrate, enzyme–effector and enzyme–inhibitor complexes are in rapid equilibrium with each other.

Individual binding constants for cAMP, in the absence of inhibitors, were estimated by fitting primary data from measured rates of hydrolysis over a wide range of cAMP concentrations to the rate equation for the two-site competitive model [48]. Equilibrium dissociation constants for cAMP to the high- and low-affinity states were 2.4 ± 8 and $410 \pm 140 \ \mu M$, respectively [42].

Initial rates of cAMP hydrolysis were measured in the presence of a number of methylxanthine derivatives and analyzed according to the two-site competitive model; equilibrium dissociation constants of the various inhibitors to high-affinity (BK_I) and low-affinity (K_I) sites were estimated. Methylxanthines were also classified by graphical analysis of classical inhibition kinetics at saturating cAMP [42]. This approach yielded K_m/K_I ratios which estimated the relative effectiveness of the binding of substrates and inhibitors to the activated or high-affinity state without establishing individual binding constants to high- and low-affinity states [42]. In earlier studies, Egrie and Siegel also utilized this approach, i.e. at substrate concentrations above $100 \ \mu M$, to determine apparent K_I values for cAMP and cGMP [22]. Couchie determined K_I values for a number of cyclic nucleotide analogs by measuring inhibition of cAMP hydrolysis in assays containing 10% ethanol which activated the PDE and 'desensitized' it to further stimulation by cGMP [46]. Thus, with the enzymes activated by either saturation of regulatory sites by substrate or by conformational changes induced in the absence of substrate, competitive inhibitors presumably reduce activity by competing with substrates at catalytic, not regulatory, sites; such data can be analyzed by standard Dixon linear transformations.

Analysis of kinetic data in the presence of methylxanthine inhibitors according to the two-site competitive model allowed estimation of K_s/BK_I ratios for each inhibitor, a direct comparison of equilibrium constants for binding of substrate and inhibitors to the high-affinity state. The rank order of effectiveness of the different inhibitors for binding to the high-affinity state was very similar whether determined from analysis of the data according to the two-site competitive model (K_s/BK_I) or from graphical analyses of inhibition at cAMP concentrations above its apparent K_m (K_m/K_i), i.e. IBMX $>$ 1,3-dipropylxanthine $>$ 1,3-dimethylxanthine (theophylline) $>$ 1,3,7-trimethylxanthine (caffeine) $>$ 1,3,9-trimethylxanthine (isocaffeine) [42].

From estimations of individual binding constants for inhibitors to high- and low-affinity sites, some of the topographical features at these sites were inferred [42]. Several of the derivatives were as effective or more effective than cAMP in binding to the low-affinity state. The presence of short hydrocarbon chain substitutions at the 1- and 3-positions of the purine ring (in IBMX; 1-methyl-3-isopropylxanthine; 1,3-dipropylxanthine; 1-ethyl-3-methylxanthine) promoted binding to the low-affinity state. Substitutions at the N-7 position (1,7-dimethylxanthine; 1,3,7-trimethylxanthine (caffeine); 1,3-dipropyl-7-methylxanthine) reduced binding to the low-affinity state, as did bulky hydrocarbon side chain (hexyl or isoamyl) substitutions at the 1- and 3-positions. No derivative, however, was more effective than cAMP in binding to the high-affinity state. IBMX was nearly as effective, as were 1-methyl-3-isopropylxanthine and 1,3-

dipropylxanthine. Thus structural requirements and determinants were much more stringent for binding to the high-affinity state than to the low-affinity state [42]. These findings suggested that allosteric transitions might alter topography of specific hydrophobic domains at binding sites. In this regard, short-chain alcohols such as ethanol, isopropanol and butanol, but not methanol, were reported to induce allosteric transitions [39]; with increasing concentrations of isopropanol, hydrolysis of low cAMP concentrations was increased with decreased apparent K_m and Hill coefficient for cAMP and no decrease in V_{max} for cAMP [39].

3.4 REGULATION AND PHYSIOLOGICAL ROLE

Little is known concerning the regulation of the cGMP-stimulated PDE or its physiological role in the control of cyclic nucleotide metabolism in intact cells. Several systems have been described, however, which may provide useful information [50–55]; one study in particular suggests an important role for cGMP stimulation of cAMP hydrolysis in the regulation of Ca^{2+} current in frog ventricular tissue [54,55].

Cultured rat hepatoma (HTC) cells contain at least three soluble PDEs analogous to those found in rat liver supernatants, including a cGMP-stimulated PDE. Incubation of HTC cells with dexamethasone for 48–72 h reduced cGMP-stimulated PDE activity with little or no effect on 'low-K_m' cAMP-PDE activity. Incubation with dibutyryl cAMP increased cAMP-PDE activity with little or no effect on cGMP-stimulated PDE activity [50]. Another cell line, MDCK$_T$ (Maden Darby Canine Kidney cells transformed by Harvey Murine Sarcoma virus) contains several PDEs, including cGMP-stimulated PDE. Incubation of MDCK$_T$ cells with butyrate, which induces several differentiated functions in these cells [52], increased both soluble and particulate cGMP-stimulated PDE activity [51]. Thus, these two cell lines (HTC and MDCK$_T$) may be especially useful in studying the regulation of both soluble and particulate forms of cGMP-stimulated PDE.

In examining various tissues, Beavo and co-workers demonstrated that adrenal glandular tissue contains the highest amount of cGMP-stimulated PDE activity; in the gland itself, virtually all the cGMP-stimulated PDE is confined to the outer 1–2 mm layer of the adrenal gland [53]. This remarkable distribution suggests an important role for the cGMP-stimulated PDE in zona glomerulosa function, i.e. aldosterone production, adrenal cell growth, etc. [53]. These workers have suggested that agents which might increase cGMP in the zona glomerulosa, i.e. atrial natriuretic factor, might bring about some of their physiological responses by increasing cAMP hydrolysis and inhibiting cAMP-mediated processes.

Perhaps the most interesting system to date which suggests a physiological role for the cGMP-stimulated PDE has been described by Hartzell and Fischmeister

[54]. In their patch clamp studies of slow inward Ca^{2+} current in frog ventricle, they demonstrated that cGMP inhibited slow inward Ca^{2+} current induced by agents which increased cAMP. Isolated frog cardiac myocytes contained cGMP-stimulated PDE activity, most of which was associated with particulate (20 000 g, 10 min) fractions [53]. The dose–response curves for the effects of cGMP on Ca^{2+} current and for activation of the cGMP-stimulated PDE were quite similar [55]. The effects of cGMP on Ca^{2+} current were blocked by PDE inhibitors, especially IBMX, and were not reproduced by 8 Br cGMP, which does not activate PDE but is a potent activator of cGMP-dependent protein kinase [54,55]. In view of these findings, Hartzell and associates have suggested that in frog myocardium, agents which increase cGMP content, by activating the cGMP-stimulated PDE, reduce cAMP and thus inhibit cAMP-induced slow Ca^{2+} inward current [54,55].

It has also been suggested that activation of this enzyme might provide a route through which insulin lowers intracellular cAMP concentrations in hepatocytes (see Chapter 8) and that such an effect could be achieved by a soluble 'mediator' [33].

3.5 CONCLUSIONS

Of the major classes of PDEs, the cGMP-stimulated PDE is unique in exhibiting positively co-operative kinetics for hydrolysis of both cAMP and cGMP [1–7]. By virtue of these positive homotropic responses for cAMP and cGMP, one cyclic nucleotide can stimulate hydrolysis of the other; cGMP is, however, preferred as substrate and effector. At physiological temperatures, the cGMP-stimulated PDE may exist in a 'low-affinity' state [39], with apparent K_m for both cAMP and cGMP well above presumed cellular concentrations of either cAMP (in the micromolar range) or cGMP (usually at least 10-fold less than cAMP) [24–29]. Under these conditions, where substrate concentrations are below saturation and do not fully occupy regulatory sites on the PDE, sigmoidal responses in substrate–velocity relationships exhibited by the cGMP-stimulated PDE could provide a sensitive, 'on–off' rapid regulation of cyclic nucleotide hydrolysis in situations where either cAMP or cGMP is increased. In addition, and perhaps more important, although the apparent K_m of the PDE for cGMP is about two orders of magnitude above concentrations of cGMP in many tissues, cGMP at 10^{-7} M (within the physiological range) can stimulate hydrolysis of physiological concentrations of cAMP (in the micromolar range). Thus under physiological conditions in the presence of the cGMP-stimulated PDE, cGMP may very likely function as an allosteric regulator of cAMP degradation. Activation of the cGMP-stimulated PDE by cGMP could thus provide a molecular basis for the opposing actions of cAMP and cGMP described in some biological systems [53–55]. It is not surprising that the allosteric properties and catalytic activity of the cGMP-stimulated PDE can be regulated by a variety of circumstances and conditions that alter the cellular macro- as well as microenvironment, e.g. hormonal balance [50,51], cellular differentiation [51,53], pH [40] and temperature [39], as well as by potential physiological and therapeutic effectors, substrates [23,43–47], fatty acids [33,49], short-chain alcohols [21,23,31,37], sulfhydryl agents [15,21,22], competitive inhibitors [38,41,45], etc. Cellular distribution and subcellular localization may also be important. The cGMP-stimulated PDE seems to be highly concentrated in the outermost cells of the bovine adrenal zona glomerulosa [53], and in association with particulate fractions in rabbit [34] and bovine cerebral cortex [13] and perhaps bovine brain coated vesicles [56]. As the distinctive characteristics of the cGMP-stimulated PDEs and the other major PDE types become more appreciated and understood, future challenges to researchers will involve understanding those mechanisms for integrating the regulation and functions of individual PDEs in intact cells and delineating their roles in the metabolism of intracellular cyclic nucleotides in general and in control or functional compartmentalization of discrete processes mediated via cAMP and cGMP.

Powerful molecular and cellular biological approaches should soon answer some of the interesting questions related to the cGMP-stimulated PDE. Understanding the protein amino acid sequence and nucleotide sequences at the level of the gene will begin to shed light on potential relationships between the cGMP-stimulated PDE and other PDEs, their appearance during development, regulation by hormones, cellular distribution, etc. Site-directed mutagenesis coupled with efficient expression/transfection systems will allow identification of allosteric and catalytic domains, and help in understanding molecular mechanisms for allosteric regulation of the PDE in the test tube as well as intact cells. Computer-assisted modeling techniques will allow synthesis of appropriate drug and peptide inhibitors (and effectors), and, coupled with sophisticated delivery and cell targeting systems, may make it possible to focus, for both research and therapeutic purposes, on the functions of the cGMP-stimulated PDE in individual intact cells.

REFERENCES

1. Appleman, M. M., Thompson, W. J., and Russell, T. R. (1973) *Adv. Cyclic Nucleotide Res.*, **3**, 66–98.
2. Wells, J. N., and Hardman, J. (1977) *Adv. Cyclic Nucleotide Res.*, **8**, 119–143.
3. Strada, S., and Thompson, W. J. (1978) *Adv. Cyclic Nucleotide Res.*, **9**, 265–283.
4. Vaughan, M., Danello, M. A., Manganiello, V. C., and Strewler, G. J. (1981) *Adv. Cyclic Nucleotide Res.*, **14**, 263–271.
5. Beavo, J. A., Hansen, R. S., Harrison, S. A., Hurwitz, R. L., Martino, T., and Mumby, M. D. (1982) *Mol. Cell. Endocrinol.*, **28**, 387–410.

6. Manganiello, V. C., Yamamoto, T., Elks, M., Lin, M. C., and Vaughan, M. (1984) *Adv. Cyclic Nucleotide Protein Phosphorylation Res.*, **16**, 291–301.
7. Beavo, J. A. (1988) *Adv. Second Messenger Phosphoprot. Res.*, **22**, 1–39.
8. Charbonneau, H., Beier, N., Walsh, K. A., and Beavo, J. (1986) *Proc. Natl Acad. Sci. USA,* **83**, 9308–9312.
9. Sharma, R. K., and Wang, J. H. (1985) *Proc. Natl Acad. Sci. USA,* **82**, 2603–2607.
10. Sharma, R. K., and Wang, J. H. (1986) *J. Biol. Chem.,* **261**, 14160–14166.
11. Geremia, R., Rossi, P., Mocini, D., Pezzotti, R., and Conti, M. (1984) *Biochem. J.,* **217**, 693–670.
12. Rossi, P., Georgi, M., Geremia, R., and Kincaid, R. (1988) *J. Biol. Chem.,* **263**, 15521–15527.
13. Murashima, S., Tanaka, T., Hockman, S., and Manganiello, V. C. (1989) *Biochemistry* (in press).
14. Beavo, J. A., Hardman, J. G., and Sutherland, E. W. (1970) *J. Biol. Chem.,* **245**, 5649–5655.
15. Beavo, J. A., Hardman, J. G., and Sutherland, E. W. (1971) *J. Biol. Chem.,* **246**, 3841–3846.
16. Franks, D., and MacManus, J. B. (1971) *Biochem. Biophys. Res. Commun.,* **42**, 844–849.
17. Manganiello, V. C., and Vaughan, M. (1972) *Proc. Natl Acad. Sci. USA,* **69**, 269–273.
18. Klotz, W., and Stock, K. (1972) *Naunyn-Schmeidebergs Arch. Pharmacol.,* **274**, 54–62.
19. Hidaka, H., and Asano, T. (1976) *Biochim. Biophys. Acta,* **429**, 485–497.
20. Russell, T. R., Terasaki, W. L., and Appleman, M. M. (1973) *J. Biol. Chem.,* **248**, 1334–1340.
21. Terasaki, W. L., and Appleman, M. M. (1975) *Metabolism,* **24**, 311–319.
22. Egrie, J. C., and Siegel, F. L. (1977) *Biochim. Biophys. Acta,* **483**, 348–366.
23. Moss, J., Manganiello, V. C., and Vaughan, M. (1977) *J. Biol. Chem.,* **252**, 5211–5215.
24. Gilman, A. G. (1972) *Adv. Cyclic Nucleotide Res.,* **2**, 9–25.
25. Kuo, J.-F., and Greengard, P. (1972) *Adv. Cyclic Nucleotide Res.,* **2**, 41–50.
26. Steiner, A. L., Wehmann, R. E., Parker, C. W., and Kipnis, D. M. (1972) *Adv. Cyclic Nucleotide Res.,* **2**, 51–63.
27. Goldberg, N. D., O'Dea, R., and Haddox, M. K. (1973) *Adv. Cyclic Nucleotide Res.,* **3**, 155–223.
28. Goldberg, N. D., and Haddox, M. K. (1977) *Annu. Rev. Biochem.,* **46**, 823–846.
29. Murad, F., Arnold, W. P., Mittal, C. K., and Braughler, J. M. (1979) *Adv. Cyclic Nucleotide Res.,* **11**, 176–204.
30. Martins, T. J., Mumby, M. C., and Beavo, J. A. (1982) *J. Biol. Chem.,* **257**, 1973–1979.
31. Miot, F., Van Haastert, P., Erneux, C. (1985) *Eur. J. Biochem.,* **149**, 59–65.
32. Yamamoto, T., Manganiello, V. C., and Vaughan, M. (1983) *J. Biol. Chem.,* **258**, 12526–12533.
33. Pyne, N. J., Cooper, M. E., and Houslay, M. D. (1986) *Biochem. J.,* **235**, 325–334.
34. Whalin, M. E., Strada, S. J., and Thompson, W. J. (1988) *Biochim. Biophys. Acta,* **972**, 79–94.
35. Hurwitz, R., Hansen, R. S., Harrison, S. A., Martins, T. J., Mumby, M. C., and Beavo, J. A. (1984) *Adv. Cyclic Nucleotide Res. Protein Phosphorylation Res.,* **16**, 89–106.
36. Mumby, M. C., Martins, T. J., Chang, M. L., and Beavo, J. A. (1982) *J. Biol. Chem.,* **257**, 13283–13290.
37. Pyne, N. J., and Houslay, M. D. (1988) *Biochem. Biophys. Res. Commun.,* **156**, 290–296.

38. Erneux, C., Miot, F., Van Haastert, P. J. M., and Jastorff, B. (1985) *J. Cyclic Nucleotide Protein Phosphorylation Res.*, **10**, 463–472.
39. Wada, H., Osborne, J. C. Jr, and Manganiello, V. C. (1987) *J. Biol. Chem.*, **262**, 5139–5144.
40. Wada, H., Osborne, J. C. Jr, and Manganiello, V. C. (1987) *Biochemistry*, **26**, 6565–6570.
41. Yamamoto, T., Yamamoto, S., Osborne, J. C. Jr, and Manganiello, V. C. (1983) *J. Biol. Chem.*, **258**, 14173–14177.
42. Wada, H., Manganiello, V. C., and Osborne, J. C. Jr (1987) *J. Biol. Chem.*, **262**, 13938–13945.
43. Braumann, T., Erneux, C., Petrides, G., Stohrer, W.-D., and Jastorff, B. (1986) *Biochim. Biophys. Acta*, **871,** 199–206.
44. Erneux, C., Couchie, D., Dumont, J., Baraniak, J., Stek, W. J., Garcia-Abbod, E., Perides, G., and Jastorff, B. (1981) *Eur. J. Biochem.*, **115**, 503–510.
45. Erneux, C., Miot, F., Boeynaems, J., and Dumont, J. E. (1982) *FEBS Lett.*, **142**, 251–254.
46. Couchie, D., Petrides, G., Jastorff, B., and Erneux, C. (1983) *Eur. J. Biochem.*, **136**, 571–575.
47. Erneux, C., Couchie, D., Dumont, J. E., and Jastorff, B. (1984) *Adv. Cyclic Nucleotide Protein Phosphorylation Res.*, **16**, 107–118.
48. Segal, I. (1975) *Enzyme Kinetics*, Wiley Interscience, New York, pp. 385–387.
49. Yamamoto, T., Yamamoto, S., Manganiello, V. C., and Vaughan, M. (1984) *Arch. Biochem. Biophys.*, **229**, 81–89.
50. Ross, P. S., Manganiello, V. C., and Vaughan, M. (1977) *J. Biol. Chem.*, **252**, 1448–1452.
51. Manganiello, V. C., Yamamoto, T., Elks, M., Lin, M. C., and Vaughan, M. (1984) *Adv. Cyclic Nucleotide Protein Phosphorylation Res.*, **16**, 291–301.
52. Lin, M. C., Koh, S. L., Dykman, D., and Shih, T. (1982) *Exp. Cell Res.*, **142**, 181–189.
53. MacFarland, R. T., Zeles, B. D., Novack, J. P., and Beavo, J. A. (1987) *Adv. Cyclic Nucleotide Calcium Protein Phosphorylation Res.*, **23**.
54. Hartzell, H. C., and Fischmeister, R. (1986) *Nature*, **323**, 273–275.
55. Simmons, M. A., and Hartzell, H. C. (1988) *Mol. Pharmacol.*, **33**, 664–671.
56. Silva, W. I., Schook, W., Mittag, T. M., and Puszkin, S. (1986) *J. Neurochem.*, **46**, 1263–1271.
57. Cleveland, D. W., Fischer, S. G., Kirschner, M. W., and Laemmli, U. K. (1977) *J. Biol. Chem.*, **252**, 1102–1106.

4

CYCLIC GMP-INHIBITED CYCLIC NUCLEOTIDE PHOSPHODIESTERASES

Vincent C. Manganiello,* Carolyn J. Smith,* Eva Degerman† and Per Belfrage†

*Section of Biochemical Physiology, NHLBI, NIH, Bethesda, MD 20892, USA
†Department of Medical and Physiological Chemistry, University of Lund, Lund, Sweden

4.1 HISTORICAL PERSPECTIVES

Sutherland and co-workers initially described cyclic nucleotide hydrolytic activity in tissue extracts [1], and Butcher and Sutherland partially purified what was thought to possibly be a ubiquitous enzyme responsible for the specific hydrolysis of the 3'-bond of cyclic nucleotides [2]. In the early 1970s, however, especially due to the work of Appleman and his associates, it became apparent that multiple forms of cyclic nucleotide phosphodiesterases (PDEs) could be isolated from various tissues [3–5]. We now recognize that PDEs constitute a complex group of enzymes, multiple forms of which are found in various amounts and proportions in most mammalian cells. PDEs are found in cytoplasmic fractions as well as in association with cellular membranes; these enzymes differ in their physiochemical properties, substrate affinities, kinetic characteristics, responsiveness to various effectors and pharmacologic agents (especially PDE inhibitors) and mechanisms of regulation [6–9]. At least six or seven distinct major classes of PDEs have been isolated, purified, and characterized. While details are unfolding as to the evolutionary and molecular relationships between the major classes at the level of protein structure and gene structure and regulation, experimental evidence supports the idea that each major PDE class, i.e. Ca^{2+}- and calmodulin (CaM)-sensitive PDE, cGMP-stimulated PDE, photoreceptor cGMP-PDE, cGMP-inhibited PDE etc., represents a family of isoenzymes [10]. For example, at least four or five distinct Ca^{2+}- and CaM-sensitive and two cGMP-stimulated PDEs have been identified and characterized [10–15].

With increasing appreciation of the multiple and complex molecular and cellular mechanisms for regulation of the different PDEs, several important

Cyclic Nucleotide Phosphodiesterases: Structure, Regulation and Drug Action
Edited by J. Beavo and M. D. Houslay © 1990 John Wiley and Sons Ltd

questions have begun to come into focus regarding mechanisms and functions for subcellular localization and cellular distribution of different PDEs, the roles of specific PDEs in regulation of signal transduction and of specific processes mediated by cyclic nucleotides, and in the integration of the multiple PDE activities in metabolic control within a single cell. Much of the work described in this chapter will be directed towards some of these questions, with reference to a specific PDE class, the so-called 'cGMP-inhibited (cGI) PDEs'.

4.2 CYCLIC GMP-INHIBITED PDEs IN ADIPOCYTES AND OTHER CELL TYPES

4.2.1 Introduction

Based on their order of elution from DEAE anion exchange columns, PDEs were initially classified as three major kinetic types, one of which exhibited a high affinity for cAMP and was designated as the 'low-K_m' or type III cAMP-PDE [5]. In calf liver supernatant, the type III cAMP-PDE was later demonstrated to consist of two distinct 'low-K_m' cAMP-PDEs, each with high affinity for cAMP, but with distinct physical properties, kinetic characteristics, and inhibitor specificities [16]. One was very sensitive to inhibition by cGMP and cilostamide, the other to Ro 20-1724. Enzymes analogous to these two classes have been described in a number of tissues; in some instances they are apparently associated with different intracellular compartments and seem to participate in regulation of specific cellular functions, e.g. lipolysis [16–21], myocardial contractility [22–39], platelet aggregation [40–49], and desensitization to follicle stimulating hormone [50–52].

As will be discussed below, cGI type III cAMP-PDEs (in the following referred to as cGI-PDE) have been purified from rat and bovine adipose tissues [53,54], bovine cardiac ventricle [55], human platelets [56], rat liver [57,58], and, more recently, bovine aorta smooth muscle [59]. These PDEs exhibit very characteristic and similar kinetic properties—high affinity for both cAMP and cGMP (apparent $K_m^{app} < 1\ \mu M$) as substrates, V_{max} for cAMP greater (4–10-fold) than for cGMP, potent inhibition of cAMP hydrolysis by cGMP ($K_i = K_m$), but relative insensitivity to inhibition by Ro 20-1724 and rolipram. cGI-PDEs are very sensitive to inhibition by a number of inotropic and antithrombotic agents, including cilostamide and other OPC (Otsuka Pharmaceutical Company) derivatives, imazodan, milrinone, fenoxamine, LY 195115, Y-590, anagrelide, RS 82056 [10,60–62]. Whereas inhibition of cGI-PDEs by therapeutically effective concentrations of these drugs correlates with physiological effects, it is not known if inhibition by endogenously produced cGMP is of physiological consequence and relevance.

In intact adipocytes, platelets, and hepatocytes, incubation with hormones or agents that elevate cAMP increase the activity of cGI-PDEs [19–21,46–49,

63–68]; cAMP-induced activation of cGI-PDEs may be involved in termination of the cAMP signal generated by activation of adenylate cyclase and consequently in 'feedback' regulation of cAMP content [19–21,66]. In adipocytes, insulin also increases cGI-PDE activity [19–21,63,64,66,67,69–72]; insulin-induced activation of the adipocyte cGI-PDE seems to be important in the antilipolytic action of insulin [17–21]. Hepatic cAMP-PDEs are also activated by insulin [65,68] as is discussed in detail in Chapter 8.

The apparent subcellular localization of cGI-PDEs differs in various cell types. Whereas most of the platelet cGI-PDE is cytosolic [46–49], in adipocytes and hepatocytes most of the cGI-PDE is isolated in association with particulate (centrifuged at 100 000 g) fractions [63,64,67,68,70,72]. The precise membranous organelle in the adipocyte, however, remains to be established [73,74].

Since other contributors to this volume will consider cGI-PDEs in hepatic and cardiovascular tissues, much of the work discussed in this chapter will primarily focus on the characterization of the adipocyte cGI-PDE and its importance in the regulation of lipolysis and the antilipolytic action of insulin.

4.2.2 Purification of the adipocyte cGI-PDE and other cGI-PDEs

Loten and Sneyd first reported that incubation of intact rat adipocytes with insulin increased 'low-K_m' cAMP-PDE activity (assayed at $< \mu$M cAMP) in homogenates [69]. This increment in PDE activity induced by insulin (as well as by lipolytic hormones and cAMP analogs) was found to be primarily confined to a particulate fraction (100 000 g) with little or no increase in cytosolic cAMP-PDE activity [63,64,70,72], and was inhibited by cGMP [75]. In addition, in 3T3-L1 cells, the hormone-responsive particulate cGI-PDE appeared during conversion (differentiation) of 3T3-L1 fibroblasts into adipocytes [76]. 3T3-L1 and rat adipocyte cGI-PDEs were also found to be potently and selectively inhibited by cilostamide and other OPC derivatives and several cardiotonic inotropic agents including milrinone, CI-914 and CI-930 [19,53,54,77].

Although the cellular regulation of the adipocyte cGI-PDE and some of its biochemical and pharmacologic properties had thus been studied extensively [63,78–82], it proved difficult to obtain the enzyme in a pure form. For the development of efficient purification procedures, the responsiveness to the selective inhibitors was very important. Since it had been reported that these agents were not potent inhibitors of other PDEs [10,60,77], the N-(2-isothiocyanato) ethyl derivative of cilostamide (CIT) was synthesized and coupled to aminoethyl–agarose for use as an affinity ligand to purify the cGI-PDE from rat and bovine adipose tissue [53,54], and more recently from human platelets and rat and bovine aorta (unpublished data).

Adipocyte cGI-PDE was first purified from particulate fractions (100 000 g) prepared from a large number of rat epididymal fat pads (Table 4.1). The initial

Table 4.1 Purification of cGI-PDE from adipose tissue

Purification	Total protein (mg)	Total activity[b] (nmol/min)	Specific activity[b] ((nmol/min)/mg)	Yield (%)	Fold-purification
(1) 100 000 g pellet	720	59.0	0.082	100	1
(2) Solubilized enzyme	391	98.5	0.25	167	3
(3) DEAE–Sephacel	22	42.1	1.9	71	23
(4) Concentration and Sephadex G-200	0.9	17.0	18.9	29	230
(5) Dialysis, CIT–agarose affinity chromatography and removal of cAMP by DEAE–Sephacel	0.0017	9.1	5353	18	65 280

[a] 100 000 g pellets were prepared from the epididymal fat pads (about 475 g) of 432 rats.

[b] Substrate concentration, 0.5 M [^{3}H]cAMP.

Reproduced with permission from *J. Biol. Chem.* [53].

high-speed centrifugation separated cytosolic PDEs (the Ca^{2+}- and calmodulin-sensitive and the cGMP-stimulated) from the cGI-PDE which constituted $> 90\%$ of the particulate PDE activity. The cGI-PDE exhibited the properties typical for a peripheral membrane protein and was released from membranes by chaotropic salts with and without sonication and by low concentrations of non-ionic detergent [53]. Part (20–50%) of the cGI-PDE was also released during homogenization, depending on buffer ionic strength and homogenization conditions (unpublished observations). A combination of a moderate salt concentration and sonication, low concentrations of non-ionic detergent, and 20% glycerol was found to solubilize with maximal yield and to stabilize cGI-PDE activity. The enzyme was then partially purified by two conventional fractionation steps, DEAE–Sephacel chromatography and gel filtration on Sephadex G-200.

Final purification (65 000-fold relative to the initial high-speed pellet fraction) was achieved through the use of the CIT–agarose affinity chromatography step. An apparently homogeneous 64 kDa protein (SDS-PAGE), subsequently found to be a 62/66 kDa doublet, was obtained at about 20% overall yield of enzyme activity.

Because of the low tissue abundance (0.001% of tissue proteins) of cGI-PDE in rat adipose tissue, bovine omental fat was chosen as a more easily available and less expensive source for the preparation of larger amounts of the enzyme [54]. Since the large volumes of homogenate obtained from up to 100 kg of tissue could not be readily centrifuged at high speed, the enzyme was initially enriched through isoelectric precipitation at pH 5.0. The subsequent purification required two successive DEAE–Sephacel ion exchanges and a Sephadex G-200 gel chromatography step, before the final affinity chromatography on CIT–agarose. In this scaled-up procedure, about 30 μg of enzyme protein, apparently composed of a 77 kDa and a 62 kDa polypeptide (the latter subsequently found to be a 61/63 doublet, see below), was obtained from each 25 kg batch of omental fat, at 16% total yield. More recently CIT–agarose chromatography has been used in a one-step purification procedure for the human platelet cGI-PDE (unpublished) and a two-step procedure for the enzyme from bovine aortic smooth muscle [59]. The resulting products were proteins in the 105, 80, 60, 50 and 30 kDa ranges (SDS-PAGE) (see Table 4.2, below).

cGI-PDEs have also been identified and purified from bovine cardiac tissue, human platelets, and rat liver, with methods not employing CIT–agarose affinity chromatography. The platelet enzyme was isolated by DEAE–cellulose chromatography followed by adsorption to blue dextran–Sepharose, yielding a 61 kDa protein (SDS-PAGE) [56]. A purification procedure including chromatography on DEAE–cellulose, cAMP–Sepharose and cGMP–Sepharose to remove contaminating cyclic nucleotide binding proteins, blue dextran–Sepharose and HPLC–DEAE was utilized to purify the bovine cardiac cGI-PDE [55]. The final product consisted of three components, a major 80 kDa and two other, 60 and 67 kDa, proteins (SDS-PAGE). As for the platelet and cardiac enzymes, blue dextran–Sepharose was also used as an important step to purify a rat liver cGI-PDE, the so-called 'dense-vesicle' enzyme [58]. Alternatively, guanine–Sepharose was used together with hydrophobic chromatography on aminopentyl–agarose as final purification steps for this liver enzyme [58]. In both cases the final product was a 57 kDa polypeptide by SDS-PAGE (for details see Chapter 8). More recently, Boyes and Loten [57] purified a cGI-PDE from rat liver which differed in some aspects from the 57 kDa product. Mild chymotrypsin treatment of rat liver particulate fractions was used to solubilize a catalytically active fragment of the enzyme which was thereafter extensively purified using cellulose phosphate, Ecteola–cellulose, $(NH_4)_2SO_4$ precipitation, hydroxyapatite, theophylline–Sepharose and finally DEAE–HPLC. The purified cGI-PDE was identified with a 73 kDa protein (SDS-PAGE).

As will be discussed below, the polypeptides obtained after purification seem to originate from larger native forms, which are proteolytically nicked during purification due to the exquisite sensitivity of this enzyme to proteolysis.

4.2.3 Antibodies

Several batches of pure cGI-PDE from bovine adipose tissue were pooled and used to produce polyclonal rabbit antibodies which inactivated, immunoprecipitated, and crossreacted on Western immunoblots with the rat adipose tissue enzyme [54] (see below). However, on Western blots the antibodies did not recognize the CIT–agarose purified cGI-PDEs from human platelets or bovine aortic smooth muscle (see below), even if they could be used to immunoprecipitate these enzyme preparations. The monoclonal antibody, cGI-2, raised by Beavo and associates against the bovine cardiac cGI-PDE [55], recognized on Western blots the CIT–agarose-purified cGI-PDEs from bovine adipose tissue [54], human platelets (unpublished) and aortic smooth muscle [59], but not the rat adipose tissue enzyme [54]. The use of the antibodies to quickly isolate and identify native cGI-PDE in extracts from isolated rat adipocytes is described below.

4.2.4 Molecular size and proteolytic nicking during purification

The pure bovine and rat adipose tissue cGI-PDEs exhibited single protein bands which co-migrated with enzyme activity during nondenaturing gel electrophoresis, and appeared as single peaks of enzyme activity with apparent M_r of about 110 000 on Sephadex G-200 chromatography [53,54]. However, Western blot analysis showed that the purified rat cGI-PDE was a 62/66 kDa doublet whereas the bovine enzyme consisted of a 61/63 kDa doublet and a 77 kDa polypeptide (Table 4.2). In contrast, as will be discussed below, cGI-PDE, rapidly partially

Table 4.2. Molecular sizes of cGI-PDEs from various tissues (kDa)

	Adipose tissue		Smooth muscle, bovine aortic[c]	Platelets, human[d]	Platelets, human/ bovine[e]	Heart, bovine[f]	Liver	
	Rat[a]	Bovine[b]					Rat[g]	Rat[h]
Apparent 'native' enzyme—rapid isolation/SDS-PAGE/Western blot	135	ND	105	105	110	110		63
Purified enzyme—gel chromatography	110	110	ND	ND	140	230	130	112
Purified enzyme after SDS-PAGE/Western blot (major polypeptides)	66/62	77, 63/61	77/74, 53, 30	83, 60	110, 61	80, 67, 60	73[i]	57

Rat and bovine adipose tissue, bovine aortic smooth muscle and human platelet enzymes were affinity purified with CIT-agarose.
[a] From ref. 53; [b] from ref. 54; [c] from ref. 59; [d] E. Degerman, unpublished; [e] from refs. 46, 48, 49, 56, 83; [f] from ref. 55; [g] from ref. 57; [h] from refs. 58, 85; [i] Proteolytic fragment obtained after proteolysis/solubilization with chymotrypsin.

purified from rat fat cell extracts, was identified as a 135 kDa protein species by Western immunoblots and after ^{32}P phosphorylation in intact fat cells (see below, Figure 4.2), and was accompanied by minor and more variable 90 kDa and 44 kDa components [84]. These findings demonstrated that during large-scale purification the native (135 kDa protein) adipose tissue cGI-PDE was proteo-lytically nicked but retained catalytic activity. Comparable results have been obtained for analogous cGI-PDEs from other tissues (Table 4.2). Antibodies towards the 61 kDa (SDS-PAGE) cGI-PDE purified from human platelets recognized a 110 kDa platelet protein with a faintly staining 61 kDa protein on Western immunoblots [48]. Also, antibodies raised against the bovine cardiac cGI-PDE polypeptides (80, 67 and 60 kDa) identified a single larger 110 kDa form of the enzyme in fresh extracts of bovine cardiac muscle and human and bovine platelet lysates [46,49,55]. When the bovine platelet lysate was incubated at room temperature for 3–6 h prior to SDS-PAGE, the enzyme was proteolyzed to 80 kDa and 60 kDa peptides [46].

Antibodies towards the 57 kDa rat liver 'dense-vesicle' cGI-PDE recognized a 63 kDa polypeptide from fresh liver homogenate which was proposed to be the native form of the enzyme [85]. However, the fact that chymotryptic solu-bilization of a rat liver cGI-PDE resulted in a 73 kDa product after ex-tensive purification [57] suggests that a larger native form exists also in this tissue.

4.2.5   Molecular and catalytic properties

As discussed above and summarized in Table 4.2, the cGI-PDEs from all tissues so far examined appear to be polypeptides in the 105–135 kDa range which are extensively proteolyzed to produce 30–80 kDa fragments (SDS-PAGE), the only exception possibly being the rat liver 'dense-vesicle' enzyme [58,85]. Gel chromatography of purified adipose tissue cGI-PDE preparations give apparent molecular sizes in the 100–140 kDa range, indicating that the enzyme may appear as a monomer in its native state. The bovine cardiac enzyme had an apparent M_r of 230 000 by gel chromatography [10] and may represent a dimer, as possibly also does the rat liver 'dense-vesicle' enzyme [58].

As shown in Table 4.3, all enzyme preparations (except the rat liver 'dense-vesicle' enzyme) exhibited linear Michaelis–Menten kinetics, with K_m for cAMP and cGMP in the 0.1–0.8 μM range in most cases, and V_{max} for cAMP between 2 and 9 (μmol/min)/mg and for cGMP between 0.3 and 2 (μmol/min)/mg. The inhibitory properties were also similar for the enzymes from the different tissues, with IC_{50} generally below 0.1 μM for OPC 3911 or cilostamide, and about 0.5 μM for milrinone and CI-930, whereas Ro 20-1724 was not an effective inhibitor ($IC_{50} > 25$ μM). As expected, cGMP induced a competitive inhibition with an IC_{50} in the 0.1–0.6 μM range, i.e. the same as the K_m for cGMP.

Table 4.3. Catalytic properties of and inhibitors for cGI-PDEs from various tissues

	Adipose tissue		Smooth muscle, bovine aortic[c]	Platelets, human[d]	Platelets, bovine/human[e]	Heart, bovine[f]	Liver	
	Rat[a]	Bovine[b]					Rat[g]	Rat[h]
Catalytic properties								
K_m (μM)								
cAMP	0.4	0.3	0.2	0.1	0.2	0.2	0.2	0.3; 28
cGMP	0.3	0.8	0.1	0.03	0.02	0.1	0.2	10
V_{max} ((μmol/min)/mg)								
cAMP	8.5	2.5	3.1	2.0	3.0	6.0	6.2	0.11; 0.63
cGMP	2.0	1.6	0.3	0.3	0.3	0.6	2.1	0.004
Inhibitors								
IC_{50} (μM)								
OPC-3911 or cilostamide	0.04	0.06	0.05	0.09	0.04	0.01[i]	—	—
Milrinone	0.6	2.2	0.4	0.7	0.5	0.3[i]	—	1
CI 930	0.4	0.6	0.3	0.2	—	—	—	—
Ro 20-1724	190	1100	>30	25	220	62[i]	—	50
cGMP	0.2	0.6	0.3	0.4	0.1	0.1[i]	0.2[i]	2

Rat and bovine adipose tissue, bovine aortic smooth muscle and human platelet enzymes were affinity purified with CIT–agarose.
[a] From ref. 53; [b] from ref. 54; [c] from ref. 59; [d] E. Degerman, unpublished; [e] from refs. 46, 56; [f] from ref. 55; [g] from ref. 57; [h] from refs. 58, 85; [i] K_i values.

4.3 SHORT-TERM REGULATION OF THE ADIPOCYTE cGI-PDE AND ITS IMPORTANCE IN THE ANTILIPOLYTIC ACTION OF INSULIN

4.3.1 Hormonal control of adipose tissue lipolysis

Cyclic AMP is the important intracellular second messenger mediating the hormonal regulation of triacylglycerol hydrolysis. In adipocytes, lipolytic hormones such as catecholamines, ACTH and glucagon interact with specific cell surface receptors and activate adenylate cyclase, leading to increases in cAMP and activation of cAMP-dependent protein kinase (cAMP-PrK) [86–90]. The 'activated' protein kinase in turn phosphorylates the hormone-sensitive lipase on a specific serine residue (Ser-563) in a regulatory domain [91,92] and thereby increases its activity, leading to the hydrolysis of stored triacylglycerol, with release of glycerol and free fatty acid. This sequence of events involved in activation of hormone-sensitive lipase can be completely or partially prevented by a number of antilipolytic agents, including insulin. The antilipolytic agents adenosine and prostaglandin E_1 interact with specific cell surface receptors and transmit inhibitory signals to the catalytic unit of adenylate cyclase via inhibitory guanyl nucleotide binding (G) proteins [90,93], causing inhibition of adenylate cyclase and of cAMP formation. Evidence suggests that rat adipocytes are subject to tonic inhibitory regulation by endogenously produced adenosine, since lipolysis can be activated by either removal of endogenously produced adenosine by adenosine deaminase (ADA) [89,94] or inactivation of the inhibitory G-proteins by pertussis toxin [95].

Insulin is a physiologically potent and important inhibitor of lipolysis [86–88, 91,96]. Although the precise mechanism(s) for its antilipolytic action is unknown, insulin has been reported to decrease rat adipocyte cAMP content [97], and its antilipolytic action can be related to a reduction in hormone-activated cAMP-PrK [98], which presumably reflects a decrease in physiologically relevant cell cAMP content [99]. This decrease in cAMP could result from either inhibition of adenylate cyclase and/or activation of cAMP-PDE. Whereas insulin inhibition of adenylate cyclase has been reported by some workers [100,101] but not by others [102,103] and is somewhat controversial, a number of investigators have demonstrated that activation of the cGI-PDE in 3T3-L1, rat, and human adipocytes may play an important part in the antilipolytic action of insulin [17–21].

4.3.2 Activation of cGI-PDE by agents which increase cAMP

In addition to insulin, incubation of intact adipocytes with lipolytic hormones which activate adenylate cyclase, such as catecholamines, glucagon, and ACTH, as well as cAMP analogs which activate cAMP-PrK and lipolysis, increases particulate cGI-PDE activity with little or no effect on soluble cAMP-PDE activity [19–21,53,63,64,66,78,80–82].

In recent studies, for example, a maximally effective concentration of isoproterenol (100–300 nM) was demonstrated to rapidly activate adenylate cyclase, resulting in activation of both cAMP-PrK (6–8-fold) and cGI-PDE (two-fold), which were both maximal within 2 min and remained relatively constant for the next 20–25 min [21]. Following a variable delay (1–2 min), isoproterenol-induced activation of hormone-sensitive lipase was maximal and glycerol production remained constant for 20–25 min. The concentration dependency for isoproterenol-induced activation of cGI-PDE was identical to that for activation of hormone-sensitive lipase (lipolysis); the K_{act} for isoproterenol was related to, and increased by, increases in PIA (N^6-(phenylisopropyl)-adenosine) concentrations. Isoproterenol produced maximal activation of cGI-PDE and hormone-sensitive lipase at the same cAMP-PrK ratios of < 0.5, suggesting that activation of cGI-PDE and hormone-sensitive lipase occurred over the same concentration range of cAMP and was maximal at intracellular cAMP concentrations well below those required to fully saturate and activate cAMP-PrK [21].

Catecholamine-induced activation of the cGI-PDE could be specifically blocked or reversed by the β-antagonist propranolol [63,64]. Corbin and co-workers found that the concentration dependency for the 8-chloro-cAMP-induced activation of cGI-PDE was identical to that for stimulation of lipolysis [66].

In rat adipocytes incubated with adenosine, addition of ADA removes exogenous or endogenously produced adenosine and relieves inhibitory constraints on adenylate cyclase, leading to rapid activation (< 2 min) of cAMP-PrK (4–6-fold), cGI-PDE (two-fold) and lipolysis. Similar concentrations of ADA were required for activation of cGI-PDE and hormone-sensitive lipase (lipolysis) (unpublished data) [21].

Thus, lipolytic effectors, which increase cAMP and activate cAMP-PrK via activation of adenylate cyclase (catecholamines, other lipolytic hormones) [19–21,63,64,67,78,80–82], or removal of endogenous inhibitory ligands (ADA) [89,94], or directly activate cAMP-PrK (cAMP analogs) [66], rapidly increase both cGI-PDE and hormone-sensitive lipase (lipolysis) activities. The activation of both cGI-PDE and hormone-sensitive lipase appears to be secondary to changes in cAMP and activation of cAMP-PrK [21].

Taken together, these observations suggest a close, presumably physiologically important, functional coupling of the regulation of adenylate cyclase, cGI-PDE and cAMP-PrK in adipocytes, since lipolytic effectors activate, within the same time-frame and over the identical concentration range, both synthesis (adenylate cyclase) and degradation (cGI-PDE) of cAMP. In addition, the fact that maximal activation of both the cGI-PDE and lipolysis occurs at cAMP concentrations well below that required to fully saturate and maximally activate cAMP-PrK suggests that cAMP-induced activation of cGI-PDE is not merely a pharmacologic 'feedback' mechanism for metabolism of excess cAMP generated during hormonal activation of adenylyl cyclase [21,66]. Activation of the cGI-PDE by cAMP may also be involved in regulation of cAMP 'turnover' in adipocytes [21]. This concept of regulation of cyclic nucleotide turnover as serving a regulatory function in signal transduction mechanisms in platelets, retina and pancreas has been emphasized by Goldberg and his associates [104–106].

4.3.3 Activation of cGI-PDE by insulin

Incubation of 3T3-L1 or rat adipocytes with insulin results in activation of particulate cGI-PDE with little or no effect on supernatant PDE activity [19–21, 53,63,64,67,69–72]. The time-course for activation of the cGI-PDE with insulin was somewhat slower than with isoproterenol. In the presence of a maximally effective concentration of insulin (0.1–3 nM), activation was maximal within 10–12 min in contrast to 2–3 min with isoproterenol [21]. Other workers have reported maximally effective concentrations for insulin of 0.1 nM in 3T3-L1 adipocytes and 2–3 nM in rat fat cells [19,64]. The extent of insulin-induced activation of the particulate cGI-PDE has been somewhat variable, ranging from increases of 50% [21] to 200–400% [71]; the reasons for this variation are not obvious. The concentration dependency for insulin activation of cGI-PDE ($K_{act} \approx 10$ pM) was similar to that for insulin inhibition of lipolysis. In adipocytes incubated in the presence of adenosine (or ADA plus PIA), insulin increased cGI-PDE activity without altering basal cAMP-PrK or lipolysis [21].

Although insulin clearly activates cGI-PDE in the absence of lipolytic effectors, the functional significance of this activation in terms of the cellular actions of insulin is not understood. As will be discussed below, however, in a 'physiological setting', i.e. in the presence of lipolytic effectors, insulin activation of cGI-PDE is important in the antilipolytic action of insulin.

4.3.4 Dual activation of cGI-PDE by insulin and lipolytic effectors (cAMP)

As described above, in adipocytes cGI-PDE can be rapidly activated by opposing classes of effectors—the antilipolytic agent insulin as well as lipolytic effectors which increase cAMP and activate cAMP-PrK. Since most of the physiological effects of lipolytic effectors and insulin are antagonistic, this apparent paradoxical dual regulation of cGI-PDE by lipolytic hormones and insulin has raised questions concerning the importance of activation of cGI-PDE in the antilipolytic action of insulin. Much of our recent work bears on what we suggest as a resolution of this apparent paradox.

Several lines of evidence strongly suggest that insulin and isoproterenol activate the same (or closely related) cGI-PDE in isolated adipocytes. The adipocyte cGI-PDE was purified from rat adipose particulate fractions in good yield with little evidence of substantial heterogeneity [53]. In adipocytes treated with isoproterenol and insulin, partial purification of particulate fractions by sucrose density gradient centrifugation [80–82] or DEAE chromatography [67,78–82] did not separate basal from insulin- and isoproterenol-stimulated activities; insulin- and catecholamine-stimulated activities exhibited almost identical substrate affinities, pH optima, and sensitivity to temperature, salts, sulfhydryl blocking reagents and detergents [82]. Cilostamide, the selective and specific cGI-PDE inhibitor, and antibodies to the purified bovine cGI-PDE, inhibited basal and hormone-stimulated activities (unpublished observations). Most importantly, as will be discussed below, incubation of ^{32}P-labeled adipocytes either with insulin or with isoproterenol results in phosphorylation of a 135 kDa protein which was immunoprecipitated from solubilized particulate fractions by the specific anti-cGI-PDE and identified as the native enzyme [84].

Thus from available data, it is apparent that insulin and lipolytic hormones activate the same (or closely related) cGI-PDE, but by different mechanisms. Activation by isoproterenol is clearly related to changes in adipocyte cAMP and presumably secondary to activation of cAMP-PrK [21,66,107]. Although the post-receptor molecular events regulated by insulin have not been defined with certainty, insulin-induced activation was not associated with changes in cAMP-PrK and was somewhat slower than isoproterenol-induced activation [21].

Despite these apparent differences in mechanisms of activation, insulin and isoproterenol may interact in regulation of adipocyte particulate cGI-PDE and lipolysis (Figure 4.1). Incubation of rat adipocytes with isoproterenol and insulin was associated with a decrease in isoproterenol-activated cAMP-PrK to a new steady state and a reduction in lipolysis [21]. With insulin plus isoproterenol there was a rapid, transient, synergistic activation of the cGI-PDE which was maximal at 9–10 min and correlated temporally with the reduction in cAMP-PrK [21] (Figure 4.1).

In previously published reports, there has not been agreement as to the additive effects of insulin and lipolytic effectors [82]. In contrast to our recent studies, however, neither adipocyte concentrations nor production of endogenous adenosine were rigorously controlled in these earlier studies, and time-courses of cAMP-PrK activation, cGI-PDE activity, and glycerol production were not simultaneously assessed.

Thus, in adipocytes exposed to both insulin and lipolytic effectors, synergistic activation of the cGI-PDE may result in increased cAMP hydrolysis which may be very important in (if not responsible for) the reduction in cAMP and cAMP-PrK which is associated with the antilipolytic action of insulin [21,97,98]. This mechanism could also account for the failure of insulin to inhibit lipolysis at high concentrations of lipolytic effectors [98], i.e. conditions wherein cAMP concentrations and cAMP-PrK activities are sufficiently elevated so that insulin activation of cGI-PDE is not sufficient to reduce cAMP and cAMP-PrK to the extent that hormone-sensitive lipase phosphorylation/activation is inhibited.

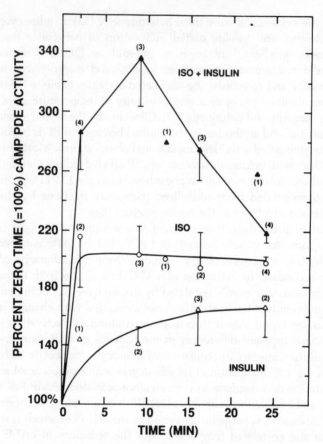

Figure 4.1   Time-course of insulin activation of cGI-PDE in the presence and absence of isoproterenol. Adipocytes were incubated in duplicate with 3 nM PIA plus 1 unit ADA/ml and 100 nM isoproterenol (O), 0.1 nM insulin (△), or isoproterenol plus insulin (▲) for the indicated times. Complete time-courses (five time-points) were compared for at least two different conditions (e.g. insulin versus insulin plus isoproterenol) in each experiment. Particulate cGI-PDE activity was normalized relative to the specific activities of time-zero values. ISO, isoproterenol. (Reproduced with permission from *Mol. Pharmacol.* [21])

We suggest that the apparently paradoxical dual regulation of cGI-PDE by lipolytic effectors and the antilipolytic agent insulin may actually reflect mechanisms whereby both hormones regulate cAMP-PrK. In adipocytes incubated with isoproterenol, the functional coupling of adenylate cyclase, cGI-PDE and cAMP-PrK activities may rapidly regulate steady-state concentrations of cAMP, and consequently the activation state of cAMP-PrK and ultimately hormone-sensitive lipase. With insulin plus isoproterenol, synergistic activation of the cGI-PDE (perhaps in combination with some inhibition of adenylate cyclase) might decrease cAMP content so as to reduce cAMP-PrK to a new steady state with reduced hormone-sensitive lipase activity and lipolysis.

In addition to these kinetic studies which correlate insulin-induced activation of particulate cGI-PDE with a reduction in isoproterenol-induced activation of cAMP-PrK and lipolysis, several indirect studies also support the hypothesis that activation of the particulate cGI-PDE is important in the antilipolytic action of insulin. In rat and 3T3-L1 adipocytes, the concentration dependency for insulin activation of cGI-PDE was similar to that for insulin inhibition of isoproterenol- and ADA-stimulated lipolysis [19,21]. In studies utilizing cAMP analogs, Corbin and his associates have reported that, of a number of cAMP derivatives, all of which directly activated cAMP-PrK and stimulated lipolysis in intact fat cells, only some were hydrolyzed by and competitively inhibited cGI-PDE isolated from adipocyte homogenates [17]. Insulin effectively inhibited lipolysis stimulated by those derivatives which were substrates for the cGI-PDE; insulin did not effectively inhibit lipolysis stimulated by analogs which were poor substrates for the cGI-PDE [17], and which therefore would not be metabolized by insulin activation of the PDE. Similar findings have been reported in human adipocytes [18].

In 3T3-L1 adipocytes, cGI-PDE is selectively and potently inhibited by cilostamide, whereas cytosolic cAMP-PDE activity is very sensitive to inhibition by Ro 20-1724 [19]. IBMX (3-isobutyl-1-methylxanthine) inhibits both 3T3-L1 adipocyte cytosolic and particulate cAMP-PDE activities. Insulin inhibited lipolysis stimulated in the presence of Ro 20-1724 alone or in the presence of isoproterenol, but did not inhibit lipolysis stimulated by cilostamide or IBMX alone or in the presence of isoproterenol [19]. Similarly, in rat adipocytes 1-3 μM OPC 3911 (a cilostamide derivative) almost completely prevented the antilipo-lytic activity of insulin (H. Eriksson, E. Degerman and C. J. Smith, unpublished observations). In adipocyte preparations from hypothyroid rats, Goswami and Rosenberg have demonstrated a marked increase in cGI-PDE and a reduction in catecholamine-stimulated lipolysis. Incubation of rat adipocytes from hypo-thyroid animals with cGMP restores the catecholamine-stimulated lipolytic response to control values [16]. Taken together, these findings point to an important role for the cGI-PDE in control of a cAMP pool involved in regulation of lipolysis, and suggest that activation of the cGI-PDE is essential in the antilipolytic action of insulin.

4.3.5   Mechanisms for the dual activation of cGI-PDE in the antilipolytic action of insulin

Until recently little was known regarding the mechanisms whereby the cAMP-enhancing agents and insulin activated the cGI-PDE. However, the recent identification/purification of the enzyme protein from several tissues and the preparation of specific antibodies have provided tools through which new insight regarding the activation mechanisms has been obtained. In the following, the discussion will be focused on the mechanisms in rat adipocytes, with occasional comparisons with those believed to exist in other cell types. Regulation of the enzyme in heart, platelets, and liver will be briefly treated in Section 4.5 and also in Chapters 8 and 12.

The fact that the activation of cGI-PDE by cAMP-enhancing agents and insulin was a stable modification suggested that it might be mediated through a phosphorylation–dephosphorylation reaction [19,53,67,78–82]. Since activation obtained with cAMP analogs correlated with the cAMP-PrK activity state [21,66], it was natural to assume that it involved phosphorylation catalyzed by cAMP-PrK. Recent work in our laboratories has shown that the pure adipose tissue cGI-PDE (rat, bovine) is indeed a good substrate for cAMP-PrK, comparable to hormone-sensitive lipase [108]. Phosphopeptide mapping demonstrated that a single serine residue (obtained in a 10–20 amino acid proteolytic peptide) was phosphorylated, with a stoichiometry of up to about 0.6 moles of phosphate per mole of PDE. The enzyme was readily dephosphorylated with purified preparations of protein phosphatase type 1, 2A and 2C [108], the latter two being the most active, as in the case of hormone-sensitive lipase [109]. The predominance of protein phosphatase 2A in the adipocyte may indicate that cGI-PDE dephosphorylation by this phosphatase would be the most important in the intact cell.

The adipose tissue enzyme is thus a good substrate for cAMP-PrK and the reversibility of the phosphorylation can be demonstrated. Enzyme preparations from bovine heart and human platelets have previously been shown to be phosphorylated in the same way, although phosphorylation site and stoichiometry have not been reported [46,48,49,55]. Recent work in our laboratories indicates that also the bovine aortic smooth muscle enzyme is phosphorylated by cAMP-PrK [59]. The phosphorylation/activation of the corresponding 'dense-vesicle' liver enzyme is discussed in Chapter 8.

It has not been possible to demonstrate activation as a result of the cAMP-PrK-mediated phosphorylation of the purified adipose tissue enzyme, in spite of a great deal of work in our laboratories. Presumably, this reflects the fact that the pure enzyme has been proteolytically nicked and/or removed from its association with membranes. The adipose tissue enzyme in a crude, membrane-associated form has been activated less than two-fold by incubation with cAMP-PrK, indicating that the activation involved a phosphorylation catalyzed by this kinase, although the PDE was not definitively identified as the substrate [107]. The human platelet enzyme, rapidly immunoisolated to avoid proteolytic nicking, was activated about 40% by phosphorylation with cAMP-PrK [49,83].

Phosphorylation of the cGI-PDE has also been shown to be stimulated by hormones in the intact cell, which verifies its physiological significance. In ^{32}P-labeled rat adipocytes, isoproterenol and insulin stimulate serine

[32]P-phosphorylation of the immunoisolated 135 kDa enzyme protein under conditions during which the hormones activate the PDE by 100% and 50%, respectively [84] (Figure 4.2). In the absence of hormones little phosphorylation occurred (usually < 10% of that with hormones). Identification of the 135 kDa ^{32}P-phosphorylated polypeptide with the enzyme protein was based on the correlation of enzyme inactivation with its precipitation, the selective blocking of its immunoprecipitation by preincubation of the antibody with pure rat or bovine adipose tissue enzyme, the recognition of a 135 kDa protein in Western

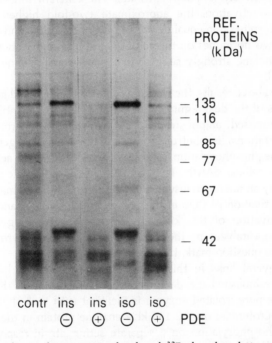

REF.
PROTEINS
(kDa)

— 135
— 116

— 85

— 77

— 67

— 42

contr   ins   ins   iso   iso
  ⊖   ⊕   ⊖   ⊕   PDE

Figure 4.2 Insulin- and isoproterenol-induced ^{32}P-phosphorylation of cGI-PDE. Rat fat cells, labeled with [32]P$_i$, were incubated at 37°C with vehicle (contr), insulin (ins) (1 nM, 12 min) or isoproterenol (iso) (300 nM, 3 min). In four experiments cGI-PDE activity was increased 90–100% by isoproterenol and 40–50% by insulin. Particulate fractions were solubilized and subjected to immunoprecipitation with anti-cGI-PDE (−), or with antibody which had been blocked by preincubation with pure rat cGI-PDE (+). The immunoprecipitates were subjected to SDS-PAGE followed by autoradiography of the dried slab gels. The data from the insulin- and isoproterenol-treated cells are not directly comparable; approximately 40% less material from insulin-treated cells was applied to the SDS-PAGE gel in this particular experiment. Values to the right are mobilities for reference proteins with the indicated molecular mass; in the case of the 135 kDa value, this has been calculated from comparison with the reference proteins. (Reproduced with permission from *Proc. Natl Acad. Sci. USA* [84])

immunoblots, its co-purification with enzyme activity over DEAE–Sephacel chromatography and the isolation of the 135 kDa phosphoprotein by CIT–agarose affinity chromatography [84]. Isoproterenol and insulin were found to induce the same extent of phosphorylation of the PDE [84] with, from preliminary estimates, a stoichiometry of 0.6 moles of phosphate per mole of cGI-PDE (unpublished data).

In more recent experiments the time-course and concentration dependency of both the isoproterenol- and insulin-induced phosphorylation were found to correlate well with the activation of the enzyme, with $t_{1/2}$ values of about 1 min and 5 min, respectively (unpublished data). The different time-courses for the phosphorylation/activation, the approximate two-fold higher activation by isoproterenol in spite of an equal extent of phosphorylation, and the synergistic effects of the two hormones on the activity (and phosphorylation (unpublished data)) of the enzyme strongly suggest that two separate serine residues were phosphorylated.

Based on the above results, the working hypothesis illustrated in Figure 4.3 for the mechanisms of the dual regulation of the cGI-PDE in the antilipolytic action of insulin is proposed: adipocyte cGI-PDE (which both reflects and regulates cAMP concentrations) is controlled by cAMP-enhancing agents through a single-site serine phosphorylation, catalyzed by cAMP-PrK, the activity of which directly reflects cellular cAMP. The cGI-PDE is phosphorylated at a different serine residue by an unknown insulin-activated serine protein kinase, resulting in a synergistic activation of PDE in the presence of cAMP-enhancing agents. In Figure 4.3, activation of this kinase is depicted as resulting from tyrosine phosphorylation, catalyzed by the activated insulin receptor tyrosine kinase; as indicated by the question mark, this is entirely speculative.

Naturally, several links in this hypothetical signal network remain to be experimentally established, e.g. demonstration that the cAMP-PrK phosphorylation site on the pure isolated enzyme is identical with that phosphorylated in response to isoproterenol on the 135 kDa enzyme protein in the intact fat cell. Furthermore, phosphorylation on a separate serine site in response to insulin must be verified.

Other mechanisms have been proposed to contribute to the antilipolytic action of insulin. Insulin has been postulated to alter the sensitivity of cAMP-PrK [110] and to induce net dephosphorylation of hormone-sensitive lipase through activation of one of the protein phosphatases (type 1 or 2A), which act on the lipase [98,111]. Insulin inhibition of adenylate cyclase has been reported to occur [101,102] although this is controversial [103,104]. A novel phosphatidylinositol glycan (one of the so-called 'mediators' of insulin action [112–114] has been proposed to mediate insulin activation of cAMP-PDE [119], as one of several insulin-induced effects. However, little substantial experimental evidence has appeared to further support the original observations and this mechanism remains hypothetical.

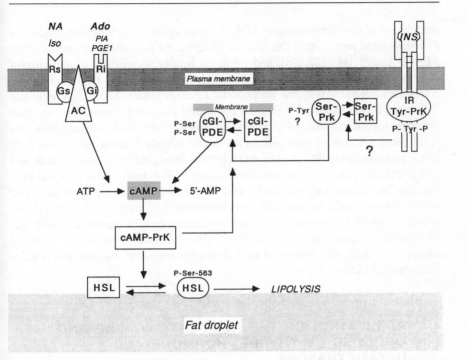

Figure 4.3 Major mechanisms for hormonal control of cGI-PDE and adipose tissue lipolysis—working hypothesis. Adenylate cyclase (AC) modulators and insulin (INS) determine adipocyte cAMP through control of its synthesis and degradation. cAMP-increasing agents, e.g. catecholamines, induce lipolysis through cAMP-PrK-catalyzed phosphorylation/activation of hormone-sensitive lipase (HSL Ser-563). The activation of lipolysis is functionally coupled to feedback control of cAMP through cAMP-PrK-catalyzed phosphorylation/activation of membrane-associated cGI-PDE. Insulin antagonizes lipolysis, induced by cAMP-enhancers, through serine phosphorylation of cGI-PDE. The dual serine phosphorylation of cGI-PDE results in synergistic activation of the PDE and sufficient cAMP reduction to prevent cAMP-PrK-catalyzed phosphorylation of hormone-sensitive lipase. The initial insulin signal is coupled to the insulin-induced phosphorylation of the cGI-PDE through activation of a serine protein kinase (Ser-PrK). The mechanism of this activation is unknown; one possibility is tyrosine phosphorylation of the kinase catalyzed by the insulin receptor tyrosine kinase (IR Tyr-PrK). NA, noradrenaline; Iso, isoproterenol; Ado, adenosine; PIA, N^6-(phenylisopropyl)-adenosine; PGE_1, prostaglandin E_1

Recent studies suggest that some aspects of insulin signalling mechanisms—including activation of PDE—may be linked to the G-proteins. Guanine nucleotides have been reported to activate insulin-stimulated PDE activity in liver membranes [120] or PIA-stimulated low-K_m PDE in rat brain [121]. Pertussis toxin treatment of intact cells, which inactivates G_i [93], inhibited insulin stimulation of the 'dense-vesicle' PDE in hepatocytes [122], inhibited insulin- or PIA-stimulated particulate PDE without effect on the β-agonist-activated PDE in 3T3-L1 adipocytes [123,124], and inhibited insulin stimulation of supernatant PDE in guinea pig ureter [125]. Although pertussis toxin treatment of rat adipocytes interfered with the antilipolytic action of insulin [126,127], Weber and colleagues found that such treatment prevented β-agonist-dependent activation rather than insulin activation of particulate 'low K_m' cAMP-PDE in rat adipocytes [128]. Since removal of endogenous adenosine inhibition in rat adipocytes with either pertussis toxin [95] or with ADA [89,94] dramatically increases cAMP and cAMP-PrK activity, and since insulin-sensitive PDE in both adipocytes and liver can also be regulated by elevated cAMP, it seems possible that these effects of pertussis toxin could be secondary to elevated cAMP levels. In this regard, the effects observed by Weber et al. [128] are consistent with increased cAMP and cAMP-PrK-mediated activation of PDE, since cholera toxin (which elevated adipocyte cAMP) also interfered with β-agonist activation but not with insulin activation of PDE.

4.4 REGULATION OF THE ADIPOCYTE cGI-PDE AND LIPOLYSIS BY THYROID HORMONE AND DEXAMETHASONE

'Permissive' hormones, such as glucocorticoids and thyroid hormones, alter responsiveness of cells to insulin as well as to agents whose effects are mediated via cAMP. These hormones usually exert their effects over a long period of time (hours or days), presumably by altering the amounts of one or more of the components of signal transduction systems, sometimes in a co-ordinate fashion.

Incubation of differentiated 3T3-L1 adipocytes in hypothyroid medium reduced isoproterenol-stimulated cAMP accumulation, lipolysis and adenylate cyclase activity, with an increase in particulate cGI-PDE as well as soluble cAMP-PDE [129]. Incubation of 3T3-L1 adipocytes in hyperthyroid medium resulted in decreased particulate cGI-PDE and soluble cAMP-PDE activities, with an increase in isoproterenol-induced cAMP accumulation, adenylate cyclase and lipolysis [129]. Thyroid status did not affect thyroid receptor number or affinity for iododihydroxybenzylpindolol in 3T3-L1 adipocytes [129]. Similar alterations in lipolysis, cAMP accumulation, and adenylate cyclase and PDE activities, without alterations in β-adrenoceptor number, have also been documented in adipocyte preparations isolated from hypo- or hyperthyroid animals [130–134]. In rat adipocytes from hypothyroid animals, catecholamine-stimulated lipolysis was decreased and cGI-PDE activity was increased approximately 2.5-fold; incubation of hypothyroid adipocytes with cGMP restored catecholamine-stimulated lipolysis to control values [16].

Glucocorticoid hormones also alter responsiveness of tissues to both insulin and agents whose actions are mediated by cAMP. These hormones can both induce a state of 'insulin resistance' as well as enhance effects of agents that act via cAMP by both receptor and post-receptor mechanisms [67,135,136].

Glucocorticoids have been reported to enhance the effects of catecholamines on lipolysis in isolated rat adipocytes [137]. In 3T3-L1 adipocytes, incubation of differentiated adipocytes with dexamethasone resulted in an increase in both the number of β_2-adrenergic receptors and the sensitivity of adenylate cyclase to catecholamines [135]. Incubation of 3T3-L1 adipocytes with dexamethasone (10 nM) for 48–72 h had no effect on basal cGI-PDE activity, but reduced both the maximal increase in cGI-PDE produced by insulin and catecholamines and the sensitivity to these agents; higher concentrations of dexamethasone also reduced basal cGI-PDE activity [67]. After exposure to dexamethasone, the initial increment in cAMP accumulation (maximal within 5 min) was not altered, but the decline in cAMP to basal concentrations was delayed [138].

In general, effects of dexamethasone and thyroid hormone on cGI-PDE represent one component of the co-ordinate regulatory mechanisms (including alterations in receptors, adenylate cyclase, and receptor–cyclase coupling via G-proteins) by which these permissive hormones alter tissue responsiveness and amplify cellular responsiveness to agents which act via cAMP-mediated processes. Specifically, effects of dexamethasone on hormonal regulation of cGI-PDE could account in part for not only some of the 'permissive' effects of glucocorticoids on enhancing cAMP-mediated processes but also the 'anti-insulin' effects of glucocorticoids.

4.5 REGULATION OF OTHER cGI-PDEs

4.5.1 Heart

The major PDE classes in cardiac tissue include Ca^{2+}- and CaM-sensitive, cGMP-stimulated, and cGMP- and Ro 20-1724-inhibited forms [10]. A cGI-PDE has been purified from bovine ventricular muscle [55]. The subcellular distribution of the cardiac cGI-PDE exhibits species variation, being predominantly soluble in guinea pig, particulate in canine and both particulate and soluble in simian and human tissues [27,31,38,39]. This PDE has been identified as a relatively specific intracellular target site of action for a number of inotropic/vasodilator drugs, some of which are being tested as therapeutic agents for the treatment of congestive heart failure in human subjects [25,60–62]. The purified PDE can serve as a substrate for cAMP-PrK [55]; whether physiologically relevant phosphorylation occurs in intact cells has not been established.

Studies in cardiac tissue, like adipocytes, may provide new insights into the roles of specific PDEs in the regulation of discrete cAMP pools and specific processes mediated by cAMP, as well as in the compartmentalization of cAMP-mediated processes. Whereas both cGMP-inhibited and Ro 20-1724-inhibited cAMP-PDEs are found in cardiac tissue from most species, inhibition of the Ro 20-1724-sensitive cAMP-PDE can increase cAMP content without exerting a direct inotropic effect [33,36,138]. On the other hand, the potency of various cardiotonic agents in producing positive inotropic responses parallels the potency of these drugs in inhibiting the cGI-PDEs [27,31,35,38,39,60–62]. Furthermore, as suggested by Weishaar, species variation in responsiveness to these inotropic agents may be related to the subcellular localization of the cGI-PDE [27,35]. In those species (i.e. rhesus monkey and dog) in which inotropes exhibit potent effects, the cGI-PDE is associated with particulate elements, presumably the sarcoplasmic reticulum [31]; in species in which inotropes are less potent (i.e. in guinea pig, rat, hamster) the cGI-PDE is predominantly soluble. In rat heart, PDEs have also been implicated in the differential effects of PGE_1 and isoproterenol on post-receptor activation of cAMP-PrK and regulation of the enzymes of glycogen metabolism [140].

4.5.2 Platelets

Human platelets contain three types of PDEs [40–49], including a cGI-PDE which accounts for virtually all of the hydrolysis of low substrate concentrations ($< 1 \mu M$ cAMP) in platelet homogenates/lysates [46]. Increases in platelet cAMP are associated with inhibition of platelet aggregation and secretion, and a number of the same potent and selective inhibitors of the cardiac cGI-PDE which elicit positive inotropic effects also inhibit platelet cGI-PDE as well as platelet aggregation [40–49]. The platelet cGI-PDE has been purified from human platelets; its kinetic properties and inhibitor specificities are similar to those of other cGI-PDEs [56].

Incubation of intact human platelets with agents that activate adenylate cyclase (i.e. prostacyclin (PGI_2), the stable prostacyclin analog iloprost, forskolin) and increase cAMP resulted in a rapid activation of the cGI-PDE [48,49]. In platelets labeled with ^{32}P, activation was associated with incorporation of ^{32}P into the cGI-PDE. A phosphorylated protein of 110 kDa was identified by chromatography on Blue Dextran–Sepharose, Western immunoblots or immunoprecipitation with polyclonal or monoclonal antibodies against platelet or cardiac cGI-PDEs [48,49]. Beavo and co-workers demonstrated that after adsorption to a specific monoclonal antibody preparation, the platelet cGI-PDE could be phosphorylated and activated by cAMP-PrK, suggesting that its activation in intact platelets was also mediated via the cAMP-PrK [49]. The increases in activity associated with phosphorylation in these broken cell systems were much smaller than the increases produced by incubation of intact platelets with PGI_2 (two-fold) or forskolin (11-fold) [48]. By ^{18}O labeling of adenine nucleotides in intact platelets, Goldberg and co-workers have also reported an increase in platelet cAMP turnover, i.e. increases in both synthesis and hydrolysis of cAMP [105]. There has been no speculation as to whether this increased turnover is involved in the regulation of the activation 'state' of platelet cAMP-PrK, as has been suggested in rat adipocytes [21].

4.5.3 Liver

Agents that increase cAMP (i.e. glucagon) as well as insulin increase particulate 'low-K_m' cAMP-PDE activity in rat hepatocytes [65]. Houslay and co-workers have distinguished two types of hormone-sensitive cAMP-PDE activity in hepatocytes, a peripheral plasma membrane cAMP-PDE activated by insulin alone and a 'dense-vesicle' (microsomal-like)-associated cAMP-PDE which can be activated by both cAMP (via cAMP-PrK) and insulin. These workers further suggest that insulin regulates the plasma membrane and 'dense-vesicle' PDEs by distinct mechanisms (see Chapter 8). The 'dense-vesicle' enzyme can be inhibited by cGMP and several cardiotonic drugs and the plasma membrane cAMP-PDE was apparently more sensitive to Ro 20-1724, suggesting that in liver insulin can regulate both cGMP-inhibited and Ro 20-1724-inhibited cAMP-PDEs, perhaps by different mechanisms [58]. It has also been suggested that activation of the plasma membrane cAMP-PDE in intact rat hepatocytes by vasopressin may be related to inhibition of glucagon-stimulated cAMP accumulation by vasopressin [141].

4.6 CONCLUSIONS AND PERSPECTIVES

As discussed above and in Chapter 8, the recent development of highly efficient purification procedures, e.g. CIT–agarose affinity chromatography, has made it possible to isolate members of the cGI-PDE family from a number of tissues and cell types, in spite of their exceedingly low tissue abundance. Several of the enzyme preparations have been at least partially characterized, monoclonal and polyclonal antibodies have been produced and highly selective inhibitors identified.

Striking similarities in molecular and catalytic properties of cGI-PDE preparations from most tissues have been found. During purification, the native enzyme is exquisitely sensitive to proteolytic nicking by various endogenous proteases, but retains catalytic activity; apparently only very rapid isolation by immunoprecipitation allows recovery of a substantial proportion of the native enzyme (105–135 kDa subunits), whereas larger-scale preparations contain mainly nicked components in the 80–30 kDa range.

The use of the selective antibodies and inhibitors has allowed rapid progress in the knowledge of the function and regulation of the enzyme to be made. It appears likely that cGI-PDEs have a critical role in the control of cAMP pools essential for the regulation of such diverse and biologically important processes as adipose tissue lipolysis, platelet aggregation, vascular smooth muscle relaxation, and myocardial contraction. The cardiac cGI-PDE is believed to be the target for a number of the new cardiotonic inotropic drugs, e.g. milrinone, CI-914 and CI-930. The apparent role of the enzyme in the other tissues suggests additional, potential uses for inhibitors as antithrombotic and vasodilatory drugs. In fact, this is the proposed function of cilostazol, a cilostamide derivative recently marketed by Otsuka Pharmaceuticals, Osaka, Japan.

The most important new information derived from the work with the adipocyte cGI-PDE has been the elucidation of the mechanism whereby cAMP enhancers and insulin regulate cGI-PDE activity and the role of the enzyme in the control of adipose tissue lipolysis, and especially in the antilipolytic action of insulin. In the rat adipocyte there seems to be a close functional coupling between hormone-induced activation of adenylate cyclase, cAMP-PrK and cGI-PDE; activation of cGI-PDE under these circumstances may be a key factor in feedback regulation of cAMP-mediated activation of cAMP-PrK, hormone-sensitive lipase and consequently the rate of lipolysis. In the absence of lipolytic effectors, insulin also activates cGI-PDE; the functional consequences of this insulin-induced activation of cGI-PDE in 'basal' cells has not yet been determined. However, in the presence of insulin and lipolytic effectors, which is the 'physiological condition', there is a transient and synergistic activation of cGI-PDE, associated with a reduction in hormone-stimulated cAMP-PrK and hormone-stimulated lipolysis. Cyclic AMP enhancers and insulin, by activating discrete serine protein kinases (i.e. cAMP-PrK and an unidentified intracellular serine protein kinase, respectively), induce serine phosphorylation and activation of the cGI-PDE. The enzyme thus appears to be a true 'physiological substrate' for insulin-mediated serine phosphorylation, directly coupling insulin signaling to the cAMP pathway. Clearly this may represent a major step forward towards an understanding of the molecular basis of insulin's antilipolytic action, one important biological effect of the hormone. Further studies will have to substantiate the working hypothesis proposed in Figure 4.3 for the early steps coupling insulin receptor activation to the insulin-induced serine phosphorylation/activation of the cGI-PDE. Current work in our laboratories involves additional structure–function studies, including peptide mapping of the phosphorylation sites and identification of the insulin-activated serine protein kinase. Access to pure cGI-PDE as substrate for this kinase will be useful in its isolation and identification, and in elucidation of mechanisms of activation of both the kinase and the cGI-PDE by insulin.

ACKNOWLEDGEMENTS

We thank Drs Ana Rascon and Valeria Vasta for permission to present unpublished results, Dr Hans Tornqvist for helpful discussions and computer-

CYCLIC GMP-INHIBITED PDEs

drawn illustrations, and Mrs C. Kosh and Mrs Ruth Lovén for excellent secretarial assistance. Our original work reviewed in this chapter has been supported by grants from the following foundations: A, Påhlsson, Malmö; Nordisk Insulin, Copenhagen; Swedish Diabetes, M. Bergvall, Swedish Hoechst and Swedish Medical Association, Stockholm; the faculty of Medicine, Lund; the Diabetes Research Education Foundation, NJ, USA. It was also supported from the Swedish Medical Research Council (Project 3362) and by stipends to Eva Degerman from the Carl Jönsson Memorial Foundation, Malmö.

REFERENCES

1. Sutherland, E. W., and Rall, T. W. (1960) *J. Biol. Chem.*, **232**, 1027–1091.
2. Butcher, R. W., and Sutherland, E. W. (1962) *J. Biol. Chem.*, **237**, 1244–1250.
3. Thompson, W. J., and Appleman, M. M. (1971) *Biochemistry*, **10**, 311–316.
4. Thompson, W. J., and Appleman, M. M. (1971) *J. Biol. Chem.*, **246**, 3145–3150.
5. Appleman, M. M., and Terasaki, W. L. (1975) *Adv. Cyclic Nucleotide Res.*, **5**, 153–163.
6. Wells, J. N., and Hardman, J. G. (1977) *Adv. Cyclic Nucleotide Res.*, **8**, 119–143.
7. Strada, S. S., and Thompson, W. J. (1978) *Adv. Cyclic Nucleotide Res.*, **9**, 265–283.
8. Vaughan, M., Danello, M. A., Manganiello, V. C., and Strewler, G. J. (1981) *Adv. Cyclic Nucleotide Res.*, **14**, 263–271.
9. Beavo, J. A., Hansen, R. S., Harrison, S. A., Hurwitz, R. L., Martins, T. I., and Mumby, M. D. (1982) *Mol. Cell. Endocrinol.*, **28**, 387–410.
10. Beavo, J. (1988) *Adv. Second Messenger Phosphoprotein Res.*, **22**, 1–38.
11. Sharma, R. K., Adachi, A. M., Adachi, K., and Wang, J. H. (1984) *J. Biol. Chem.*, **259**, 9248–9254.
12. Sharma, R. K., and Wang, J. H. (1986) *J. Biol. Chem.*, **261**, 14160–14165.
13. Shenolikar, S., Thompson, W. J., and Strada, S. J. (1985) *Biochemistry*, **24**, 672–678.
14. Ho, H. C., Wirch, E., Stevens, F. C., and Wang, J. H. (1977) *J. Biol. Chem.*, **252**, 43–50.
15. Yamamoto, T., Lieberman, F., Osborne, J. C. Jr, Manganiello, V. C., Vaughan, M., and Hidaka, H. (1984) *Biochemistry*, **23**, 670–675.
16. Goswami, A., and Rosenberg, I. (1985) *J. Biol. Chem.*, **260**, 82–85.
17. Beebe, S. J., Redman, J. B., Blackmore, P. W., and Corbin, J. (1985) *J. Biol. Chem.*, **260**, 15781–15788.
18. Lönnroth, P., and Smith, U. (1986) *Biochem. Biophys. Res, Commun.*, **141**, 1157–1161.
19. Manganiello, V. C., and Elks, M. (1986) In *Mechanisms of Insulin Action* (Belfrage, P., Donnér, J. and Strålfors, P., eds) Elsevier/North Holland Press, pp. 123–133.
20. Manganiello, V. C., Smith, C., Newman, A. H., Rice, K., Degerman, E., and Belfrage, P. (1987) *J. Cyclic Nucleotide Protein Phosphorylation Res.*, **11**, 497–511.
21. Smith, C. J., and Manganiello, V. C. (1989) *Mol. Pharmacol.* **35**, 381–386.
22. Endoh, M., Yamashita, S., and Taira, N. (1982) *J. Pharmacol. Exp. Ther.*, **221**, 775–783.
23. Alousi, A. A., Cantor, J. M., Montenaro, M. J., Fort, D. J., and Ferrari, R. A. (1983) *J. Cardiovasc. Pharmacol.*, **5**, 792–803.

24. Aku, H.-S., Eardley, D., Watkins, R., and Prioli, N. (1986) *Biochem. Pharmacol.*, **35**, 1113–1121.
25. Colucci, W. S., Wright, R. F., and Braunwald, E. (1986) *N. Engl. J. Med.*, **314**, 290–299, 349–358.
26. Weishaar, R. E., Burrows, S. D., Kobylarz, D. C., Quade, M. M., and Evans, D. B. (1986) *Biochem. Pharmacol.*, **35**, 787–800.
27. Weishaar, R. E., Kobylarz-Singer, D. C., Quade, M. L., Steffen, R. P., and Kaplan, H. R. (1987) *J. Cyclic Nucleotide Protein Phosphorylation Res.*, **11**, 513–530.
28. Erhardt, P. W. (1987) *J. Med. Chem.*, **30**, 231–237.
29. Jones, G. H., Venuti, M. C., Alvarez, R., Bruno, J. J., Berks, A. H., and Prince, A. (1987) *J. Med. Chem.*, **30**, 295–302.
30. Kariya, T., Willie, L. J., and Dage, R. C. (1987) *J. Cardiovasc. Pharmacol.*, **4**, 509–514.
31. Kauffman, R. F., Crowe, V. G., Utterback, B. G., and Robertson, D. W. (1987) *Mol. Pharmacol.*, **30**, 609–616.
32. Manganiello, V. C. (1987) *J. Mol. Cell. Cardiol.*, **19**, 1037–1040.
33. Reeves, M. L., Leigh, B. K., and England, P. J. (1987) *Biochem. J.*, **241**, 535–541.
34. Venuti, M. C., Jones, G. H., Alvarez, R., and Bruno, J. J. (1987) *J. Med. Chem.*, **30**, 303–318.
35. Weishaar, R. E., Kobylarz-Singer, D. C., Steffen, R. P., and Kaplan, H. R. (1987) *Circ. Res.*, **61**, 539–547.
36. Simpson, A. W. M., Reeves, M. L., and Renk, T. J. (1988) *Biochem. Pharmacol.*, **37**, 2315–2320.
37. Silver, P. (1989) *Am. J. Cardiol.*, **63**, 2A–8A.
38. Weishaar, R. E., and Bristol, J. A. (1989) In *Comprehensive Medicinal Chemistry* (Sammes, P. G. and Taylor, J. B., eds), Pergamon Press, New York.
39. Weishaar, R. E., Kobylarz-Singer, D. C., and Klinkenfuss, B. (1989) *Cardiovasc. Drugs Ther.*, **3**, 29–39.
40. Hidaka, H., and Asano, T. (1976) *Biochim. Biophys. Acta*, **429**, 485–497.
41. Asano, T., Ochiai, Y., and Hidaka, H. (1977) *Mol. Pharmacol.*, **13**, 400–406.
42. Hidaka, H., Hayashi, H., Kohri, Y., Hosokawa, T., Igawa, T., and Saitoh, Y. (1979) *J. Pharmacol. Exp. Ther.*, **211**, 26–30.
43. Alvarez, R., Taylor, A., Fazzari, J. J., and Jacobs, J. R. (1981) *Mol. Pharmacol.*, **20**, 302–309.
44. Hidaka, H., and Endo, T. (1984) *Adv. Cyclic Nucleotide Res.*, **16**, 248–259.
45. Muggli, R., Tschopp, T. B., Mittelholzer, E., and Baumgartner, H. R. (1985) *J. Pharmacol. Exp. Ther.*, **235**, 212–220.
46. Macphee, C. H., Harrison, S. A., and Beavo, J. A. (1986) *Proc. Natl Acad. Sci. USA*, **83**, 6660–6663.
47. Alvarez, R., Banerjie, G. L., Bruno, J. J., Jones, G. L., Littschwager, K., Strossberg, A., and Venuti, M. C. (1986) *Mol. Pharmacol.*, **29**, 554–560.
48. Grant, P. G., Mannerino, A. F., and Colman, R. W. (1988) *Proc. Natl Acad. Sci. USA*, **85**, 9071–9075.
49. Macphee, C. H., Reifsnyder, D. H., Moore, T. A., Levea, K. M., and Beavo, J. A. (1988) *J. Biol. Chem.*, **263**, 10353–10358.
50. Conti, M., Toscano, M. V., Petrelli, L., Geremia, R., and Stefanini, M. (1982) *Endocrinology*, **110**, 1189–1196.
51. Conti, M., Toscano, M. V., Petrelli, L., Geremia, R., and Stefanini, M. (1983) *Endocrinology*, **113**, 1845–1853.

52. Conti, M., Monaco, L., Geremia, R., and Stefanini, M. (1986) *Endocrinology*, **118**, 901–908.
53. Degerman, E., Belfrage, P., Newman, A. H., Rice, K. C., and Manganiello, V. C. (1987) *J. Biol. Chem.*, **262**, 5797–5807.
54. Degerman, E., Manganiello, V. C., Newman, A. H., Rice, K. C., and Belfrage, P. (1988) *Adv. Second Messenger Phosphoprot. Res.*, **12**, 171–182.
55. Harrison, S. A., Reifsnyder, D. H., Gallis, B., Cadd, C. G., and Beavo, J. A. (1986) *Mol. Pharmacol.*, **25**, 506–514.
56. Grant, P., and Colman, R. W. (1984) *Biochemistry*, **23**, 1801–1807.
57. Boyes, S., and Loten, E. G. (1988) *Eur. J. Biochem.*, **174**, 303–309.
58. Pyne, N., Cooper, M., and Houslay, M. D. (1987) *Biochem. J.*, **242**, 33–42.
59. Rascon, A., Belfrage, P., Degerman, E., Lindgren, S., Andersson, K. E., Newman, A., Manganiello, V. C., and Degerman, E. (1989) In *Purine Nucleosides and Nucleotides in Cell Signalling: Targets for New Drugs* (Jacobsen, K., Daly, J. and Manganiello, V. C., eds), Springer-Verlag, New York, pp. 353–358.
60. Weishaar, R. E., Carn, M. H., and Bristol, J. A. (1985) *J. Med. Chem.*, **28**, 537–545.
61. Pang, D. (1988) *Drug Dev. Res.*, **12**, 85–92.
62. Pang, D., Cantor, E., Hagedorn, A., Erhardt, P., and Wiggens, I. (1988) *Drug Dev. Res.*, **14**, 141–149.
63. Zinman, B., and Hollenberg, C. H. (1974) *J. Biol. Chem.*, **249**, 2182–2187.
64. Pawlson, L. G., Lovell-Smith, C. J., Manganiello, V. C., and Vaughan, M. (1974) *Proc. Natl Acad. Sci. USA*, **71**, 1639–1642.
65. Loten, E. G., Assimacopoulos-Jeannet, F. D., Exton, J. H., and Park, C. R. (1978) *J. Biol. Chem.*, **253**, 746–757.
66. Gettys, T. W., Blackmore, P. F., Redman, J. E., Beebe, S. J., and Corbin, J. D. (1987) *J. Biol. Chem.*, **262**, 333–339.
67. Elks, M. E., Manganiello, V. C., and Vaughan, M. (1983) *J. Biol. Chem.*, **258**, 8582–8587.
68. Heyworth, C. H., Wallace, A. V., and Houslay, M. D. (1983) *Biochem. J.*, **214**, 99–110.
69. Loten, E. G., and Sneyd, J. G. T. (1970) *Biochem. J.*, **120**, 187–193.
70. Manganiello, V. C., and Vaughan, M. (1973) *J. Biol. Chem.*, **248**, 7164–7170.
71. Kono, T., Robinson, F. W., Sarver, J. A., Vega, F. V., and Pointer, R. H. (1977) *J. Biol. Chem.*, **252**, 2226–2233.
72. Sakai, T., Thompson, W. J., Lavis, U. R., and Williams, R. H. (1974) *Arch. Biochem. Biophys.*, **162**, 331–339.
73. Kono, T., Robinson, F. W., and Sarver, J. A. (1975) *J. Biol. Chem.*, **250**, 7826–7835.
74. Macauley, S. L., Kiechle, F., and Jarrett, L. (1983) *Biochim. Biophys. Acta*, **760**, 293–299.
75. Weber, H. W., and Appleman, M. M. (1982) *J. Biol. Chem.*, **257**, 5339–5341.
76. Manganiello, V. C., Reed, B. R., Lieberman, F., Moss, J., Lane, D. M., and Vaughan, M. (1983) *J. Cyclic Nucleotide Protein Phosphorylation Res.*, **9**, 143–154.
77. Manganiello, V. C., Degerman, E., and Elks, M. C. (1988) *Methods Enzymol.*, **154**, 504–520.
78. Lovell-Smith, C. J., Manganiello, V. C., and Vaughan, M. (1977) *Biochim. Biophys. Acta*, **497**, 447–458.
79. Loten, E. G., Francis, S. H., and Corbin, J. D. (1980) *J. Biol. Chem.*, **255**, 7838–7844.
80. Makino, H., and Kono, T. (1980) *J. Biol. Chem.*, **255**, 7850–7854.

81. Makino, H., deBuschiazzo, P. M., Pointer, R. H., Jordan, J. E., and Kono, T. (1980) *J. Biol. Chem.*, **255**, 7845–7849.

82. Francis, S. H., and Kono, T. (1982) *Mol. Cell. Biochem.*, **42**, 109–116.

83. Macphee, C. H., Reifsnyder, D. H., Moore, T. A., and Beavo, J. A. (1987) *J. Cyclic Nucleotide Protein Phosphorylation Res.*, **11**, 487–496.

84. Degerman, E., Smith, C. J., Tornqvist, H., Vasta, V., Belfrage, P., and Manganiello, V. C. (1989) *Proc. Natl Acad. Sci. USA* (in press).

85. Pyne, N. I., Anderson, N., Lavan, B., Milligan, G., Nimmo, H. G., and Houslay, M. D. (1987) *Biochem. J.*, **248**, 897–201.

86. Fain, J. N. (1980) *Biochem. Actions of Hormones*, **VII**, 119–204.

87. Steinberg, D., Mayer, S. E., Khoo, J. C., Miller, E. A., Miller, R. E., Fredholm, B., and Eichner, R. (1975) *Adv. Cyclic Nucleotide Res.*, **5**, 549–568.

88. Belfrage, P., Fredrikson, G., Olsson, H., and Strålfors, P. (1983) *The Adipocyte and Obesity: Cellular and ¯Molecular Mechanisms*, Raven Press, New York, pp. 217–224.

89. Honnor, R. C., Dhillon, G. S., and Londos, C. (1985) *J. Biol. Chem.*, **260**, 15122–15129.

90. Honnor, R. C., Dhillon, G. S., and Londos, C. (1985) *J. Biol. Chem.*, **260**, 15130–15138.

91. Strålfors, P., Björgell, P., and Belfrage, P. (1984) *Proc. Natl Acad. Sci. USA*, **81**, 3317–3321.

92. Holm, C., Kirchgessner, T. G., Svenson, K. L., Fredrikson, G., Nilsson, S., Miller, C. G., Shively, J. E., Heinzmann, C., Sparkes, R. S., Mohandas, T., Lusis, A. J., Belfrage, P., and Schotz, M. C. (1988) *Science*, **241**, 1503–1506.

93. Gilman, A. G. (1987) *Annu. Rev. Biochem.*, **56**, 615–649.

94. Schwabe, U., Ebert, R., and H. C. Erbler (1975) *Adv. Cyclic Nucleotide Res.*, **5**, 569–584.

95. Olansky, L., Myers, G. A., Pohl, S. L., and Hewlett, E. L. (1983) *Proc. Natl Acad. Sci. USA*, **80**, 6547–6551.

96. Jungas, R. L., and Ball, E. G. (1963) *Biochemistry*, **2**, 383–388.

97. Butcher, R. W., Sneyd, J., Park, C. R., and Sutherland, E. W. (1966) *J. Biol. Chem.*, **241**, 1651–1653.

98. Londos, C., Honnor, R. C., and Dhillon, G. S. (1985) *J. Biol. Chem.*, **260**, 15139–15145.

99. Soderling, T. R., Corbin, J. D., and Park, C. R. (1973) *J. Biol. Chem.*, **248**, 1822–1829.

100. Iliano, G., and Cuatrecasas, P. (1972) *Science*, **175**, 906–908.

101. Hepp, K. O., and Renner, R. (1972) *FEBS Lett.*, **20**, 191–194.

102. House, P., Poulis, P., and Weidemann, M. J. (1972) *Eur. J. Biochem.*, **24**, 429–437.

103. Vaughan, M., and Murad, F. (1969) *Biochemistry*, **8**, 3092–3099.

104. Goldberg, N. D., Ames, A. III, Gander, J. E., and Walseth, T. F. (1983) *J. Biol. Chem.*, **258**, 9213–9219.

105. Goldberg, N. D., Walseth, T., Eide, J., Krich, T. P., Kuehn, B. L., and Gander, J. E. (1984) *Adv. Cyclic Nucleotide Protein Phosphorylation Res.*, **16**, 363–379.

106. Deez, M. A., Graeff, R. M., Walseth, T. F., and Goldberg, N. D. (1988) *Proc. Natl Acad. Sci. USA*, **85**, 7867–7871.

107. Gettys, T. W., Vine, A. J., Simonds, M. F., and Corbin, J. D. (1988) *J. Biol. Chem.*, **263**, 10359–10363.

108. Degerman, E. (1988) 'Purification, characterization and regulation of the hormone-sensitive cAMP-phosphodiesterase from adipose tissue.' Thesis, Lund University.

109. Olsson, H., and Belfrage, P. (1987) In *Advances in Protein Phosphatases* (W. Merlevede and J. Di Salvo, eds), Leuven University Press, Leuven, pp. 409–429.

110. Ashby, C. D., and Walsh, D. A. (1972) *J. Biol. Chem.*, **247**, 6638–6642.

111. Nemenoff, R. A., Blackshear, P. J., and Avruch, J. (1983) *J. Biol. Chem.*, **258**, 9437–9443.

112. Larner, J. (1982) *J. Cyclic Nucleotide Res.*, **8**, 289–296.

113. Low, M. G., and Saltiel, A. R. (1988) *Science*, **239**, 268–275.

114. Saltiel, A. R., and Cuatrecasas, P. (1986) *Proc. Natl Acad. Sci. USA*, **83**, 5793–5797.

115. Alemany, S., Mato, S. J., and Strålfors, P. (1987) *Nature*, **330**, 77–79.

116. Fox, J. A., Soliz, N. M., and Saltiel, A. R. (1987) *Proc. Natl Acad. Sci. USA*, **84**, 2663–2667.

117. Kelly, K. I., Mato, J. M., Mereda, P., and Jarett, R. (1987) *Proc. Natl Acad. Sci. USA*, **84**, 6404–6407.

118. Romero, G. L., Luttrell, L., Rogol, A., Zeller, K., Hewlett, E., and Larner, J. (1988) *Science*, **240**, 509–511.

119. Saltiel, A. R. (1987) *Endocrinology*, **120**, 967–972.

120. Heyworth, C. M., Rawal, S., and Houslay, M. (1983) *FEBS Lett.*, **154**, 87–91.

121. De Mazancourt, P., and Giudicelli, Y. (1984) *FEBS Lett.*, **167**, 142–146.

122. Heyworth, C. M., Grey, A.-M., Wilson, S. R., Hanski, E., and Houslay, M. D. (1986) *Biochem. J.*, **235**, 145–149.

123. Elks, M. L., Watkins, P. A., Manganiello, V. C., Moss, J., Hewlett, E., and Vaughan, M. (1983) *Biochem. Biophys. Res. Commun.*, **116**, 593–598.

124. Elks, M. L., Jackson, M., Manganiello, V. C., and Vaughan, M. (1987) *Am. J. Cell. Physiol.*, **252** (Cell Physiol. 21), C342–C348.

125. Weiss, R. M., and Wheeler, M. A. (1988) *J. Pharmacol. Exp. Ther.*, **247**, 630–634.

126. Goren, H. J., Northup, J. K., and Hollenberg, M. (1985) *Can. J. Physiol. Pharmacol.*, **63**, 1017–1022.

127. Mills, I., and Fain, J. N. (1985) *Biochem. Biophys. Res. Commun.*, **130**, 1059–1065.

128. Weber, H. W., Chung, F.-Z., Day, K., and Appleman, M. M. (1987) *J. Cyclic Nucleotide Protein Phosphorylation Res.*, **11**, 345–354.

129. Elks, M. L., and Manganiello, V. C. (1985) *Endocrinology*, **117**, 947–953.

130. Armstrong, K. J., Stouffer, J. E., Van Inwegen, R. G., Thompson, W. J., and Robison, G. A. (1974) *J. Biol. Chem.*, **249**, 4226–4231.

131. Correze, C., Laudat, M. Y., Laudat, P. L., and Nunez, J. (1974) *Mol. Cell. Endocrinol.*, **1**, 309–327.

132. Van Inwegen, R. G., Robison, G. A., Thompson, W. J., Armstrong, K. J., and Stouffer, J. E. (1975) *J. Biol. Chem.*, **250**, 2452–2456.

133. Stiles, G., and Lefkowitz, R. J. (1984) *Thyroid Today*, **7**, 1–6.

134. Malbon, C. C., Graziano, M. P., and Johnson, G. L. (1984) *J. Biol. Chem.*, **259**, 3254–3260.

135. Lai, E., Rosen, O. M., and Ruben, C. S. (1982) *J. Biol. Chem.*, **257**, 6691–6696.

136. Grunfeld, C., Baird, K., Van Oberghen, E., and Kahn, R. (1981) *Endocrinology*, **109**, 1723–1790.

137. Fain, J. W. (1968) *Endocrinology*, **82**, 825–830.

138. Elks, M. L., Manganiello, V. C., and Vaughan, M. (1984) *Endocrinology*, **115**, 1350–1356.
139. Gristwood, R. W., Owen, D. A. A., and Reeves, M. L. (1985) *Br. J. Pharmacol.*, **85**, 22P.
140. Bode, D. C., and Brunton, L. L. (1988) *Mol. Cell. Biochem.*, **82**, 13–18.
141. Miot, F., Kepper, S., Erneux, C., Wells, J., and DeWulf, H. (1988) *Biochem. Pharmacol.*, **37**, 3447–3453.

5

CYCLIC GMP-BINDING CYCLIC GMP-SPECIFIC PHOSPHODIESTERASE FROM LUNG

Sharron H. Francis, Melissa K. Thomas and Jackie D. Corbin

Department of Molecular Physiology and Biophysics and the Howard Hughes Medical Institute, Vanderbilt University, Nashville, TN 37232-0295, USA

5.1 INTRODUCTION

Throughout the history of research into the action of cyclic nucleotides in eukaryotes, attention has been largely focused on cyclic nucleotide-dependent protein kinases (cAMP and cGMP kinases) as mediators of cAMP and cGMP action. However, other receptors for these nucleotides may also serve important roles. The cGMP-binding cyclic nucleotide phosphodiesterases (PDEs) are one such group of cGMP receptors, being distinguished by the presence of two distinct types of sites for cGMP; one site functions as the catalytic site where the cyclic nucleotide is enzymatically cleaved, and the other site is a specific binding site for cGMP but exhibits no hydrolytic activity. This is a heterogeneous family of enzymes consisting of at least several subgroups, including the widely distributed cGMP-stimulated cyclic nucleotide phosphodiesterase (cGMPs-PDE) [1,2], the retinal cGMP phosphodiesterases (rod outer segment phosphodiesterase (ROS-PDE) [3,4] and cone phosphodiesterases) [5], a cGMP-stimulated cGMP-specific phosphodiesterase from *Dictyostelium discoideum* [6–9] and the cGMP-binding cGMP-specific phosphodiesterase (cG-BPDE) which has been described in rat and bovine lung [9,10] and in rat and human platelets [9,11].

The cGMPs-PDEs represent a major portion of PDE activity in mammalian liver, heart, adipose tissue and adrenal gland [1,12–17]. By binding to a cGMP-specific allosteric binding site, micromolar concentrations of cGMP stimulate hydrolysis of cAMP at the catalytic site of this enzyme [18]. Antibodies generated to this enzyme do not crossreact with the ROS-PDE or the cG-BPDE from rat lung [15,19].

Cyclic Nucleotide Phosphodiesterases: Structure, Regulation and Drug Action
Edited by J. Beavo and M. D. Houslay © 1990 John Wiley and Sons Ltd

The ROS-PDE is a key element in the proposed cyclic nucleotide cascade of vision, although the function of the cGMP binding site in its mechanism of action has not been elucidated. The catalytic site is highly specific for cGMP and catalytic function is modulated by the association of the low molecular weight protein, transducin [20]. This enzyme has been purified from mammalian [21] and amphibian [22,23] rod outer segments. Many properties of the ROS-PDE, such as subunit molecular weight, cGMP specificity for catalysis and binding and 3-isobutyl-1-methyl-xanthine (IBMX) stimulation of cGMP binding, closely resemble those of the cG-BPDE and suggest some degree of homology between the two enzymes. However, the two enzymes have differences in numerous properties [3,4,10], including distinct immunologic features, marked differences in the K_m and V_{max} values, and responses to protease treatment.

The cGMP-stimulated cGMP-specific PDE from *Dictyostelium discoideum* also has two distinct cGMP sites: a cGMP-specific catalytic site and a cGMP-specific allosteric site [24]. Cyclic GMP binding to the allosteric site alters the catalytic site by lowering the K_m for cGMP hydrolysis while leaving the V_{max} unchanged. Analog studies have demonstrated distinctive analog specificities for the two sites, neither of which interacts with cAMP, and catalytic activity is unaffected by Ca^{2+} [25,26]. This intracellular PDE appears to be involved in lowering cGMP levels which have been elevated by chemoattractant signals [25,27–29].

The cG-BPDE from lung was first noted by Lincoln et al. [30] in supernatants of lung extracts as a heterogeneous cGMP binding activity which is not associated with cGMP-dependent protein kinase activity. Subsequently, this cGMP binding activity in rat lung [9,10] and platelets [9,11] has been shown to co-migrate with a highly specific cGMP PDE activity, which differs from most PDEs in its nearly complete inability to hydrolyze cAMP. The cG-BPDE differs from cGMP-dependent protein kinase not only in chromatographic behavior and molecular size, but also in cyclic nucleotide analog specificity [10]. The rat lung enzyme is clearly distinct from the Ca^{2+}–calmodulin (CaM)-stimulated PDE, since the hydrolytic activity is unaffected by the presence of calcium chloride and CaM or by pretreatment with trifluoperazine [31,32]. The cG-BPDE and the CaM-dependent 'high-K_m' enzyme chromatograph similarly on DEAE–cellulose, and proteolyzed forms of the latter enzyme are reportedly insensitive to CaM [33]. Therefore, caution must be exercised in identifying the cG-BPDE in various tissues. The cG-BPDE can be more clearly distinguished by assessment of its native molecular mass (177 kDa), monomer size (93 kDa). PDE inhibitor specificity and IBMX-stimulated cGMP-specific binding [10,34,35].

The cG-BPDE has been purified to homogeneity from rat [35] and now from bovine lung [36]. The presence of this or very similar enzymes has been noted in several tissues and species, including rat and bovine lung [9,10,35,36], rat and human platelets [9,11], rat spleen [37], guinea pig lung [38], vascular smooth muscle [11,37] and sea urchin sperm [10]. However, the physiological function of the cGMP binding site of this enzyme is still unclear, although communication between the binding and catalytic sites has been demonstrated [10,36,39]. An important role for the cG-BPDE in modulating the effects of cGMP is supported by the fact that significant levels of cG-BPDE relative to cGMP-dependent protein kinase are present in many of these tissues. Furthermore, guanylate cyclase and cGMP levels are relatively high in these tissues, emphasizing the possibility that cGMP and the enzymes responsible for its synthesis and degradation play an important physiological role in modulating the cellular effects of cGMP [9–11].

5.2 PURIFICATION

The cG-BPDE has been purified from rat lung [35] and bovine lung [36]. The purification steps include sequential chromatography on DEAE–cellulose, Blue Sepharose CL-6B, zinc chelate adsorption, and HPLC–TSK DEAE 5PW. Steps used in the purification of the bovine and rat lung cG-BPDE are the same except for the use of a sizing column as the final step in the purification of the bovine enzyme. The cGMP binding activity and the cGMP PDE activity co-purify through all steps and recoveries for the two activities are similar [35]. For either enzyme, the purification yields predominantly a protein band of 93 kDa on SDS-PAGE. The distribution of this band in chromatographic fractions from a HPLC gel filtration column correlates directly with cGMP binding and cGMP PDE activities (see Figure 5.1 for bovine lung cG-BPDE).

5.3 PHYSICAL PROPERTIES

The physical characteristics are shown in Table 5.1 for the rat lung, bovine lung, and rat platelet cG-BPDE. Although the cG-BPDEs from the three sources have similar calculated native molecular masses, there is some variation in the sedimentation coefficients and Stokes radii. These variations in physical parameters could be due to the use of different standards, columns, etc., in the determination of the values, or they could reflect true differences in the proteins. Since the platelet enzyme has not yet been completely purified and specific antibodies are not available for either the lung or platelet enzymes, one cannot definitively conclude that the enzymes are exactly the same species of PDE. Using the purified rat or bovine lung enzymes, the 93 kDa protein band described above has been identified as the cG-BPDE subunit by specific [^{32}P]cGMP binding in photoaffinity labeling experiments. These results suggest that the native enzyme is a homodimer [35,36,39]. In studies of platelet extracts, Walseth and Goldberg also reported specific [^{32}P]cGMP photoaffinity labeling of a platelet protein of similar size on SDS-PAGE [40]. The N-terminus of the purified bovine cG-BPDE does not yield any amino acid sequence when subjected to gas phase sequence analysis and is presumed to be blocked [41].

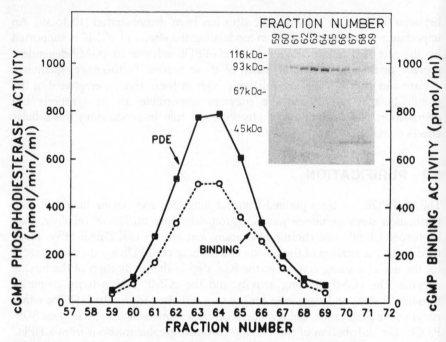

Figure 5.1 Chromatographic profile of purified bovine lung cG-BPDE. The partially purified cG-BPDE was applied to a HPLC Bio-Sil TSK-250 column (600 × 21.5 mm) equilibrated in 100 mM sodium phosphate, pH 6.8, 2 mM EDTA, 25 mM 2-mercaptoethanol. Fractions were assayed for cGMP hydrolytic activity at 30°C in the presence of 20 μM cGMP, 10 mM $MgCl_2$, 0.33 mg/ml bovine serum albumin, and 50 mM TRIS, pH 7.5. [³H]cGMP binding activity was measured at 4°C in the presence of 0.2 mM IBMX and 0.5 mg/ml histone VIII-S. Samples were then filtered onto Millipore HAWP filters (0.45 μM) and rinsed with cold 10 mM KH_2PO_4, pH 6.8, 1 mM EDTA (3 × 2 ml). Inset: Aliquots from fractions exhibiting activity were analyzed by SDS-PAGE on 8% minislab gels and stained with 0.1% Coomassie Brilliant Blue. The migration of the molecular weight markers are indicated at the left margin of the inset

Table 5.1 Physical properties of cG-BPDE from rat and bovine lung, and rat platelets

	Rat lung [10,35]	Bovine lung [41]	Rat platelets [11,58]
Native molecular mass (kDa)	177	178	176
Monomer molecular mass (kDa)	93	93	ND
Sedimentation coefficient (S)	7.8	9.35	6.44
Stokes radius (nm)	5.5	4.4	5.6
K_m for cGMP hydrolysis	4 μM	5 μM	ND

5.4   CATALYTIC FUNCTION

The PDE activity of the cG-BPDE is highly specific for cGMP and hydrolyzes cAMP at a rate which is approximately 100 times slower than the hydrolytic rate for cGMP. The purified enzymes from rat and bovine lung exhibit similar specific activities of cGMP hydrolysis, i.e 3–10 μmol/min/mg with a K_m of $\sim 5\ \mu$M [10,36,41], and the pH optimum for catalysis is ~ 8.0–8.4. Catalysis is unaffected by a wide concentration range of Ca^{2+}, CaM or pretreatment with trifluoperazine. The hydrolytic process is completely dependent on the presence of either Mn^{2+} or Mg^{2+}. The rat lung enzyme has a much higher affinity for Mn^{2+} as the cation supporting catalysis, with half-maximal catalytic activity being supported by 50 μM Mn^{2+} as compared to 5 mM Mg^{2+}. This marked difference in cation affinity suggests that the catalytic function of the enzyme in vivo may be primarily dependent on Mn^{2+}. The catalytic function of the cG-BPDE is potently inhibited by Zn^{2+} and is unaffected by Fe^{2+} and Fe^{3+} [34]. Hematin is also a potent inhibitor of catalysis, and the inhibition does not appear to be competitive. The cGMP PDE hydrolytic activity is rapidly inactivated by heat or trypsin and by treatment with the arginine-specific protein modifier, 2,3-butanedione [34]. Coquil and co-workers have reported the reversible inhibition of the catalytic activity of the partially purified rat lung enzyme by unsaturated 18- and 20-carbon fatty acids, whereas saturated fatty acids have no effect [37].

5.5   PDE INHIBITOR SPECIFICITY

The PDE activity of the cG-BPDE from both bovine and rat lung is inhibited by a number of PDE inhibitors, as shown in Figure 5.2. IBMX has been used routinely in the studies of this enzyme, since it was originally observed that this compound is a competitive inhibitor [10] of the PDE catalytic activity, thereby providing strong evidence that the stimulation of cGMP binding is due to an interaction between the catalytic site and the allosteric site. In Figure 5.2 it is evident that PDE inhibitors exhibit a broad range of potencies for inhibition of the catalytic activity. Papaverine and 8-methoxymethyl-IBMX have IC_{50} values of $\sim 6.0\ \mu$M and are equipotent with IBMX ($IC_{50} \sim 8\ \mu$M). Inhibitors exhibiting 10–20 times greater potencies than IBMX include zaprinast and dipyridamole with IC_{50}s of 300 nM and 830 nM, respectively; cilostamide, theophylline, Cl-914, Ro-1724 and rolipram are much weaker by comparison. The abilities of these inhibitors to inhibit the cG-BPDE clearly differ from the inhibition patterns observed for the Ca^{2+}–CaM-sensitive PDEs, the cGMP-stimulated cyclic nucleotide PDE and the 'low-K_m' cAMP-PDE [42–45].

  Even the most potent PDE inhibitors of the cG-BPDE catalytic activity have very diverse molecular structures, making interpretation of the most important sites of interaction difficult. In studies with specific cyclic nucleotide analogs, the

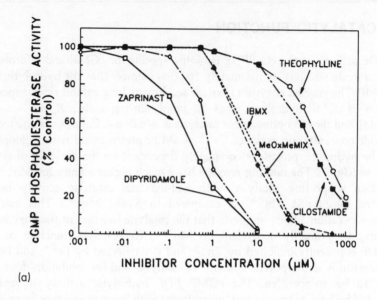

(a)

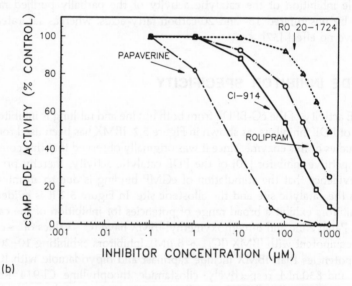

(b)

Figure 5.2 Sensitivity of cGMP-PDE catalytic activity of the cG-BPDE to a variety of inhibitors. Purified bovine cG-BPDE was assayed for 10 min at 30°C in the presence of 20 μM [³H]cGMP, 10 mM MgCl₂, 50 mM Tris-Cl, pH 7.5, 0.33 mg/ml bovine serum albumin with the addition of various concentrations of inhibitors. In the absence of added inhibitors approximately 15–20% of the substrate was hydrolyzed during the course of the experiment

relative importance of various structural features can be more systematically evaluated. In Figure 5.3 the potencies of five cyclic nucleotide analogs have been studied. In this group, the most potent analog to inhibit cGMP catalysis is β-phenyl-1-N^2-etheno-cGMP (PET-cGMP). The inhibition by this compound is remarkable, since it strongly interacts with the catalytic site despite the large substitution involving two positions in the pyrimidine portion of the purine ring, i.e. positions 1 and 2. Subsequent studies have shown, however, that this analog is not hydrolyzed. Thus, despite a strong interaction at the catalytic site, the phosphodiester bond is not positioned appropriately for catalysis to proceed. The strong interaction of this analog indicates that when cyclic nucleotides bind to the catalytic site, there is ample space available near the N-1 and N^2 positions to accommodate a large chemical grouping as is found in PET-cGMP. Likewise, N^2-hexyl-cGMP (not shown) potently interacts with the catalytic site, and this analog is hydrolyzed at a rate comparable to that of cGMP. The inhibition of catalysis by N^2-hexyl-cGMP is ~ 15 times less potent than that by PET-cGMP. The analogs, 2'-deoxy-cGMP, cIMP, and 1-methyl-cGMP also interact with the catalytic site with reasonable potencies, indicating the relative unimportance of the 2'-hydroxyl in the ribose moiety, and re-emphasizing the tolerance for analog modification at N-1 and N^2. In contrast, derivatives such as 8-(4-chlorophenyl-thio)-cGMP (8pCPT-cG) and 8-bromo-cGMP (not shown) interact weakly with the cG-BPDE. This is presumably due to either chemical or steric constraints in that region of the catalytic site or to preference for the *anti* conformation of

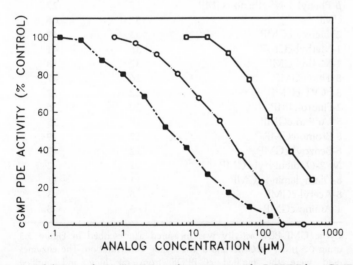

Figure 5.3 Inhibition of cGMP-PDE catalytic activity of cG-BPDE by cGMP analogs. The ability of various concentrations of five analogs of cGMP to inhibit hydrolysis of 20 μM [^{3}H]cGMP was determined as described in Figure 5.2. ■, PET-cGMP; ○, cIMP, 1-CH$_3$-cGMP, or 2'-deoxy-cGMP; □, 8-pCPT-cGMP

cGMP instead of the *syn* conformation that would result from substitutions at C-8 [46].

A more extensive study of the cyclic nucleotide analog specificity of the catalytic site of the cG-BPDE has been conducted for the rat lung enzyme and is shown in Table 5.2. The pattern of analog inhibition potencies for the rat lung enzyme is similar to that for the bovine lung cG-BPDE (data not shown). Despite many types of chemical substitutions at C-8, these analogs are generally poor inhibitors of catalytic activity. The apparent lack of specific requirements in the pyrimidine portion of the cGMP molecule make it difficult to understand the poor interaction of cAMP with the catalytic site of this enzyme. Since cIMP is a moderately potent inhibitor, the oxygen at the C-6 position is likely to be an important determinant for the binding site interaction with the potential for hydrogen bond formation at this position. It is possible that features which are

Table 5.2 Analog inhibition of cGMP-PDE activity of cG-BPDE and Ca^{2+}–CaM-sensitive PDE

Analog	cG-BPDE inhibition %	Ca^{2+}–CaM-sensitive PDE inhibition %
N^2-Hexyl-cGMP	> 90	84
β-Phenyl-1-N^2-etheno-cGMP	67	27
cIMP	57	68
2′-Deoxy-cGMP	43	47
1-Methyl-cGMP	38	41
1-Methyl-cIMP	32	18
8-Thio-cGMP	26	47
8-pCPT-cGMP	22	12
2-Fluoro-cIMP	20	6
5′-Amino-cGMP	18	35
8-Bromo-cGMP	12	19
8-Benzoyl-cGMP	12	17
N^2-2-O'-dibutyryl-cGMP	11	2
8-Diethylamino-cGMP	11	11
8-Acetyl-cGMP	8	3
1-Amino-cGMP	3	8

Cyclic GMP-PDE activity was measured as described in Figure 5.3 using 0.5 μM [³H]cGMP and 5 μM analog concentration. The enzymes used were the purified cG-BPDE from rat lung and purified Ca^{2+}–CaM-sensitive PDE from porcine cerebral cortex [49]. The latter frozen enzyme was thawed and assayed in the presence of 10 μM $CaCl_2$, and 0.5 μg/ml purified CaM.

not easily identified in analog studies, such as electron distributions in the purine ring, provide for the strong selectivity for cGMP over cAMP in this enzyme.

It is of further interest to compare the analog specificities of the catalytic site of the cG-BPDE with sites in other proteins which interact with cGMP. For instance, PET-cGMP interacts potently with the cGMP-dependent protein kinase binding sites; the kinase is also activated reasonably well by 1-methyl-cGMP [47,48]. Studies of analogs modified at C-2 suggest that the amino group at this position is highly important for binding of cGMP to the kinase, possibly by forming a hydrogen bond in the binding site. As noted above, this is apparently not the case for the cG-BPDE. Very marked differences between the kinase sites and the cG-BPDE catalytic site are also evident with analogs in which modifications have been introduced at the 2'-OH or C-8 position. The 2'-OH group in the ribose moiety is particularly critical for high-affinity interaction with the kinase. This group could be important for hydrogen bonding to the cGMP kinase binding site, but not to the catalytic site of cG-BPDE. Analogs derivatized at C-8, such as 8-bromo-cGMP and 8-pCPT-cGMP, are potent activators of the kinase, suggesting a *syn* conformation for kinase-bound cyclic nucleotide. Although the results of the cyclic nucleotide analog studies are suggestive, further work is needed to establish if cGMP binds in an *anti* conformation to the cG-BPDE catalytic site but in a *syn* conformation to the cGMP kinase binding sites.

The analog specificity of the rat lung cG-BPDE catalytic site can also be compared with the specificity exhibited by the catalytic site of the Ca^{2+}–CaM-stimulated 'high-K_m' PDE from porcine cerebral cortex [49] (Table 5.2). This comparison reveals a more similar pattern than that observed with the cGMP kinase. Like the cG-BPDE, the CaM-dependent enzyme has a marked preference for cGMP at the hydrolytic site, but, unlike the cG-BPDE, cleaves cAMP reasonably well [49–51]. It is clear from the analog studies presented in Table 5.2 that the hydrolytic sites of these two enzymes possess many similarities in their chemical specificities. Derivatives with modification of N^2, namely N^2-hexyl-cGMP and PET-cGMP, are potent inhibitors, whereas derivatives modified at C-8 are relatively ineffective as inhibitors of the catalytic activities of both PDEs. In addition, cIMP, 1-methyl-cGMP, and 2'-deoxy-cGMP interact well with the catalytic site. More thorough chemical and structural analyses of the catalytic sites in these two enzymes will establish features responsible for the similarities in the analog specificities for the respective catalytic sites. Even though the results suggest evolutionary relatedness, the molecular weights of these proteins differ dramatically, as does the regulation of catalysis. Thus, it would appear that many of the critical features of the catalytic sites have been preserved in these enzymes despite the evolution of varied physical and regulatory structures. It is likely that these similarities in catalytic site structures may be present in other PDEs which show a preference for cGMP.

5.6   CYCLIC GMP BINDING FUNCTION

Multiple lines of evidence indicate that the cGMP binding site(s) in the rat and bovine lung cG-BPDE is distinct from the site of cGMP hydrolysis [10,34–36,41]. In the absence of metal, competitive inhibitors of the catalytic activity [10,41], such as IBMX, and other PDE inhibitors, such as papaverine, dipyridamole and zaprinast, enhance cGMP binding to the cG-BPDE. This clearly indicates a communication between the two sites, since the stimulation of binding occurs through a mechanism independent of substrate protection. Some cGMP analogs, notably 2'-O-monobutyryl-cGMP, N^2-hexyl-cGMP, and PET-cGMP, potently inhibit cGMP PDE activity (Table 5.2), but not cGMP binding (not shown). As shown in Figure 5.4, analogs such as IBMX and N^2-hexyl-cGMP,

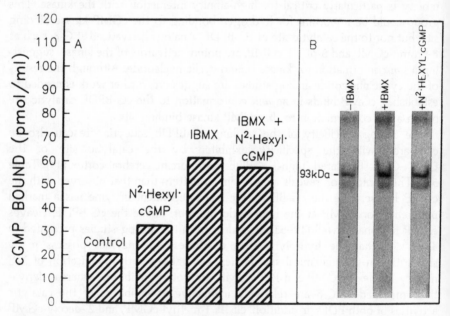

Figure 5.4   Effect of catalytic site-specific agents on the allosteric cGMP binding site in cG-BPDE. (A) Stimulation of [^{3}H]cGMP binding by N^2-hexyl-cGMP and IBMX. Purified rat lung cG-BPDE was incubated with [^{3}H]cGMP (0.1 μM), 10 mM NaH_2PO_4, pH 6.8, 1 mM EDTA and 12 mM 2-mercaptoethanol for 60 min at 4°C in the absence and presence of 0.2 mM IBMX and/or 5 μM N^2-hexyl-cGMP in a total volume of 100 μl. Samples were then filtered as described in Figure 5.1. (B) Autoradiograph of [^{32}P]cGMP photoaffinity labeling of cG-BPDE. Purified bovine cG-BPDE was preincubated in quartz tubes with [^{32}P]cGMP (0.06 μM) for 60 min at 4°C in the presence of 10 mM NaH_2PO_4, pH 6.8, 2 mM EDTA, 12 mM 2-mercaptoethanol with the addition of 0.2 mM IBMX or 10 μM N^2-hexyl-cGMP where indicated. Samples were then irradiated at 254 nm in a Rayonet photochemical reactor (5 × 10 s), treated with 2M 2-mercaptoethanol, and 2% SDS, and subjected to SDS-PAGE on 8% gels

which interact with the catalytic site, stimulate cG-BPDE binding of [³H]cGMP or photoaffinity labeling with [³²P]cGMP under assay conditions in which no cGMP hydrolytic activity is detectable, further indicating a communication between the catalytic site(s) and the cGMP binding site(s). Two important conclusions can be drawn concerning these observations. First, at least for this PDE, the cyclic nucleotide can bind to the catalytic site in the absence of metal. This would suggest that the cation is primarily involved in the catalytic process rather than in the binding of cGMP at the catalytic site. Second, the binding of cyclic nucleotide at the catalytic site, in the absence of catalysis, is sufficient to generate the conformational change required to elicit increased binding at the allosteric site.

Several lines of evidence have suggested that the cG-BPDE could contain at least two functionally, if not structurally, distinct allosteric cGMP binding sites. In studies of the association of cGMP with the enzyme, cGMP binding exhibits a biphasic pattern (Figure 5.5) with increasing concentrations of cGMP in the presence of IBMX. A curvilinear pattern is also observed when the cGMP dissociation rate from the enzyme is studied. Using conditions that permit saturation of the binding site with the labeled cGMP, a chase with excess

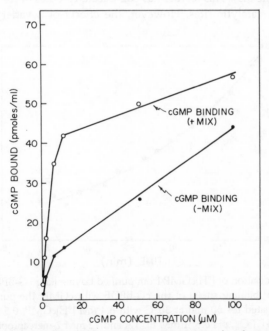

Figure 5.5 cGMP binding by the rat lung cG-BPDE in the absence and presence of IBMX. [³H]cGMP binding by the partially purified rat lung cG-BPDE was assessed at the indicated concentration of cGMP in a reaction mix containing 10 mM NaH$_2$PO$_4$, pH 7.0, 1 mM EDTA, 12 mM 2-mercaptoethanol, ± 0.2 mM IBMX. Samples were incubated at 4°C for 1 h followed by filtration as described in Figure 5.4A

unlabeled cGMP causes displacement of labeled cGMP in a time-dependent manner (Figure 5.6). The dissociation curve thus generated implicates the presence of multiple kinetic binding sites. However, since the stoichiometry obtained for [³H] cGMP binding to the enzyme has been consistently slightly less than 1 mol per subunit, the kinetic evidence suggesting multiple sites must be viewed with caution. Kinetically distinct species of cGMP binding sites could be (1) the result of multiple structurally distinct binding sites in each monomer, (2) a product of asymmetric kinetic interaction between structurally identical sites in the subunits of the dimeric enzyme, or (3) the influence of covalent modification(s) which could generate microheterogeneity in the purified preparations of cG-BPDE.

The specificity for cGMP binding at the allosteric site has been shown to be quite stringent [10,41]. Interaction of cAMP with the binding site is very poor. To date, no cyclic nucleotide analog has been found which will interact preferentially with the binding site(s) although several analogs exhibit preference for the catalytic site [10,36,39,41]. In fact, only a very few analogs (cIMP, 8-methyl-cGMP, 8-thio-cGMP) will compete effectively with cGMP for the binding site (Table 5.3). This feature has complicated studies of the interaction of the binding and catalytic sites. However, the effects of a variety of chemical

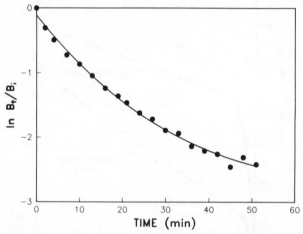

Figure 5.6 Dissociation of [³H]cGMP from purified bovine lung cG-BPDE. In order to saturate the cGMP binding site(s) of the cG-BPDE with cGMP, the purified cG-BPDE (60 nM) was incubated for 60 min at 30°C with 6.4 μM [³H]cGMP, 0.5 mg/ml histone VIII-S, 50 mM NaH_2PO_4, pH 6.8, 1 mM EDTA and 12 mM 2-mercaptoethanol. At time 0 (*i*) aliquots were withdrawn and subjected to rapid Millipore filtration in order to assess the amount of [³H]cGMP bound. Unlabeled cGMP (100 μM) was immediately added and aliquots were removed at various time points for Millipore filtration. Data is plotted as ln B_t/B_i versus time t where B_t is the amount of [³H]cGMP bound at any given time and B_i is the amount of [³H]cGMP bound at 0 time

substitutions on the interaction of cGMP with the allosteric binding site reveal important distinctions between this site and the catalytic site. In general, any alteration of the cGMP molecule appears to have deleterious effects on the interaction of cGMP with the binding site. The binding site exhibits high specificity for cGMP over cAMP, whereas cIMP competes relatively well. This

Table 5.3 Effects of analogs on [^{3}H]cGMP binding by bovine lung cG-BPDE

Analog	Analog : cGMP	
	20 : 1 (%)	200 : 1 (%)
Control	100	100
8-pCPT-cAMP	100	106
8-Acetyl-cGMP	95	104
N^2-2-O'-Dibutyryl-cGMP	102	106
1-Methyl-cGMP	127	99
8-pCPT-cGMP	102	102
8-Benzyl-cGMP	99	93
8-Neopentyl-cGMP	94	98
5'-cGMP	ND	87
8-Benzylthio-cGMP	103	95
8-Butyryl-cGMP	97	86
2-Fluoro-cIMP	87	77
8-(1-Hydroxyethyl)-cGMP	100	69
8-Carbamoyl-cGMP	86	66
5'-Amino-cGMP	94	64
8-Hydroxy-cGMP	100	64
2'-Deoxy-cGMP	84	49
N^2-Monobutyryl-cGMP	82	56
8-Methyl-cGMP	76	38
β-Phenyl-1-N^2-etheno-cGMP	100	41
8-Bromo-cGMP	71	32
2'-O-Monosuccinyl-cGMP	63	30
cAMP	82	18
8-Thio-cGMP	51	11
cIMP	46	11

ND = not determined.
Purified bovine lung cG-BPDE was incubated at 0–4°C in the presence of histone VIII-S (0.5 mg/ml), IBMX (0.2 mM), [^{3}H]cGMP (0.15 μM), 10 mM KH$_2$PO$_4$, 1 mM EDTA, 12 mM 2-mercaptoethanol in the absence or presence of added cyclic nucleotide analogs at a ratio of analog to [^{3}H]cGMP of either 20 : 1 or 200 : 1. Activity is expressed as percentage of [^{3}H]cGMP binding in the control tubes in the absence of added analogs. Filtration of binding assay was performed as described in Figure 5.1

suggests that the amino group at C-2 in cGMP is not a pivotal feature of binding, whereas the C-6 keto oxygen is likely to be an important element, possibly via formation of a hydrogen bond at this position. Other results substantiate this interpretation. An analog modified by a chloro group in the C-6 position (6-chloropurineriboside 3',5'-cyclic monophosphate, data not shown) interacts very poorly with the binding site. Substitution of an aliphatic group on the nitrogen at C-2, as in N^2-hexyl-cGMP (not shown), or aryl C-2 substitutions as in β-phenyl-1-N^2-etheno-cGMP, decreases inhibitory potency and suggests steric constraints that are not present in the catalytic site. Likewise, substitution of a fluoro group at C-2 as in 2-fluoro-cIMP dramatically reduces interaction of cIMP with the binding site. The introduction of a fluoro group at C-2 in cIMP increases the electronegativity of the purine ring [52], so that alteration of the electron distribution pattern could also be responsible for the affinity change. When the N-1 position is modified by methylation as in 1-methyl-cGMP, or 1-methyl-cIMP, interaction with the binding site is virtually eliminated, suggesting that a hydrogen bond is formed between the N-1 proton and the binding site. Substitution of an amino group for the proton at N-1 also interferes with the binding but to a lesser extent than that of a methyl group. The weak interactions of 2'-deoxy-cGMP, 5'-amino-cGMP and 2'-O-monobutyryl-cGMP with the allosteric binding site suggests the importance of hydrogen bonding through the 2'- and 5'-positions in the ribose moiety. However, studies of analogs with other modifications (such as 2'-O-monosuccinyl-cGMP and 2'-O-monosuccinyl-cGMP tyrosyl methyl ester) raise the possibility that these side groups interact with structures external to the binding site, and serve to stabilize the binding of these analogs.

All substitutions at C-8 reduce the affinity for the binding site. Large aliphatic or aromatic substitutions at the C-8 positions such as those present in 8-benzylthio-cGMP, 8-neopentyl-cGMP, 8-pCPT-cGMP and 8-butyryl-cGMP almost completely eliminate apparent binding potency. Although the chemical nature of the side group should be considered, this pattern may be due to the bulk of the substitutions. In this regard, it is apparent that introduction of other groups at C-8 as in 8-bromo-cGMP, 8-thio-cGMP, and 8-methyl-cGMP are weakly accommodated. All substitutions at C-8 are likely to affect the distribution of the cGMP molecule between the *syn* and *anti* conformations [46] and could suggest that the allosteric binding site binds cGMP in the *anti* conformation. However, interpretation of analog effects are complicated by the fact that the differences in the affinity of the binding site for a given analog are likely to be a product of a number of factors, including conformation, steric constraints, hydrophobicity, and electron distribution. Furthermore, if the cG-BPDE is subsequently shown to contain two structurally distinct cGMP binding sites, the above pattern of analog specificity for the cGMP binding function could represent features of both sites.

The function of the cGMP binding site of the cG-BPDE is unknown. Our laboratory has proposed that the allosteric site regulates the activity of the cGMP hydrolytic site [9]. Although conclusive proof is not yet available, several observations provide support for this proposal. As noted above, [³H]cGMP binding is stimulated by catalytic site-specific analogs, and photoaffinity labeling of the binding site by [³²P]cGMP is enhanced by these same analogs and by IBMX (Figure 5.4) [35,36,39]. Further, since ligands interacting with the catalytic site induce an activity change in the allosteric site function, presumably by an effect on the conformation of the enzyme, the principle of reciprocity predicts that the converse will also be true [53]. When cGMP is bound to the allosteric site, the enzyme exhibits a charge shift on DEAE chromatography. Under these experimental conditions, the cG-BPDE elutes from the DEAE at a higher ionic strength than does the free enzyme, suggesting a significant increase in surface electronegativity (due to a change in enzyme conformation) when cGMP is bound to the enzyme. A conformational change induced by cGMP binding is also supported by the recently discovered cGMP-dependent phosphorylation of this enzyme by the purified catalytic subunit of cAMP-dependent protein kinase [36,41]. If cGMP analogs can be found which are specific for the allosteric site as compared to the catalytic site, then this question can be addressed more directly. Hamet and co-workers have suggested that the cGMP binding site in this enzyme might serve as a significant buffer for intracellular cGMP [54]. The allosteric binding site of the enzyme may serve both functions, i.e. to buffer intracellular cGMP concentrations via direct cGMP binding to the enzyme while modulating catalytic function as a result of that interaction.

Cyclic GMP binding affinity for the lung enzyme varies significantly with assay conditions. In this laboratory, a number of factors have been shown to enhance cGMP binding in the lung enzyme. These include highly charged proteins, such as histones, phosvitin and polyarginine, neutral salts, short-chain aliphatic alcohols, limited proteolysis, and heat treatment [10,34,35,37,55]. Both heat and trypsin treatment cause a rapid loss of cGMP hydrolytic activity, but enhance overall cGMP binding [10,54]. In studies with the rat lung cG-BPDE, Coquil [37] reported that the cGMP binding is increased in the presence of 18- and 20-carbon unsaturated fatty acids. It is possible that the effects of ligands such as these in altering cGMP binding are mimicked by intracellular factors, which mediate the physiological regulation of the cGMP binding.

In the cell, cGMP binding affinity may vary depending on the presence of specific modulators or on localized changes in the microenvironment. In the absence of added factors, optimal cGMP binding occurs at pH 8.0–8.4 with little binding evident below pH 6.5; maximum sensitivity to stimulation of cGMP binding by agents such as IBMX is most evident at pH 7.0–7.8. With more alkaline assay conditions, basal cGMP binding increases, and the IBMX stimulation is less, suggesting that titration of specific ionizable groups in this pH range is responsible for an induced conformational change in the cG-BPDE to increase cGMP binding. At pH 7.0–8.0, the presence of histone VIII-S and IBMX greatly increases the binding affinity so that the association of cGMP with the binding site occurs well within the physiological concentrations of cGMP. Studies with the bovine lung enzyme have determined the K_d for cGMP to be ~ 200 nM at pH 6.8 in the presence of IBMX and histone VIII-S [41]. In the rat lung enzyme, half-maximal saturation in the presence of IBMX is ~ 500 nM [10]. Hamet et al have reported the K_d for cGMP binding as 353 nM and 13 nM in the absence and presence of IBMX for the platelet enzyme [9,53,55]. The higher affinity for cGMP binding reported for the partially purified platelet enzyme is a very significant difference between this enzyme and the rat and bovine lung cG-BPDE. The difference in affinities may reflect unique structural features in the enzymes from the two tissue types, or, alternatively, the presence of endogenous modulators of enzyme function may be present in the partially purified platelet enzyme. With either the rat or bovine lung enzyme, the rate, affinity and capacity for cGMP binding are not greatly affected by temperature changes (4°C versus 37°C).

5.7 SENSITIVITY TO SULFHYDRYL REAGENTS

The cG-BPDE purification scheme for the lung enzymes requires the continual presence of sulfhydryl compounds, such as 2-mercaptoethanol, in order to preserve the binding and catalytic activity. However, the addition of specific sulfhydryl modifying agents such as iodoacetate, iodoacetamide and hydrogen peroxide to the rat lung enzyme causes little change in either PDE or binding activity. However, at high concentrations of N-ethylmaleimide (5 mM), the binding of cGMP in the presence of IBMX is significantly decreased. Tremblay et al have reported that oxygenation of platelet extracts causes a time-dependent activation of both the PDE and cGMP binding activities of the cG-BPDE [54]. These workers have also found that Mn^{2+} increases the activating effect of oxygenation and that H_2O_2 is a potent inhibitor of the PDE activity [54].

Hematin has been shown to be a potent inhibitor of PDE activity of the partially purified rat lung enzyme with an $I_{50} \simeq 0.5$ μM, and to increase cGMP binding to the allosteric site [30]. Similar inhibition of the PDE activity of the platelet enzyme has also been noted [54,55]. Since these studies have all been conducted with partially purified enzymes, it is difficult to interpret these observations, and the role of sulfhydryl residues in this protein remains enigmatic.

5.8 ENZYME PHOSPHORYLATION

In the presence of cGMP, but not in its absence, the purified cG-BPDE is phosphorylated by the purified catalytic subunit of the cAMP-dependent protein kinase (Figures 5.7 and 5.8) [36]. Since there is no direct effect of cGMP on the catalytic subunit, the stimulation of phosphorylation by cGMP implies a substrate-directed modification. The cGMP effect on phosphorylation is not mimicked by IBMX, which would interact with the active site of the PDE, indicating that the phosphorylation site in the enzyme is exposed only when cGMP is bound to the allosteric site. The phosphorylation event is a rapid catalysis which occurs at physiological concentrations of catalytic subunit, and the phosphorylated residue(s) in the cG-BPDE is a serine (Figure 5.7). The

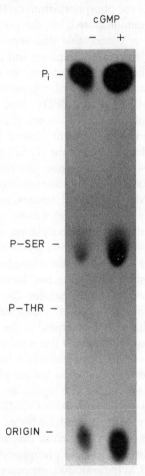

Figure 5.7 Autoradiograph of the cGMP-dependent phosphorylation of a serine residue in cG-BPDE by the catalytic subunit of cAMP-dependent protein kinase. Purified bovine lung cG-BPDE was incubated for 20 min at 30°C with 100 nM purified catalytic subunit of cAMP-dependent protein kinase, 80 μM [γ-^{32}P]ATP, and 8 mM MgCl$_2$, $\pm$ 10 μM cGMP (as noted). Samples were acid-precipitated and partially hydrolyzed in 6 M HCl. Unlabeled phosphoamino acid standards were added prior to high-voltage paper electrophoresis of the samples. The migration positions of phosphothreonine and phosphoserine standards determined by ninhydrin staining, of unhydrolyzed material at the electrophoretic origin, and of inorganic phosphate released during hydrolysis are indicated

phosphorylated peptide has been isolated from a tryptic digestion and sequenced (KISASEFDRPLR) [41]. All of the radiolabel is released with the serine at the third cycle in the sequence as would be expected from a consensus sequence for a phosphorylation site for the catalytic subunit; such a sequence contains a trypsin-sensitive basic residue two or three positions N-terminal to the phosphorylated serine. To date, no effects of the phosphorylation on the functional characteristics of the enzyme have been demonstrated, but the possibilities for modulation of function are intriguing. It is conceivable that regulation of catalytic function occurs by an interplay between cGMP binding and phosphorylation.

In addition to the purified catalytic subunit, several other purified protein kinases have been compared at equimolar concentrations for their abilities to catalyze phosphorylation of cG-BPDE [36]. Protein kinase C and Ca^{2+}–CaM-dependent protein kinase II do not phosphorylate cG-BPDE effectively, while cGMP kinase is a very potent catalyst for phosphorylation of the enzyme (Figure 5.8). Tryptic digestion of the cG-BPDE phosphorylated by either the cGMP or cAMP kinase yields the same phosphorylated peptide and with either kinase the ^{32}P is covalently linked to serine (not shown). When equimolar concentrations of the cAMP and cGMP kinases are compared, phosphate is incorporated into cG-BPDE at least ten times faster using the cGMP kinase. As shown in Figure 5.8, the state of phosphorylation is similar when high levels (2 μM) of the kinases are used, but at limiting levels of kinase (10 nM) the cGMP kinase is clearly the better catalyst for the phosphorylation of the cG-BPDE. Most in vitro protein substrates for these two kinases are known to be much better substrates for the cAMP kinase than for the cGMP kinase [56,57]. The findings that the cG-BPDE is a potent cGMP kinase substrate in vitro and that the phosphorylation requires that cGMP be bound to the cG-BPDE suggest that this PDE is a natural substrate for this kinase. If this is subsequently established, it will be the first example of a cGMP kinase substrate for which there is a defined catalytic function. It would also suggest a unique physiological control which has not been previously observed, i.e. the regulation of phosphorylation events can occur by binding of a cyclic nucleotide (cGMP) to the enzyme substrate (cG-BPDE) as well as to a protein kinase (cGMP- or cAMP-dependent).

Based on studies of intact platelets and crude preparations of cG-BPDE, Hamet and co-workers have suggested a role for phosphorylation of platelet cG-BPDE by the cAMP kinase [54,55]. In these studies, exposure of intact platelets to either forskolin or prostaglandin E_1, in the presence of IBMX, causes an activation of the platelet cG-BPDE binding and catalytic activities which persists after the cells are lysed. The activation is stable to DEAE chromatography and can be mimicked by incubation of crude platelet extracts in the presence of catalytic subunit, ATP and Mg^{2+}. Since the activation thus induced cannot be duplicated with partially purified preparations of the platelet cG-BPDE [54], it has been proposed that the activation might proceed through a regulatory subunit which is removed by further purification.

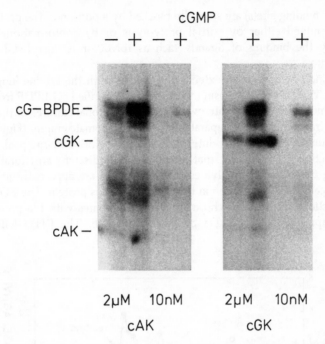

Figure 5.8 The preferential phosphorylation of the cG-BPDE by cGMP-dependent protein kinase as compared with the cAMP-dependent protein kinase. Purified bovine lung cG-BPDE was incubated in the absence or presence of 10 μM cGMP as noted for 10 min at 30°C in the presence of 8 mM $MgCl_2$, 80 μM $[\gamma\text{-}^{32}P]$ATP, and 2 μM or 10 nM of either the purified catalytic subunit of cAMP-dependent protein kinase (cAK) or the purified cGMP-dependent protein kinase (cGK). Reactions were terminated by boiling in 2 M β-mercaptoethanol and 2% SDS and analyzed by SDS-PAGE on an 8% minislab gel and subjected to autoradiography

5.9 FUNCTIONAL DOMAINS AND THEIR INTERACTIONS

In order to relate the various biochemical functions of the cG-BPDE to specific regions of the protein structure, limited proteolysis has been used to generate large fragments which retain functional features. Trypsin treatment of the rat lung cG-BPDE converts the enzyme to a lower molecular weight form (molecular mass = 105 kDa) with a concomitant increase in cGMP binding to the enzyme and marked loss of PDE catalytic activity [10]. The cGMP binding affinity of the fragment thus generated is similar to that of the intact enzyme and retains the same analog specificity. The binding fragment produced by the trypsin treatment is more electronegative than the intact enzyme, since it elutes at a higher salt concentration from DEAE columns. These results suggest that in the basal state the cGMP binding site(s) are sterically blocked by a portion of the protein which can be removed either by partial proteolysis or by conformational changes elicited by the binding of ligands such as IBMX or N^2-hexyl-cGMP at the catalytic site.

These findings have been extended in studies with the bovine lung enzyme [36,39,41]. With α-chymotrypsin treatment of the purified cG-BPDE from bovine lung, fragments can be generated which retain functional features of the enzyme, and these fragments can be separated by DEAE chromatography (Figure 5.9). A small amount of PDE activity elutes in a peak just prior to a large peak of cGMP binding activity, suggesting that there are two discrete structural domains containing the respective activities. These domains are apparently joined by a proteolytically sensitive region in the structure of this protein. The PDE activity in the fractions from DEAE chromatography correlates with the presence of a band of approximately 55 kDa on 8% SDS-PAGE. The [^{3}H]cGMP binding

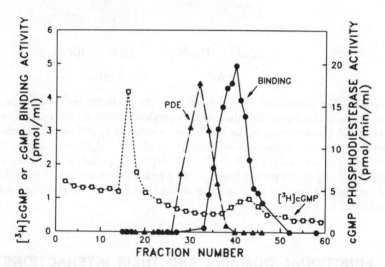

Figure 5.9 Separation of binding and catalytic domains of bovine cG-BPDE by HPLC–DEAE chromatography of an α-chymotryptic digest. The purified cG-BPDE in 20 mM NaH$_2$PO$_4$, 2 mM EDTA and 25 mM 2-mercaptoethanol was digested with α-chymotrypsin (10:1) for 30 min at 30°C, after which trace amounts of [^{3}H]cGMP were added, and the incubation proceeded at 4°C for another 30 min. The digest was then applied to a HPLC TSK-DEAE-5PW column and eluted with a linear NaCl gradient (0–200 mM) in the above buffer. cGMP binding activity ($\bullet$) was measured by incubation of aliquots of the fraction with 0.5 μM [^{3}H]cGMP followed by filtration as described in Figure 5.1. cGMP PDE activity ($\blacktriangle$) was measured at 30°C for 120 min using 20 μM [^{3}H]cGMP as substrate. Aliquots of the fractions were added to aqueous scintillant and counted ($\square$) to assess whether [^{3}H]cGMP was retained by the fragments generated by the proteolytic treatment

activity is correlated with two comigrating major bands of approximately 35 kDa and 45 kDa, and each had the amino-terminal sequence TSPRFDNDEG. Since the molecular mass of the cGMP binding fragment under non-dissociating conditions is ~ 106 kDa, these data suggest that two binding domains remain tightly associated after proteolysis and that this region of the molecule contains the structure(s) responsible for dimerization of the monomers of the native enzyme. Cyclic GMP binding fragments similar to those produced with α-chymotrypsin are also generated by proteolysis of the bovine lung cG-BPDE by V-8 protease and trypsin, lending further support to the existence of a proteolytically sensitive 'hinge' region.

Based on the current information regarding the structure of cG-BPDE, a working model of the enzyme has been constructed as shown in Figure 5.10. In this model, the 93 kDa monomers are associated through a dimerization zone located in the cGMP binding domain to form a homodimeric enzyme of molecular mass 178 kDa. Each monomer is composed of two functional domains of relatively similar sizes, i.e. a cGMP binding domain and a cGMP catalytic domain. According to this model, the enzyme undergoes several conformational changes. Interaction of cGMP (or IBMX) with the catalytic site induces a conformational change in which the allosteric cGMP binding site(s) becomes available for association with cGMP. Although only one cGMP binding site is depicted in the cGMP binding domain in Figure 5.10, kinetic evidence suggests

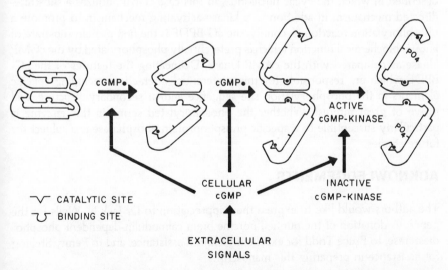

Figure 5.10 Working model for cGMP-BPDE. Hypothetical model depicting structural events which may occur when cGMP interacts with the hydrolytic site of the cG-BPDE enzyme to allow for increased binding at the allosteric site. cGMP binding at the allosteric site induces a conformational change which then allows for the phosphorylation of a specific serine residue by the cGMP-dependent protein kinase

more than one type of site. However, resolution of the total number of binding sites on the enzyme will require further study. The binding and catalytic domains appear to be joined by a length of protein which is exposed at the surface of the molecule and which is easily cleaved by a number of proteases in the presence or absence of cGMP. Upon proteolytic cleavage in the hinge region, the two functional domains can dissociate, thus removing the constraints placed on the binding site by the physical presence of the catalytic domain. Subsequent to this partial proteolysis, the cG-BPDE exhibits a marked increase in cGMP binding which is no longer sensitive to IBMX stimulation. Although this working model depicts the interaction of the catalytic and binding domain as being largely localized within a single subunit, it is possible that the regulation is effected through interactions between subunits.

Upon the association of cGMP with the allosteric site, another portion of the protein containing a specific serine residue which can be phosphorylated by either the cAMP or cGMP kinase is revealed. This phosphorylation event is strictly dependent on the conformational change elicited by the interaction of cGMP with the allosteric site and cannot be mimicked by the catalytic site-specific agent, IBMX. The phosphorylation site is a far better substrate for the cGMP kinase than for the cAMP kinase. The consequences of this phosphorylation event as it pertains to the enzyme activities have not been elucidated but two points are mentioned. Firstly, this phosphorylation event is the initial one described in which the cyclic nucleotide, in this case cGMP, utilizes a substrate-directed mechanism in addition to a kinase-activating mechanism to promote a phosphorylation reaction. Secondly, the cG-BPDE is the first protein substrate of known biochemical function which is preferentially phosphorylated by the cGMP kinase as compared with the cAMP kinase. Delineating the features of the cG-BPDE which are responsible for this preferential interaction, including specific elements in the phosphorylation site sequence and/or secondary structure, will prove of great interest. Whether the phosphorylated serine in this enzyme is particularly susceptible to a specific phosphoprotein phosphatase is a subject for future studies.

ACKNOWLEDGEMENTS

The authors would like to express their appreciation to Dr Jack N. Wells for the generous donation of the purified porcine brain calmodulin-dependent phosphodiesterase, to Bruce Todd for excellent technical assistance, and to Penny Stelling for assistance in preparing this manuscript.

REFERENCES

1. Beavo, J. A., Hardman, J. G., and Sutherland, E. W. (1971) *J. Biol. Chem.*, **46**, 3841–3846.

2. Erneaux, C., Couchie, D., Dumont, J. E., Baraniak, J., Stec, W. J., Garcia Abbad, S., Petridis, G., and Jastorff, B. (1981) *Eur. J. Biochem.*, **115**, 503–510.
3. Yamazaki, A., Sen, I., Bitensky, M. W., Casnellie, J. E., and Greengard, P. (1980) *J. Biol. Chem.*, **255**, 11619–11624.
4. Miki, N., Keirns, J. J., Marcus, F. R., Freeman, J., and Bitensky, M. W. (1973) *Proc. Natl Acad. Sci. USA*, **70**, 3820–3824.
5. Gillespie, P. G., and Beavo, J. A. (1988) *J. Biol. Chem.*, **263**, 8133–8141.
6. Dicou, E. L., and Brachet, P. (1980) *Eur. J. Biochem.*, **109**, 507–514.
7. Bulgakov, R., and Van Haastert, P. J. M. (1982) *Biochim. Biophys. Acta*, **756**, 56–66.
8. Van Haastert, P. J. M., Van Walsum, J., Van der Meer, R., Bulgakov, R., and Konijn, T. M. (1982) *Mol. Cell. Endocrinol.*, **25**, 171–182.
9. Hamet, P., and Coquil, J.-F. (1978) *J. Cyclic Nucleotide Res.*, **4**, 281–290.
10. Francis, S. H., Lincoln, T. M., and Corbin, J. D. (1980) *J. Biol. Chem.*, **255**, 620–626.
11. Hamet, P., Coquil, J.-F., Bousseau-Lafortune, S., Franks, D. J., and Tremblay, J. (1984) *Adv. Cyclic Nucleotide Protein Phosphorylation Res.*, **16**, 119–136.
12. Beavo, J. A. (1988) *Adv. Cyclic Nucleotide Res.*, **22**, 1–38.
13. Erneaux, C., Couchie, D., and Dumont, J. E. (1980) *Eur. J. Biochem.*, **104**, 297–304.
14. Russell, T. R., Terasaki, W. L., and Appleman, M. M. (1973) *J. Biol. Chem.*, **248**, 1334–1340.
15. Martins, T. J., Mumby, M. C., and Beavo, J. A. (1982) *J. Biol. Chem.*, **257**, 1973–1979.
16. Yamamoto, T., Manganiello, V. C., and Vaughan, M. (1983) *J. Biol. Chem.*, **258**, 12526–12533.
17. Klotz, U., and Stock, K. (1972) *Naunyn-Schmiedebergs Arch. Pharmacol.*, **274**, 54–62.
18. Miot, F., Van Haastert, P. J. M., and Erneaux, C. (1985) *Eur. J. Biochem.*, **149**, 59–65.
19. Hurwitz, R. L., Hansen, R. S., Harrison, S. A., Martins, T. J., Mumby, M. C., and Beavo, J. A. (1984) *Adv. Cyclic Nucleotide Protein Phosphorylation Res.*, **16**, 89–106.
20. Moss, J., Manganiello, V. C., and Vaughan, M. (1977) *Methods Enzymol.*, **252**, 5211–5215.
21. Baehr, W., Devlin, M. J., and Applebury, M. L. (1979) *J. Biol. Chem.*, **254**, 11669–11677.
22. Miki, N., Baraban, J. M., Keirns, J. J., Boyce, J. J., and Bitensky, M. W. (1975) *J. Biol. Chem.*, **250**, 6320–6327.
23. Yamazaki, A., Tatsumi, M., and Bitensky, M. W. (1988) *Methods Enzymol.*, **159**, 702–710.
24. Van Haastert, P. J. M., Van Walsum, H., Van der Meer, R. C., Bulgakov, R., and Konijn, T. M. (1982) *Mol. Cell. Endocrinol.*, **25**, 171–182.
25. Van Haastert, P. J. M., and Van Lookeren Campagne, M. M. (1984) *J. Cell Biol.*, **98**, 709–716.
26. Kesbeke, F., Baraniak, J., Bulgakov, R., Jastorff, B., Morr, M., Petridis, G., Stec, W. J., Seela, F., and Van Haastert, P. J. M. (1985) *Eur. J. Biochem.*, **149**, 179–186.
27. Mato, J. M., Woelders, H., and Konijn, T. M. (1979) *J. Bacteriol.*, **137**, 169–172.
28. Van Haastert, P. J. M., Van Walsum, H., and Pasveer, F. J. (1982) *J. Cell Biol.*, **94**, 271–278.
29. Mato, J. M., Krens, F. A., Van Haastert, P. J. M., and Konijn, T. M. (1977) *Proc. Natl Acad. Sci. USA*, **74**, 2348–2351.
30. Lincoln, T. M., Hall, C. L., Park, C. R., and Corbin, J. D. (1976) *Proc. Natl Acad. Sci. USA*, **73**, 2559–2563.

31. Ho, H. C., Wirch, E., Stevens, F. C., and Wang, J. H. (1979) *J. Biol. Chem.*, **252**, 43–50.
32. Beavo, J. A., Hansen, R. S., Harrison, S. A., Hurwitz, R. L., Martins, T. J., and Mumby, M. C. (1982) *Mol. Cell. Endocrinol.*, **28**, 387–410.
33. Moss, J., Manganiello, V. C., and Vaughan, M. (1979) *Biochim. Biophys. Acta*, **541**, 279–287.
34. Francis, S. H. (1985) *Curr. Top. Cell. Regul.*, **26**, 247–262.
35. Francis, S. H., and Corbin, J. D. (1988) *Methods Enzymol.*, **159**, 722–729.
36. Thomas, M. K., Francis, S. H., and Corbin, J. D. (1988) *14th International Congress of Biochemistry*, MO256, p. 122.
37. Coquil, J.-F. (1983) *Biochim. Biophys. Acta*, **743**, 359–369.
38. Davis, C. W., and Kuo, J. F. (1977) *J. Biol. Chem.*, **252**, 4078–4084.
39. Thomas, M. K., Francis, S. H., Todd, B. W., and Corbin, J. D. (1988) *FASEB J.*, **2**, A596.
40. Walseth, T. F., Yuen, P. S. T., Panter, S. S., and Goldberg, N. D. (1985) *Fed. Proc.*, **44**, 1854.
41. Thomas, M. K., Francis, S. H., and Corbin, J. D. In preparation.
42. Weishaar, R. E., Cain, M. H., and Bristol, J. A. (1985) *J. Med. Chem.*, **28**, 537–545.
43. Kauffman, R., Schenk, K. W., Utterback, B. G., Crowe, V. G., and Cohen, M. L. (1987) *J. Pharmacol. Exp. Ther.*, **242**, 864–872.
44. Kramer, G. L., and Wells, J. W. (1979) *Mol. Pharmacol.*, **16**, 813–822.
45. Harrison, S. A., Beier, N., Martins, T. J., and Beavo, J. A. (1988) *Methods Enzymol.*, **159**, 685–702.
46. Hoppe, J., and Wagner, K. G. (1974) *Eur. J. Biochem.*, **48**, 519–525.
47. Corbin, J. D., Ogreid, D., Miller, J. P., Suva, R. H., Jastorff, B., and Døskeland, S. O. (1986) *J. Biol. Chem.*, **261**, 1208–1214.
48. Døskeland, S. O., Vintermyr, O. K., Corbin, J. D., and Ogreid, D. (1987) *J. Biol. Chem.*, **262**, 3534–3540.
49. Keravis, T. M., Duemier, B. H., and Wells, J. N. (1987) *J. Cyclic Nucleotide Res.*, **11**, 365–372.
50. Kakiuchi, S., and Yamazaki, R. (1970) *Proc. Jpn. Acad.*, **46**, 387–392.
51. Krinks, M. H., Haiech, J., Rhoads, A., and Klee, C. B. (1984) *Adv. Cyclic Nucleotide Protein Phosphorylation Res.*, **16**, 31–47.
52. Michael, G., Muhlegger, K., Nelboeck, M., Thiessen, C., and Weimann, G. (1974) *Pharmacol. Res. Commun.*, **6**, 203–252.
53. Weber, G. (1975) *Adv. Protein Chem.*, **29**, 1–83.
54. Hamet, P., and Tremblay, J. (1988) *Methods Enzymol.*, **159**, 710–722.
55. Tremblay, J., LaChance, B., and Hamet, P. (1985) *J. Cyclic Nucleotide Protein Phosphorylation Res.*, **10**, 397–411.
56. Lincoln, T. M., and Corbin, J. D. (1983) *Adv. Cyclic Nucleotide Res.*, **15**, 139–192.
57. Glass, D. B., and Krebs, E. G. (1982) *J. Biol. Chem.*, **257**, 1196–1200.
58. Coquil, J.-F., Franks, D. J., Wells, J. N., Dupuis, M., and Hamet, P. (1980) *Biochim. Biophys. Acta*, **631**, 148–165.

6

CYCLIC CMP-SPECIFIC PHOSPHODIESTERASE ACTIVITY

R. P. Newton, S. G. Salih and J. A. Khan

Biochemistry Research Group, School of Biological Sciences, University College of Swansea, Swansea SA2 8PP, UK

6.1 INTRODUCTION

While the natural occurrence of cAMP and cGMP is firmly established, the existence of cCMP as an endogenous component of mammalian tissues has been a contentious issue. Evidence for the natural occurrence of cCMP was first described by Bloch [1], who isolated a compound from Leukemia L-1210 cell extracts which co-chromatographed with chemically synthesized cCMP and which possessed the same UV absorption spectrum and mass spectrum. Supporting evidence of the identity of the same isolated compound from normal tissues as cCMP was later produced by radioimmunoassay [2–5] and enyzme immunoassay [6]. Such claims were challenged by Wikberg and co-workers [7,8], who reported the absence of cCMP in more stringently purified extracts; the natural occurrence of cCMP has now, however, been established by means of large-scale purification of tissue extracts followed by fast atom bombardment mass spectrometry and mass analysed ion kinetic energy spectrum scanning of the resultant analyte (FAB/MIKES) [9–11].

Similar debate has existed over the occurrence of cytidylate cyclase; Ignarro and co-workers [12–15] reported the existence of an enzyme capable of cCMP biosynthesis, claims disputed by Gaion and Krishna [16,17]. Again, the identities of the major product of the putative cytidylate cyclase and of four additional novel products as 2'-O-aspartylcytidine-3',5'-cyclic monophosphate, 2'-O-glutamylcytidine-3',5'-cyclic monophosphate, cytidine 2'-monophosphate-3',5'-cyclic monophosphate and cytidine 3',5'-cyclic pyrophosphate have now been confirmed by FAB/MIKES [18,19].

The report of a phosphodiesterase (PDE) capable of cCMP hydrolysis [20] suggested that a third component of a cyclic nucleotide system analogous to those for cAMP and cGMP also existed for cCMP. Purification of this enzyme

Cyclic Nucleotide Phosphodiesterases: Structure, Regulation and Drug Action
Edited by J. Beavo and M. D. Houslay   © 1990 John Wiley and Sons Ltd

from pig liver to homogeneity, however, revealed that it was not specific for cCMP but was multifunctional, being able to hydrolyse both 3′,5′- and 2′,3′-cyclic nucleotides and having activity with both purines and pyrimidines [21–26]. As with another cCMP-PDE reported in rat liver [27], this multifunctional enzyme displayed its greatest activity with cCMP; however, its wide specificity precluded the deduction that the hydrolysis of this pyrimidine cyclic nucleotide is its primary function and that its existence is consequently evidence of a metabolic role for cCMP. Further study showed that the rat liver enzyme [27] was also similarly multifunctional [28]. Work in our laboratories indicated that rat tissues contained more than one PDE which possessed activity with cCMP as substrate; the two most active are the multifunctional enzyme and an apparent cCMP-specific PDE [29–31]. The purification, properties and potential significance of the latter enzyme are reviewed here.

6.2   DEVELOPMENT OF AN ASSAY FOR cCMP-PDE ACTIVITY

The assay of cCMP-PDE activity is based upon methods previously developed for the assay of cAMP- and cGMP-PDE, in which the radiolabelled mononucleotide product is further hydrolysed to a nucleoside by a nucleotidase [32] and the nucleoside then separated out by passage through an ion exchange column. The adaption of this method for cCMP-PDE had to be carefully evaluated, since the chromatographic behaviour of cytosine-containing compounds on ion exchange differs to a significant extent from that of adenine and guanine nucleotides; in addition, the activity of several types of nucleotidase had to be tested to ensure that an adequate coupling enzyme was identified which would utilize a pyrimidine instead of a purine substrate. Using Dowex 1 as the ion exchanger, a series of column sizes and eluents were tested; in each case the separation of cytidine from cCMP was only adequate when the column length (8 × 1.8 cm) had reduced cytidine recovery to < 80% and running time became impracticable. Combination of ion exchange with the molecular sieving effect of QAE Sephadex proved more successful. A 0.2 × 1 cm column of QAE 25 Sephadex when eluted with 20 ml water retained cCMP, CMP and cytidine; all three compounds were eluted in 5 ml by 5 mM NaCl. When 30 mM ammonium formate was used as the eluting solvent, cytidine was eluted in the first 2 ml and was satisfactorily resolved from the other compounds. The use of QAE Sephadex had a much superior recovery over Dowex, with recovery being routinely > 95% of cytidine added in control experiments, and this was adopted as the method of separation.

The choice of a coupling enzyme with the ability to hydrolyse CMP to cytidine appeared to rest between purified 5′-nucleotidase and three snake venom preparations, those from *Crotalus durissus terrificus*, *Crotalus adamanteus* and *Ophiophagus hannah*. The effectiveness of each was tested by incubating 200 μl of 10 mg/ml CMP with 100 μl of the enzyme or venom solution in the presence of 1.8 mM MgSO$_4$ in a total volume of 1 ml at pH 7.4 and 37°C. In addition to the ability to hydrolyse CMP, it was necessary to select a coupling enzyme with minimal ability to hydrolyse CMP. The above assays were therefore repeated with cCMP replacing CMP as substrate. The ratio of CMP hydrolysis/cCMP hydrolysis was less than unity for *Ophiophagus hannah* venom, 2–30 for *Crotalus durissus terrificus* venom, 2–50 for purified 5′-nucleotidase and 25–50 for *Crotalus adamanteus* venom. Although the latter two maximum ratios were similar, in view of the inconsistency observed with commercially purified nucleotidase, the crude *Crotalus adamanteus* venom was selected as the coupling enzyme despite its nucleotidase-specific activity being an apparent order of magnitude lower than that of purified nucleotidase.

The routine assay protocol utilized was thus as follows. The incubation was carried out in a final total volume of 300 μl of 50 mM Tris-HCl pH 7.4 containing 10 nmol of $FeSO_4.7H_2O$ with a final substrate concentration of 10 mM cCMP containing 40 000–80 000 d.p.m. [5-^{3}H]cCMP (15–30 Ci/mmol). Incubation was carried out at 37°C for 8 min, then the PDE was inactivated by heating at 90°C for 1 min. One thousand units of *Crotalus adamanteus* venom were added after cooling and the mixture was reincubated at 37°C for 1 h. After a second heat inactivation the supernatant was then passed through a QAE 25 Sephadex column (1 × 0.2 cm) and eluted with 30 mM ammonium formate, with the first 2 ml being retained and their radioactive content determined. No-substrate, no-enzyme and boiled enzyme controls were routinely employed.

6.3 TISSUE AND AGE DISTRIBUTION OF cCMP-PDE ACTIVITY

Analysis of cCMP-PDE activity in tissue extracts from 12-week-old adult Lister Hooded male and female rats is depicted in Table 6.1. Each tissue examined

Table 6.1 Cyclic CMP-PDE activities of rat tissues

Tissue	nmol cCMP hydrolysed per min/per g wet wt
Liver	108 ± 43
Heart	92 ± 31
Spleen	86 ± 38
Blood	20 ± 36
Lung	74 ± 38
Sketetal muscle	98 ± 32
Brain	122 ± 71
Kidney	53 ± 28
Testes	87 ± 159
Ovary	63 ± 40

For details of assay procedure see text. Each datum is the mean of at least six replicates in quadruplicate from the tissues of 12-week-old adult Lister Hooded rats.

displayed activity of a similar order; the largest variation between samples was found in the testes extracts, while no significant difference in activity was observed between equivalent tissues from male and female animals.

The variation of cCMP-PDE activity in the liver of male rats over a 60 week period is shown in Figure 6.1. There is a gradual increase in the activity over the first 30 weeks until a peak is reached, which then undergoes a gradual decay.

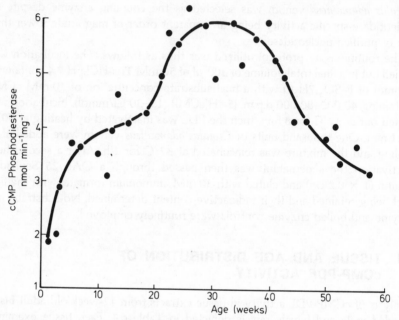

Figure 6.1 Variation of cCMP-PDE activity of rat liver with age. Activity is expressed per mg dry weight of tissue; each point is the mean of at least four replicates determined in quadruplicate

6.4 ISOLATION OF A cCMP-SPECIFIC PDE

The cCMP-PDE activity from rat liver was subjected to a sequential purification procedure of ammonium sulphate fractionation, gel filtration, two ion exchange chromatographic steps, preparative electrophoresis and two affinity chromatography stages [31]. At each stage selection was made for maximum specificity, not for maximum activity, so the cAMP- and cGMP-PDE activities of each cCMP-PDE containing fraction were also determined.

Ammonium sulphate fractionation yielded three fractions of cCMP-PDE activity, with the third fraction (60–90% sat.) showing the greatest activity and greatest specificity for cCMP as substrate (Figure 6.2a). Gel filtration of this

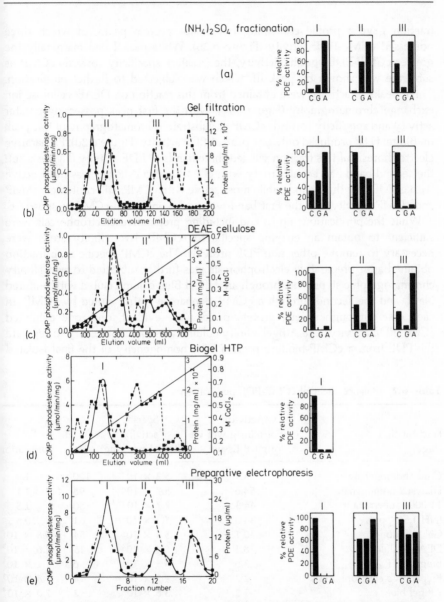

Figure 6.2 Sequential purification of cCMP-PDE. The 60–90% sat. $(NH_4)_2SO_4$ precipitate (a) was further purified by (b) Sephadex G200 gel filtration, (c) DEAE—cellulose, (d) Biogel HTP, and (e) preparative electrophoresis. ●, cCMP-PDE; ■, protein concentration. Histograms show the relative activity of each PDE peak with cCMP, cGMP and cAMP. (Reprinted with permission from 'Cyclic CMP phosphodiesterase: isolation, specificity and kinetic properties', R. P. Newton and S. G. Salih, *Int. J. Biochem.*, **18**, No. 9, 743–752, 1986, Pergamon Press plc)

fraction gave a profile containing five major protein peaks, of which three possessed cCMP-PDE activity (Figure 6.2b). While peak I had marginally the greater cCMP-PDE specific activity, the greatest specificity towards cCMP as substrate was shown by peak III, which was subjected to further purification. Three peaks of activity were obtained from this fraction on DEAE–cellulose ion exchange chromatography (Figure 6.2c), with the first peak possessing greater activity and specificity towards cCMP as substrate. Chromatography of this peak on Biogel HTP produced only one peak of PDE activity (Figure 6.2d). Preparative electrophoresis of this fraction yielded three bands of PDE activity (Figure 6.2e): the second band had equal activity with cCMP and cGMP but was most active with cAMP; the third band, while most active with cCMP, still contained cAMP and cGMP activity, but the first band was active only with cCMP as substrate.

While the purifications up to and including preparative electrophoresis were sufficient to obtain an enzyme specific for cCMP, two affinity steps were necessary to remove other non-PDE proteins. The cCMP-specific PDE fraction obtained after preparative electrophoresis was further subjected to two affinity chromatographic steps, one through an Affigel Blue column eluted with buffered NaCl, and the second through a cCMP–Sepharose matrix eluted by cCMP. In each case, although several protein peaks were evident, only one exhibited cCMP-PDE activity. The Affigel affinity column was found necessary to separate the PDE from a cCMP-binding protein. The homogeneity of the final isolated

Table 6.2 Balance sheet of cCMP-PDE purification

Fractionation stage	Activity ((nmol/min)/g wet wt original tissue)	Specific activity ((nmol/min)/mg)	Yield (%)	Fold purification
Crude homogenate	55.7	5.2×10^{-1}	100	1.0
Dialysed homogenate	54.6	5.7×10^{-1}	98	1.1
11 000g supernatant	44.3	7.7×10^{-1}	80	1.5
$(NH_4)_2SO_4$ ppt.	34.0	1.1×10^1	61	20.6
Gel filtration	30.1	5.1×10^2	54	9.8×10^2
DEAE–cellulose	8.3	8.3×10^2	15	1.6×10^3
Biogel HTP	3.6	5.5×10^3	6.5	1.1×10^4
Preparative electrophoresis	2.2	9.2×10^3	4	1.8×10^4
Affigel blue	1.2	3.6×10^4	2	7.0×10^4
cCMP–Sepharose	0.8	4.8×10^4	1.5	8.8×10^4

The cCMP-PDE activities and protein concentrations of the pooled reclaimed fractions at each stage are shown. The percentage yield and purification are expressed relative to the original total and specific activities respectively in the original tissue homogenate. For experimental detail see ref. 31. (Reprinted with permission from 'Cyclic CMP phosphodiesterase: isolation, specificity and kinetic properties', R. P. Newton and S. G. Salih, *Int. J. Biochem.* **18**, No. 9, 743–752, 1986, Pergamon Press plc)

PDE was confirmed by isoelectric focusing and HPLC. The balance sheet of the purification process showed the enzyme to have been purified 88 000-fold and to contain 1.5% of the original total activity (Table 6.2). The sequence of stages was varied; the data obtained were consistent with the cCMP-specific enzyme being non-artefactual in nature and confirmed that the sequence described was optimal.

6.5 SPECIFICITY AND OTHER PROPERTIES OF cCMP-PDE

The specificity of the enzyme was determined by using both the 2',3'- and 3',5'-isomers of cCMP, cAMP, cGMP, cUMP and cIMP at substrate concentrations between 100 nM and 50 mM. While the K_ms with each substrate were of a similar order, the values of V_{max}, V_{max}/K_m and velocity at an approximate theoretical physiological substrate concentration (5 μM) for the PDE with each of the alternative cyclic nucleotide substrates were less than 1% of the values obtained with cCMP (Table 6.3), suggesting that the enzyme has absolute specificity for the substrate.

The purified PDE was found to have a pI of 4.2–4.4, molecular mass of 28 kDa and displayed Michaelis–Menten kinetics (Figure 6.3). The catalytic activity was stable when the enzyme was stored frozen, although sensitivity to effectors decayed with storage. Examinations of the subcellular distribution by means of density gradient centrifugation in a zonal rotor revealed three sites of activity.

Table 6.3 Specificity of cCMP-PDE

Substrate	V_{max} $((\mu mol/min)/mg)$	K_m (mM)	Activity at 5 μM substrate $((nmol/min)/mg)$
3',5'-cCMP	48.6	9.0	27.1
3',5'-cAMP	2.9×10^{-1}	11.2	1.3×10^{-1}
3',5'-cGMP	1.6×10^{-1}	14.7	5.4×10^{-1}
3',5'-cUMP	9.2×10^{-2}	8.7	5.2×10^{-2}
3',5'-cIMP	1.5×10^{-1}	14.3	5.2×10^{-2}
2',3'-cCMP	2.8×10^{-1}	18.4	7.6×10^{-2}
2',3'-cAMP	4.7×10^{-1}	17.5	1.4×10^{-1}
2',3'-cGMP	3.7×10^{-1}	18.6	9.9×10^{-2}
2',3'-cUMP	3.5×10^{-1}	21.3	8.2×10^{-2}
2',3'-cIMP	1.8×10^{-1}	20.9	4.3×10^{-2}

The activity of the final, purified preparation of cCMP-PDE after affinity chromatography was determined with purine and pyrimidine 3',5'- and 2',3'-cyclic nucleotide substrates. For experimental detail see ref. 31. (Reprinted with permission from 'Cyclic CMP phosphodiesterase: isolation, specificity and kinetic properties', R. P. Newton and S. G. Salih, *Int. J. Biochem.*, **18**, No. 9, 743–752, 1986, Pergamon Press plc)

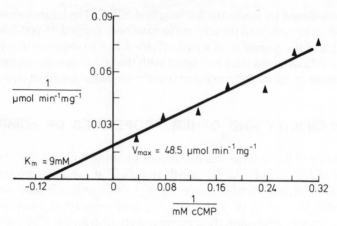

Figure 6.3   Lineweaver-Burk plot for isolated cCMP-PDE

The major site was the soluble phase, but two other active bands were the microsomal and nuclear fractions. Amino acid analysis of the enzyme revealed aspartate and glutamate as the two commonest residues (Table 6.4): the presence of glutamate and aspartate at such levels would seem to be at least partially responsible for the acidic isoelectric point of cCMP-PDE.

Table 6.4   Amino acid composition of cCMP-PDE

Asx	44	Cys	15	Tyr	7
Glx	28	Gly	15	Met	6
Pro	16	Ser	14	Phe	6
Ala	16	Thr	11	Leu	5
Try	16	Iso	8	His	3
Arg	16	Lys	8	Val	3

Total moles amino acid/mole enzyme. For experimental detail see ref. 31. (Reprinted with permission from 'Cyclic CMP phosphodiesterase: isolation, specificity and kinetic properties', R. P. Newton and S. G. Salih, *Int. J. Biochem.*, **18**, No. 9, 743–752, 1986, Pergamon Press plc)

6.6   ANTI-cCMP-PDE SERUM

Polyclonal antibodies against the cCMP-PDE raised in albino Swiss mice showed little or no binding affinity for Ca^{2+}–CaM (calmodulin)-activated, cGMP-stimulated, low-K_m cAMP-PDE or light-activated cGMP-PDE from a variety of tissue sources [33]. The multifunctional PDE possessed binding affinity for the

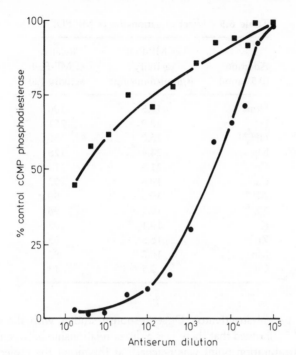

Figure 6.4   Effect of anti-cCMP-PDE serum on PDE activity. Polyclonal antibodies were raised in albino Swiss mice against cCMP-PDE; effects of a series of dilutions of this antiserum against cCMP-PDE (●) and multifunctional PDE (■) are shown

anti-cCMP-specific PDE serum greater than that of the enzymes listed above, but to attain comparable degrees of inhibition of enzyme activity the requisite concentration of antiserum was two orders of magnitude greater with the multifunctional PDE than with the cCMP-specific enzyme (Figure 6.4). This crossreactivity indicated that some structural similarity between the two types of cCMP-PDE exists, but also that there are very significant differences, as will be discussed later.

6.7   EFFECT OF CATIONS

The purified cCMP-specific PDE was active in the absence of added cations, but was stimulated by Fe^{2+}, NH_4^+, Mg^{2+} and Co^{2+} when 0.5 nmol of their salts were added to the standard incubation mixture (Table 6.5). Both chloride and sulphate salts were used and no significant difference due to anion effects was observed. No significant effect upon cCMP-PDE activity was observed with Ca^{2+}, Al^{3+}, Na^+ or K^+, while Zn^{2+} and Mn^{2+} were slightly inhibitory and

Table 6.5   Effect of cations on cCMP-PDE

Addition (0.5 nmol)	cCMP-PDE activity ($(\mu mol/min)/mg$)	Relative cCMP-PDE activity (%)
None	19.5	100
Fe^{2+}	48.9	251
NH_4^+	36.0	185
Mg^{2+}	24.9	128
Co^{2+}	21.8	112
Ca^{2+}	19.8	102
Al^{3+}	19.3	99
Na^+	19.1	98
K^+	19.1	98
Zn^{2+}	16.5	85
Mn^{2+}	16.2	83
Cu^+	2.5	13

Cu^+ was a potent inactivator. Dose–response studies with the most potent activator, Fe^{2+}, showed that there was a direct relationship between activity and Fe^{2+} until a saturation point was reached; at this point the molecular ratio of Fe^{2+} : enzyme was of the order of 10^3 : 1, while the Fe^{2+} : cCMP molecular ratio was of the order of $1 : 10^3$, suggesting that the effect of the cation was upon the enzyme rather than the substrate. (Interestingly Fe^{2+} appears to have a similar relationship with the biosynthetic enzyme, cytidylate cyclase (Newton, Salvage and Hakeem, unpublished observations).

6.8   EFFECT OF OTHER NUCLEOTIDES AND RELATED COMPOUNDS

The activity of the purified cCMP-PDE was examined in the presence of a range of other nucleotides and related compounds considered as potential effectors (Table 6.6). Classical PDE inhibitors such as theophylline and methylisobutylxanthine exerted little or no effect, nor did papavarine. Imidazole exerted a small inhibitory effect, while three nitrophenyl derivatives, the phosphate, bisphosphate and phosphonate, were inhibitory to a greater extent. Inorganic phosphate was a potent inhibitor, while pyrophosphate had a negligible effect. Both deoxy-cGMP and deoxy-cAMP were potent inhibitors, while deoxy-cCMP was the most potent inhibitor examined. Of the three dinucleotides tested only GpG had a significant inhibitory effect, while adenosine tetraphosphate was found to be a more potent inhibitor. The 3',5'-cyclic nucleotides tested each inhibited the

Table 6.6 Effect of nucleotides and related compounds on cCMP-PDE activity

Addition (10 mM)	Relative cCMP-PDE (%)	Addition (10 mM, *2 mM)	Relative cCMP-PDE (%)
None	100	2′,3′-cCMP	51
1-Methyl-3-		2′,3′-cAMP	32
isobutylxanthine	98	2′,3′-cGMP	29
Theophylline	96	2′,3′-cUMP	57
Papavarine	101	2′,3′-cIMP	57
Imidazole	86	Uracil	97
Nitrophenylphosphate	84	Uridine	98
Nitrophenylbisphosphate	78	Adenine	105
Nitrophenylphosphonate	80	Adenosine	102
Phosphate	12	Guanine	104
Pyrophosphate	97	Guanosine	103
2′-Deoxy-3′,5′-cGMP	18	Cytosine	118
		Cytidine	121
2′-Deoxy-3′,5′-cAMP	26	CTP	86
		CDP	80
2′-Deoxy-3′,5′-cCMP	8	CMP	85
		*Putrescine	109
GpG	83	*Spermine	118
CpC	101	*Spermidine	114
ApC	98	*3′,5′-cCDP	68
3′,5′-cAMP	49	*2′-O-Phospho-3′,5′-cCMP	78
3′,5′-cGMP	41		
3′,5′-cUMP	48	*2′-O-Aspartyl-3′,5′-cCMP	81
3′,5′-cIMP	73		
Dibutyryl 3′,5′-cAMP	89	*2′-O-Glutamyl-3′,5′-cCMP	80
Dibutyryl 3′,5′-cGMP	88		
Dibutyryl 3′,5′-cCMP	71		

enzyme, as did dibutyryl-cCMP and the 2′,3′-cyclic nucleotides. The presence of phosphate in these compounds for inhibition to occur appears to be essential, since uracil, uridine, adenine, adenosine, guanine and guanosine exhibited no significant effect. Cytosine and cytidine both activated the cCMP-PDE activity, while the corresponding non-cyclic nucleotides inhibited. The four cCMP derivatives produced as side products of the cytidylate cyclase reaction [18,19] which contain either an extra aminoacyl or phosphate group were also potent inhibitors. Polyamines, previously reported to be potent inhibitors of cCMP-PDE activity in total cell measurements [27], exhibited, in the form of spermidine, putrescine and spermine, activation but only to a minor degree. Glutathione (152%), cysteine (124%) and mercaptoethanol (128%) also stimulated the enzyme, suggesting a sensitivity of the enzyme system to oxidation.

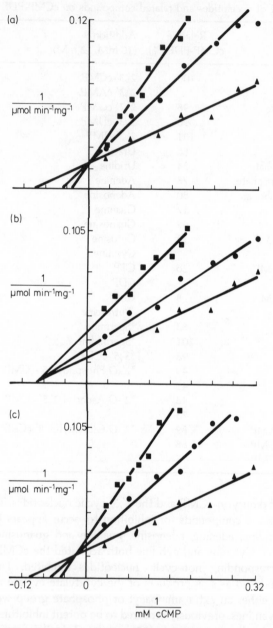

Figure 6.5 Lineweaver-Burk plots for cCMP-PDE in the presence of inhibitors. Plots are constructed in the absence of effector (▲), and in the presence of low (●) and high (■) concentrations of (a) 3′,5′-cUMP, (b) 3′,5′-cAMP, and (c) 2′,3′-cAMP

A full systematic analysis of each of the listed effects on cCMP-PDE has not yet been completed. Studies of the $3',5'$-cyclic nucleotides, $2',3'$-cyclic nucleotides, $2'$-deoxy-$3',5'$-cyclic nucleotides and the aminoacyl and phospho derivatives of $3',5'$-cCMP indicated that inhibition varied from competitive, to noncompetitive, to mixed, depending upon the nature of the nitrogenous base and the position of the cyclic phosphate ring. For example, an analogous pyrimidine isomer, $3',5'$-cUMP, behaved as a competitive inhibitor, the analogous purine isomer, $3',5'$-cAMP, displayed a noncompetitive effect, while the alternative isomer, $2',3'$-cAMP, exhibited mixed-type inhibition (Figure 6.5). The aminoacyl and phospho derivatives of cCMP were each competitive inhibitors.

6.9 OTHER FACTORS EXERTING EFFECTS UPON cCMP-PDEs

A range of hormones, each at a number of concentrations, have been tested in the standard cCMP-PDE assay. Adrenaline, thyroxine, insulin, vasopressin, oxytocin, glucagon, gonadotrophin and luteinizing hormone did not exhibit any significant effect on the enzyme's activity (Table 6.7). Aldosterone, a prostaglandin mixture and thyrocalcitonin each exerted an inhibitory effect which appeared greater than that which could be attributable to experimental error, but the response did not relate directly to doses of hormone. Diethylstilboestrol, thyroglobulin and androsterone each stimulated the PDE activity, with diethylstilboestrol being the most potent activator. However, as with the inhibitory effects of hormones, no dose responsiveness was apparent. With dihydrotestosterone not only was an activating effect obvious, but it was also an apparently dose-dependent response.

Perhaps of greater physiological significance was the observation of the effects of three endogenous proteins, which also produced dose-dependent responses. One, designated in-house as F9, had a molecular mass of 78 kDa and was activating in effect; a second, designated F13, had a molecular mass of 46 kDa and was inhibitory in effect (Figure 6.6). Neither protein had any apparent effect upon non-cCMP-related PDEs. The third endogenous protein exerting an effect on the cCMP-PDE was Ca^{2+}–CaM, which had an unexpected inhibitory effect upon the enzyme (Figure 6.6). The observed effect of F9 and F13 could be abolished by preincubation with trypsin and retained in preincubation with trypsin and trypsin inhibitor; the inhibition of cCMP-PDE by CaM was abolished by inclusion of fluphenazine in the assay mixture. Preliminary data suggest that at least part of the cellular cCMP-PDE exists in complexes also containing one or more of CaM and the two novel endogenous proteins.

Table 6.7   Effect of hormones on cCMP-specific PDE activity

Addition	Activity showing maximum change as % of control
None	100
Adrenaline	96
Thyroxine	99
Insulin	97
Vasopressin	98
Oxytocin	98
Glucagon	96
Gonadotrophin	93
Luteinizing hormone	98
Aldosterone	83
Prostaglandin mixture	88
Thyrocalcitonin	89
Diethylstilboestrol	225
Thyroglobulin	185
Androsterone	198
Progesterone	113
Dihydrotestosterone	209

Hormone preparations were added to the standard cCMP-PDE assay protocol in a concentration range from 0.1 ng/ml to 100 µg/ml; activity is expressed as that showing the maximum change as percentage of control in the absence of added hormone.

6.10   COMPARISON OF cCMP-PDE WITH MULTIFUNCTIONAL PDE

Only two enyzmes with reported cCMP-hydrolysing activity have been purified to homogeneity: the rat liver cCMP-specific PDE reviewed here [29–31], and the multifunctional PDE isolated from pig liver by Kuo's group [20–26]. Although the cCMP-specific enzyme has a slightly lower molecular mass than the multifunctional enzyme, the amino acid composition, in particular the fact that the former enzyme contains significantly more cysteine, aspartate and arginine residues per mole than the latter enzyme, suggests that the cCMP-PDE is not just a large fragment of the multifunctional PDE. While the difference in size and composition could be interpreted as merely reflecting the two enzymes' origin from two different species, our evidence and that of Cheng and Bloch [27] indicates that a multifunctional PDE, closely similar to the pig enzyme, also exists in rat tissue and is distinct from the cCMP-PDE.

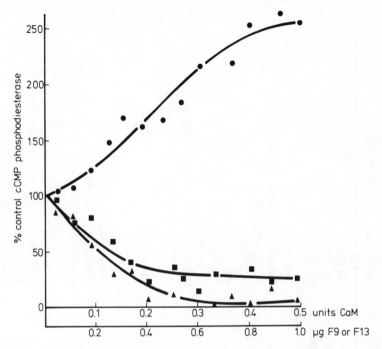

Figure 6.6 Effect of endogenous proteins on cCMP-PDE activity. The activity of isolated cCMP-PDE was determined in the presence of Ca^{2+}–CaM (■), F9 (●) and F13 (▲). For further details see text

The cCMP-specific and multifunctional PDEs have several properties in common (Table 6.8). These include similarity in pH optima and isoelectric points, insensitivity to classical PDE inhibitors such as methylisobutylxanthine, inhibition by phosphate and by 2′,3′-, 3′,5′- and 2′-deoxy-cyclic nucleotides, stimulation by cytosine and cytidine and absence of an absolute requirement for cation as cofactors. They differ in that V_{max} for cCMP-PDE is an order of magnitude greater than that of the multifunctional enzyme, while the K_m of the former is also an order of magnitude greater than that of the latter. No evidence is available to suggest that the multifunctional enzyme is sensitive to the endogenous proteins F9 and F13 which activate and inhibit respectively cCMP-PDE; the cCMP-specific enzyme differs also in that it is stimulated by antioxidants such as glutathione and the two enzymes also show significant differences in their responses to added cations. The most clear-cut difference is that one enzyme displays absolute specificity for cCMP as substrate; this enzyme is unique in that it is inhibited by CaM, and these two properties appear indicative of these two enzymes possessing different cellular functions.

Table 6.8 Comparison of cCMP-PDE and multifunctional PDE

	cCMP-PDE [29–31]	Multifunctional PDE [21–26]
M_r	28 000	33 000
Amino acid composition	cCMP-PDE enzyme contains significantly more cysteine, aspartate and arginine residues per molecule	
pH optimum	7.2	7.0
Stability	Catalytic activity stable when stored frozen	Unstable in absence of carrier protein
V_{max}	48.6 (μmol/min)/mg	4.2 (μmol/min)/mg
K_m	9 mM	182 μM
pI	4.2–4.4	4.6
Specificity	Less than 1% activity with substrate other than cCMP	Hydrolyses 3',5'- and 2',3'-cCMP, cAMP and cGMP
Metal ion	Active in absence of metal cofactor, stimulated by Fe^{2+}, Fe^{3+}, Mg^{2+}, Co^{2+}, inhibited by Zn^{2+}, Mn^{2+}, Cu^{2+}, Cu^+, Ca^{2+}	Active in absence of metal cofactor, stimulated by Mg^{2+}, Ca^{2+}, Mn^{2+},
Calmodulin	Inhibited	Insensitive
Sensitivity to effectors		
(a)	F9, F13 are proteinaceous effectors	None reported
(b)	Inhibited by phosphate	Inhibited by phosphate
(c)	Inhibited by 2',3'- and 3',5'-cyclic nucleotides and deoxycyclic nucleotides	Competitively inhibited by 2',3'- and 3',5'- cyclic nucleotides and deoxycyclic nucleotides
(d)	Stimulated by cytosine and cytidine	Stimulated by cytosine and cytidine
(e)	Insensitive to methylisobutylxanthine	Insensitive to methylisobutylxanthine
(f)	Stimulated by hormones, e.g. testosterone	None reported
(g)	Stimulated by cysteine, glutathione and mercaptoethanol	Insensitive to organic antioxidants

6.11 CYCLIC CMP-PDE AS A TARGET FOR PHARMACOLOGICAL AGENTS

The responses to effectors by cCMP-PDE suggest that it possesses great potential as a pharmacological target in that the inhibitor pattern identified differs from those reported for other PDEs elsewhere in this text. The fact that the natural substrate of the enzyme is a pyrimidine cyclic nucleotide rather than the purine substrate of other PDEs lends optimism to the view that synthetic inhibitors which will be selective for cCMP-PDE can be designed and produced. In order to justify such an exercise the physiological function of both the enzyme and its substrate needs to be elucidated and any potential role and application of developed specific inhibitors identified.

In view of the absolute specificity of the cCMP-PDE, it is reasonable to conclude that hydrolysis of this pyrimidine cyclic nucleotide is the prime function of this enzyme. However, as the K_m for cCMP of the multifunctional PDE is an order of magnitude lower than that of the cCMP-PDE, and as the former appears to be present at higher concentration in most mammalian tissues [33], then, in the absence of (1) other competing cyclic nucleotide substrates and (2) compartmentation effects, it can be deduced that the hydrolysis of cCMP to CMP would be catalysed predominantly by the multifunctional PDE. In the presence of high cellular concentrations of cAMP and cGMP, cCMP hydrolysis would on the other hand be predominantly catalysed by the cCMP-specific PDE, thereby sensitizing the cCMP turnover rate to changes in concentrations of dihydrotestosterone, CaM, the two endogenous proteinaceous effectors of cCMP-PDE and perhaps Fe^{2+} availability and redox status. The relative concentrations of cAMP and cGMP would on this basis appear to be crucial factors in determination of the rate of breakdown of cCMP.

The existence of a PDE specific for cCMP hydrolysis, together with a cyclase specific for cCMP synthesis [18,19], is evidence in support of there being a specific, significant function or functions for cCMP. As yet this role remains unelucidated; the most popular hypothesis has been a potential involvement of cCMP in the regulation of cellular proliferation rates ([34] for review) and while older data concerning changes in putative cCMP concentrations in response to, for example, luteinizing hormone release factors [35], long-acting thyroid stimulator [36] and thyroid stimulating hormone [37], remains debatable, indirect evidence of a biochemical regulatory role for cCMP also exists, including effects of exogenously administered cCMP upon DNA [38], RNA [39], protein [40] and steroid synthesis [41], and Ca^{2+} efflux [42]. Plausible interpretations of these data include a specific regulatory role for cCMP and, alternatively, that cCMP merely interferes by, for example, acting as an agonist or antagonist of cAMP and cGMP. Collectively, the data concerning cCMP potential functions remain ambiguous, but recent data [43,44] are in support of the concept of cCMP involvement in cellular proliferation, and we have now, with an unambiguous radioimmunoassay for cCMP, produced data confirming the earlier observations that cCMP concentrations increase in rapidly dividing cells [3,45]. The depressed level of cCMP-PDE activity in young tissue, described in this review, and in Morris hepatomas [46], is consistent with such a hypothesis.

6.12 CONCLUSIONS

In summary it can be said that cCMP-PDE is a PDE with some unique properties and which possesses considerable potential as a pharmacological target. At present the absence of an established function of cCMP does not justify attempts to develop specific effectors of cCMP-PDE on the basis of clearly defined potential applications. It is the view of these authors that cCMP does possess a significant, but as yet not fully elucidated, metabolic role, and that once the functions of this third cyclic nucleotide are better understood then the potential of the cCMP-specific PDE as a pharmaceutical target will be readily appreciated and developed. Indeed, the utilization of cCMP-PDE-selective effectors may well prove to be instrumental in identifying the functions of cCMP.

REFERENCES

1. Bloch, A. (1974) *Biochem. Biophys. Res. Commun.*, **58**, 652–659.
2. Cailla, H. L., Roux, D., Delaage, M., and Goridis, C. (1978) *Biochem. Biophys. Res. Commun.*, **85**, 1503–1509.
3. Murphy, B. E., and Stone, J. E. (1979) *Biochem. Biophys. Res. Commun.*, **89**, 122–128.
4. Hachiya, T., Katjita, Y., Yoshimura, M., Ijichi, H., Miyazachi, T., Ochi, Y., and Hosoda, S. (1980) *Abs. 2nd Asia and Oceania Congress of Medicine*, 137.
5. Sato, T., Kuninaka, A., and Yoshino, H. (1982) *Anal. Biochem.*, **123**, 208–218.
6. Yamamato, I., Takai, T., and Tsuji, J. (1982) *Immunopharmacology*, **4**, 331–340.
7. Wikberg, J. E. S., and Wingren, G. B. (1981) *Acta Pharmacol. Toxicol.*, **49**, 52–58.
8. Wikberg, J. E. S., Wingren, G. B., and Anderson, R. G. (1981) *Acta Pharmacol. Toxicol.*, **49**, 452–454.
9. Newton, R. P., Salvage, B. J., and Salih, S. G. (1983) *Biochem. Soc. Trans*, **11**, 354–355.
10. Newton, R. P., Salvage, B. J., and Salih, S. G. (1984) *Adv. Cyclic Nucleotide Protein Phoshorylation Res.*, **17a**, 130.
11. Newton, R. P., Salvage, B. J., Salih, S. G., and Kingston, E. E. (1984) *Biochem. J.*, **221**, 655–673.
12. Cech, S. Y., and Ignarro, L. J. (1977) *Science*, **198**, 1063–1065.
13. Cech, S. Y., and Ignarro, L. J. (1978) *Biochem. Biophys. Res. Commun.*, **80**, 119–125.
14. Ignarro, L. J. (1979) *Science*, **203**, 673.
15. Ignarro, L. J., and Cech, S. Y. (1979) In *ICN-UCLA Symposium on Molecular and Cellular Biology: Modulation of Protein Function* (Atkinson, D. E. and Fox, C. F., eds), Academic Press, New York, pp. 315–333.
16. Gaion, R. M., and Krishna, G. (1979) *Biochem. Biophys. Res. Commun.*, **86**, 105–111.

17. Gaion, R. M., and Krishna, G. (1979) *Science,* **203**, 672.
18. Newton, R. P., Kingston, E. E., Salvage, B. J., Beynon, J. H., Hakeem, N. A., and Wassenaar, G. (1986) *Biochem. Soc. Trans,* **14**, 964–965.
19. Newton, R. P., Hakeem, N. A., Salvage, B. J., Wassenaar, G., and Kingston, E. E. (1988) *Rapid Commun. Mass Spec.,* **2**, 118–126.
20. Kuo, J. F., Brackett, N. L., Shoji, M., and Tse, J. (1978) *J. Biol. Chem.,* **153**, 2518–2521.
21. Kuo, J. F., Shoji, M., Helfman, D. M., and Brackett, N. L. (1979). *ICN-UCLA Symposium on Molecular and Cellular Biology: Modulation of Protein Function* (Atkinson, D. E. and Fox, C. F., eds), Academic Press, New York, pp. 335–355.
22. Helfman, D. M., Brackett, N. L., and Kuo, J. F. (1978) *Proc. Natl Acad. Sci. USA,* **75**, 4422–4425.
23. Helfman, D. M., Shoji, M., and Kuo, J. F. (1981) *J. Biol. Chem.,* **256**, 6327–6334.
24. Helfman, D. M., and Kuo, J. F. (1981) *Biochem. Pharmacol.,* **21**, 43–47.
25. Helfman, D. M., and Kuo, J. F. (1982) *J. Biol. Chem.,* **257**, 1044–1047.
26. Helfman, D. M., Katoh, N., and Kuo, J. F. (1984) *Adv. Cyclic Nucleotide Res.,* **16**, 403–416.
27. Cheng, Y. C., and Bloch, A. (1978) *J. Biol. Chem.,* **253**, 2522–2529.
28. Conrad, D., and Bloch, A. (1980) *Adv. Cyclic Nucleotide Res.,* **11**, 1840.
29. Newton, R. P., and Salih, S. G. (1983) *Biochem. Soc. Trans,* **11**, 355–356.
30. Newton, R. P., and Salih, S. G. (1984) *Adv. Cyclic Nucleotide Protein Phosphorylation Res.,* **17a**, 54.
31. Newton, R. P., and Salih, S. G. (1986) *Int. J. Biochem.,* **18**, 743–752.
32. Thompson, W. J., and Appleman, M. M. (1971) *Biochemistry,* **10**, 311–329.
33. Salih, S. G., Khan, J. A., and Newton, R. P. (1988) *Biochem. Soc. Trans,* **16**, 774–775.
34. Anderson, T. R. (1981) *Mol. Cell. Endocrinol,* **28**, 373–385.
35. Hierowski, M. T., Warring, A. J., and Schaly, A. V. (1981) *Biochim. Biophys. Acta,* **675**, 323–328.
36. Arisawa, M., Makino, T., Lin, H., Ohno, T., and Iuzuka, R. (1982) *Endocrinol. Jpn,* **229**, 241–244.
37. Ochi, Y., Hosoda, S., Hachiya, T., Yoshimura, M., Mizazaki, T., and Kajita, Y. (1981) *Acta Endocrinol.,* **98**, 62–69.
38. Chalmers, D. A. (1975) *J. Cell Biol.,* **67**, 61a.
39. Pisarev, M. A., and Pisarev, D. L. K. (1977) *Acta Endocrinol.,* **84**, 297–301.
40. Bloch, A. (1975) *Adv. Cyclic Nucleotide Res.,* **5**, 331–338.
41. Kowal, J. (1973) *Endocrinology,* **93**, 461–466.
42. Wren, R. W., and Biddulph, D. M. (1979) *J. Cyclic Nucleotide Res.,* **5**, 239–246.
43. Sheffield, L. G. (1987) *Cell Biol. Int. Rep.,* **11**, 557–562.
44. Chan, P. J. (1987) *Experientia,* **43**, 929–931.
45. Scavennec, J., Carcassonne, Y., Gastaut, J., Blanc, A., and Cailla, H. L. (1981) *Cancer Res.,* **41**, 3222–3227.
46. Wei, J. W., and Hickie, R. A. (1982) *Int. J. Biochem.,* **15**, 789–795.

ANALYSES OF PHOSPHODIESTERASE FORMS AND THEIR FUNCTION IN SPECIFIC CELLULAR SYSTEMS

7

PHOSPHODIESTERASES IN VISUAL TRANSDUCTION BY RODS AND CONES

Peter G. Gillespie*

*Department of Pharmacology, University of Washington, Seattle,
WA 98195, USA*

7.1 INTRODUCTION

In rod photoreceptors, visual transduction commences with the absorption of a photon by rhodopsin and concludes with the closure of cGMP-sensitive channels on the plasma membrane of the outer segment of the photoreceptor. A cyclic nucleotide phosphodiesterase (PDE) plays a pivotal role in this process, hydrolyzing cGMP in response to light. Because of the readily quantified effects of light on the membrane conductance of the rod photoreceptor, the ease of isolation of rod outer segments, and the high concentrations of key enzymes, the visual transduction pathway of rods has become perhaps the most throughly characterized signal transduction system of vertebrates. The physiological function of the rod PDE—examined in this chapter—thus is thought to be understood more clearly than that of any other PDE. Because recent studies have indicated that rod photoreceptors contain both membrane-associated and soluble forms of PDE, this chapter compares their structural and functional properties. In addition, the rod PDEs are compared here with a related photoreceptor PDE, the cone PDE. Although the physiological properties of cones and rods differ, the major proteins in the rod visual transduction pathway have cone-specific isoenzymes and the cone pathway is thought to be similar. Study of the cone PDE's properties, as well as the properties of other cone-specific isoenzymes, offers an excellent opportunity to examine how the PDE has become adapted for the specific requirements of the cone photoreceptor.

Some of the properties of the membrane-associated rod PDE, the soluble rod PDE, and the cone PDE are summarized in Table 7.1. Because they have been

*Present address: Department of Cell Biology and Neuroscience, University of Texas Southwestern Medical Center, Dallas, TX 75235-9039

Cyclic Nucleotide Phosphodiesterases: Structure, Regulation and Drug Action
Edited by J. Beavo and M. D. Houslay   © 1990 John Wiley and Sons Ltd

Table 7.1 Structural and functional properties of photoreceptor PDE isoenzymes

Property	Membrane-associated rod PDE	Soluble rod PDE	Cone PDE
Structural			
Large (catalytic) subunits[a]	88,84 kDa	88,84 kDa	94 kDa
Inhibitory subunits[a]	11 kDa	11 kDa	11 kDa, 13 kDa
Other subunits	none	15 kDa	15 kDa
Native	210 kDa[b]	ND	226 kDa[c]
Kinetic			
K_m for cGMP:			
trypsin[d]	20–150 μM	20 μM	17 μM
light[e]	500–1500 μM	ND	ND
K_m for cAMP:			
trypsin	2000 μM	ND	600 μM
V_{max} for cGMP:			
trypsin	7700 s^{-1}	7000 s^{-1}	4200 s^{-1}
light	5300 s^{-1}	ND	ND
Tα-GTPγS[f]	3900 s^{-1}	ND	3800 s^{-1}
V_{max} for cAMP:			
trypsin	2800 s^{-1}	ND	ND
Noncatalytic cGMP binding			
Affinity	< 500 pM, > 100 nM (2 classes)	ND	11 nM
Stoichiometry	2 mol/mol	ND	2 mol/mol
$t_{1/2}$, dissociation	4 h[g,h]	ND	15 min[h]

[a] By SDS-PAGE.
[b] From the amino acid sequence (α and γ) and by SDS-PAGE (β).
[c] By gel filtration and sucrose density centrifugation.
[d] PDE activated by trypsin proteolysis.
[e] PDE activated by light in rod outer segments.
[f] PDE activated by purified transducin α subunit complexed with GTPγS.
[g] High-affinity site.
[h] Measured at 37°C.
ND = not determined.
Data were compiled from a variety of sources [26,34,35,42,43,47,49,50,83].

examined in the greatest detail, the bovine photoreceptor PDEs are the focus of this chapter. Where appropriate, the properties of PDEs from other species (particularly the frog) are compared with those of the bovine PDE. The physiology of photoreceptors and detailed biochemistry of visual transduction, which are not covered extensively here, are the subjects of excellent reviews [1–4].

7.2   VISUAL TRANSDUCTION BY RODS AND CONES

7.2.1   Transduction by rod photoreceptors

Low levels of illumination elicit the use of rod photoreceptors. These cells are exquisitely sensitive to light; an electrical response can be measured in response to a single photon [5]. At the threshold of human vision as few as six photons, absorbed by different rods, can be perceived as a flash of light [6]. This sensitivity is achieved by a series of amplifying enzyme reactions whereby the absorption of a single photon leads to the closure of thousands of ion channels. The following scheme for rod transduction satisfies most workers. The outer segment enzyme casade is initiated when the protein rhodopsin absorbs a photon [7]. Each activated rhodospin (Rh*) activates thousands of transducin molecules [8,9], a guanine nucleotide binding protein that is a heterotrimer of three subunits, $T\alpha$, $T\beta$ and $T\gamma$, Rh* activates transducin by catalyzing the exchange of GDP (normally bound to $T\alpha$ in the dark) for GTP [8,9]. $T\alpha$, with GTP bound, can then dissociate from the $T\beta$–$T\gamma$ complex and activate the PDE [9], which rapidly hydrolyzes cGMP. Cyclic GMP directly activates a plasma membrane cation channel (carrying primarily Na^+) in the plasma membrane [10]; a decrease in cGMP concentration, therefore, reduces the Na^+ influx [11,12]. In the dark, the constant flow of Na^+ through these channels (the dark current) keeps the cell membrane relatively depolarized [13,14]. Light thus reduces the dark current, hyperpolarizing the cell membrane [13]. The hyperpolarization spreads passively to the photoreceptor's synapse, where the signal is chemically transmitted to second-order cells of the retina.

A substantial amount of Ca^{2+} also flows through the cGMP-sensitive channel in the dark [15]; light therefore causes intracellular Ca^{2+} levels to fall because of the continued action of a Na^+/Ca^{2+}-exchanger [16]. The decrease in Ca^{2+} levels is thought to trigger light adaptation [17,18]. A guanylate cyclase, which is stimulated by low Ca^{2+} levels [19–21], is responsible for restoration of cGMP levels after illumination. This scheme is illustrated diagrammatically in Figure 7.1.

Several lines of evidence thus suggest that sequential activation of rhodopsin, transducin, and PDE leads to a reduction in the steady-state cGMP concentration, the signal sensed by the light-sensitive channel. Goldberg and his colleagues [22–24] have, however, challenged this scheme. They suggest that the conductance of the light-sensitive channel is not controlled by the steady-state levels of cGMP but rather by the *flux* of nucleotide through the PDE–cyclase system. By metabolically labeling intact rabbit or toad retinas with $[^{18}O]H_2O$, they show that the steady-state hydrolysis of cGMP in the dark is sufficiently high for the entire cGMP pool in an outer segment to be hydrolyzed in about a second, and that increasing (low) levels of light increase the turnover linearly. Total cGMP levels of the outer segment do not, however, appear to change. These authors postulate that the energy derived from hydrolysis of the cyclic phosphodiester

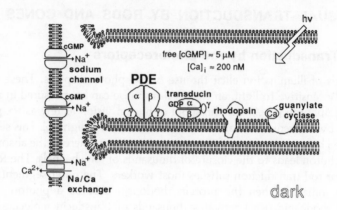

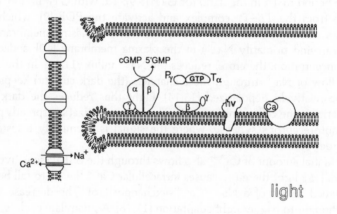

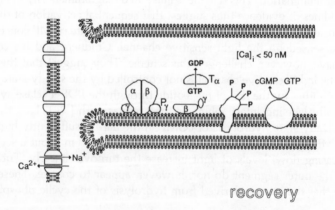

bond (or a product of the reaction)—not changes in the steady-state concentration of cGMP—is used to close the light-sensitive channel. Although the preponderance of the evidence does not support this conclusion, their measurements of cGMP metabolism in the intact retina do show that the full catalytic capability of the PDE is never achieved in the outer segment. The low intracellular concentration of cGMP (only two-fold greater than the concentration of the PDE; see below) is rapidly depleted by the PDE, so that although each activated PDE is capable of hydrolyzing thousands of molecules of cGMP per second in vitro [25], less than 11 000 cGMP molecules are hydrolyzed by the 3000 Tα–GTP–PDE complexes formed per activated rhodopsin in the intact toad retina [24]. Similar values have been estimated for salamander rods using a different technique [26]. These experiments also show that the guanylate cyclase is activated in vivo to the same level as is the PDE, so that the reduction in cGMP levels is transient and substantial amounts of cGMP are synthesized and hydrolyzed after the cGMP concentration (and thus the membrane conductance) has returned to its pre-flash level. It will be important to further relate the ^{18}O labeling measurements and the results of other methods that can assess the activity of the photoreceptor enzymes in vivo [26] to the properties of the proteins involved in visual transduction, as the in vivo methods provide restrictions on the possibilities suggested by in vitro study of these proteins.

7.2.2 Transduction by cone photoreceptors

Cone photoreceptors mediate high-luminance and color vision, and their physiological properties have been matched to these conditions. Dark-adapted cones are typically only one-hundreth as sensitive as rods [27]; furthermore, their electrical response is faster and is terminated much more quickly than that of rods

---

Figure 7.1 The rod outer segment cGMP metabolism cascade. In the dark, rhodopsin, transducin, PDE, and guanylate cyclase are inhibited, so that the turnover of cGMP is relatively low. The free concentration of cGMP is less than 10 μM, which keeps fewer than 10% of the cGMP-sensitive channels open. Sufficient Ca^{2+} enters the outer segments through these channels to keep the concentration at approximately 200 nM. Absorption of a photon by rhodopsin (light) converts it into a form (metarhodopsin II[9]) that can activate transducin. Activated transducin (with GTP bound) dissociates so that the active α subunit can activate PDE. Hydrolysis of cGMP reduces cGMP levels so that all the cGMP-sensitive channels in the vicinity of a bleached rhodopsin are closed. Recovery is initiated when (1) Ca^{2+} levels fall, because Ca^{2+} entry through light-sensitive channels is blocked while Ca^{2+} efflux through the Na^+/Ca^{2+}-exchanger is maintained transiently, so that the Ca^{2+}-dependent regulator of the guanylate cyclase dissociates, allowing cyclase activity to increase; (2) rhodopsin is phosphorylated, allowing binding of a 48 kDa protein that blocks further transducin activation; and (3) GTP is hydrolyzed by transducin, terminating its activation of the PDE

[1,27]. In addition, cones show light adaptation over a much wider range of light intensities than rods [28]. Individual cones are maximally sensitive to different wavelengths, depending on the isoform of opsin they contain [29–32]. These properties allow cones to respond rapidly under conditions where light intensity varies widely and to distinguish colors in the visual range.

The sequence of biochemical events responsible for visual transduction in cones has not been definitively established. The transduction pathway is thought to be similar to that of rods, nevertheless, in part because the key proteins in rod transduction have counterparts in cones. Thus, distinct cone opsins [32], transducins [33], PDEs [34,35], and cGMP-sensitive channels [36,37] have been characterized. Furthermore, as in rods, Ca^{2+} triggers light adaptation in cones [17,18]. It remains a challenge to elucidate the biochemical basis of the physiological differences between rods and cones.

7.3 IDENTIFICATION AND PURIFICATION OF THE PHOTORECEPTOR PDEs

7.3.1 Membrane-associated rod PDE

Pannbacker et al. [37] first noted that retinas contain a large amount of cGMP PDE activity. Subsequent work showed that light stimulates this activity when GTP is present [38–40]. Rod PDE is present in bovine and frog photoreceptors at about 30 μM [42–44]. This high concentration of the PDE and the relative ease of isolation of rod outer segments has simplified the purification of this enzyme. PDE was first purified from low-ionic-strength extracts of frog (*Rana catesbeiana*) outer segments by sucrose density gradient centrifugation and polyhistidine–agarose chromatography [45]. The initial purification of the more abundant bovine rod PDE employed low-ionic-strength elution from rod outer segment membranes, DEAE anion exchange chromatography, and gel filtration [42]. Most other preparations reported are modifications of this procedure [46–49].

The bovine rod PDE has also been purified by immunoaffinity chromatography, using an immobilized antibody (ROS-1a) [50]. The high-affinity binding of the PDE to this antibody can be reversed by a brief exposure to high pH (pH 10.7). This method is useful because very crude rod outer segment or retinal low-ionic-strength extracts can be used and only one chromatographic step is required. The rod PDE prepared in this fashion is indistinguishable from that prepared using standard methods [50].

7.3.2 Soluble rod PDE

Because it is readily lost from bovine outer segments prepared by usual methods, the soluble rod PDE was not detected until recently. Apparently, most outer-

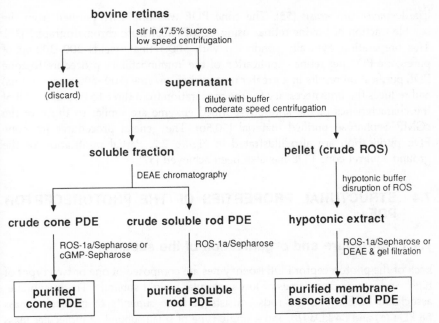

bovine retinas

stir in 47.5% sucrose
low speed centrifugation

pellet
(discard)

supernatant

dilute with buffer
moderate speed centrifugation

soluble fraction

DEAE chromatography

pellet (crude ROS)

hypotonic buffer
disruption of ROS

crude cone PDE   **crude soluble rod PDE**   **hypotonic extract**

ROS-1a/Sepharose or
cGMP-Sepharose

ROS-1a/Sepharose

ROS-1a/Sepharose or
DEAE & gel filtration

**purified
cone PDE**

**purified soluble
rod PDE**

**purified membrane-
associated rod PDE**

Figure 7.2   Purification scheme for bovine photoreceptor PDEs. The general scheme for purification of the bovine photoreceptor PDEs is illustrated. Note that the cone PDE and the soluble rod PDE are purified from a soluble fraction resulting from the initial disruption of the retina. These procedures are described in greater detail elsewhere [35,50]

segment preparations are sufficiently leaky that substantial quantities of soluble enzymes can diffuse out [51,52]. The soluble rod PDE was originally identified and purified from the soluble fraction of bovine retina fractionations [50]. The principal distinguishing feature of this isoenzyme is its δ subunit (15 kDa; see below). In preparations of sealed rod outer segments, only the PDE soluble at moderate ionic strength has the δ subunit [50]. Purification of this isoenzyme required the monoclonal antibody purification described above. The general scheme of this purification is illustrated in Figure 7.2. Use of the antibody-affinity purification is probably not essential, since the soluble rod PDE conceivably could be purified by standard methods from sealed rod outer segments [51,53].

7.3.3   Cone PDE

The bovine code PDE was first observed by Hurwitz et al. [34,54] who found that a 94 kDa polypeptide could be immunoprecipitated with the monoclonal antibody ROS-1 [54] from a small peak of PDE activity (peak I) in anion exchange separations of retinal extracts [34]. This polypeptide is enriched in bovine and human cones [34], and is the only PDE isoenzyme in the all-cone retinas of the lizard *Anolis carolinensis* [55]. The cone PDE was originally purified from the soluble fraction of bovine retinas using cGMP–Sepharose chromatography [35]. This preparation, typically conducted with 500 retinas, yields 200–300 ng of pure cone PDE per retina. Application of the immunoaffinity procedure to cone PDE purification results in a greater recovery of enzyme (500–600 ng per retina) and reduces the time necessary for the preparation from three to two days. All of the characteristics of the antibody-purified enzyme are similar to those of the cGMP–Sepharose–purified material [50,56]. The general procedures for cone PDE purification are also illustrated in Figure 7.2. Partial purification of the ground squirrel cone PDE has also been achieved [57].

7.4 STRUCTURAL PROPERTIES OF THE PHOTORECEPTOR PDEs

7.4.1 Structure and organization of the subunits

Each of the photoreceptor PDE isoenzymes are composed of one or two types of high- and one or more types of low-molecular-weight subunits. The membrane-associated PDE of bovine rods consists of large subunits of molecular mass 88 kDa (α) and 84 kDa (β), and a single type of small subunit of molecular mass 11 kDa (γ) [42,47]. The α subunit can be carboxymethylated [58]. Recent evidence suggests that this isoenzyme exists as a $\alpha\beta\gamma_2$ complex [59]; other species (e.g. $\alpha_2\gamma_2$ and $\beta_2\gamma_2$) may also exist [34]. The soluble rod PDE has three subunits (α_{sol}, β_{sol}, and γ_{sol}) of exactly the same size as the subunits of the membrane-associated isoenzyme, as well as an additional subunit of molecular mass 15 kDa (δ). Finally, the cone PDE consists of a single type of large subunit (α', molecular mass 94 kDa) [34] and three small subunits (molecular mass 11 kDa, 13 kDa and 15 kDa) [35]. While the stoichiometry of cone PDE subunits has not been established, the native molecular mass of the cone PDE has been estimated to be 226 kDa [35], which is consistent with two large subunits and several smaller ones.

Full-length cDNAs corresponding to the bovine membrane-associated rod α and γ subunits have been cloned and sequenced [62,63]. The α subunit contains 858 amino acid residues, with a total molecular mass of 99 kDa. About 70% of the β subunit has also been cloned [62]. The α and β subunits are homologous to each other and to other PDEs, including several from cow, *Drosophila*, and yeast [64–67]. The region of highest homology between these PDEs, which for the rod α subunit is the C-terminal one-third of the molecule, has been proposed to be the catalytic domain [64]. The regions of the α or β subunits that are responsible for interaction with the γ subunit or for noncatalytic cGMP binding are not known. The bovine rod γ subunit contains 87 amino acids (9700 Da) and shows no homology to any other known protein. It is characterized by a region (residues 24–45) that has an unusually large fraction of basic residues (10/22) and no acidic residues [63].

A substantial fraction of the cone α' sequence has been determined by protein sequencing, and a cDNA clone corresponding to this subunit has been isolated [103,104]. Of the sequences that can be aligned, 70% of the α' residues are identical to residues of the α and β subunits of the rod PDE. An additional 15% of the residues are identical to only α or only β, while the rest are unique to α'. The sequence shows that the α' protein is a distinct gene product that has diverged from both α and β.

Interestingly, both the membrane-associated rod PDE and the cone PDE are very asymmetric molecules, with axial ratios of at least 10:1 [35]; transducin also has a large axial ratio [60,61]. The functional significance of the elongated shape of these proteins is not known.

7.5 FUNCTIONAL PROPERTIES OF PHOTORECEPTOR PDEs

7.5.1 Activation of the PDEs

A common feature of all the photoreceptor PDEs is their activation by transducin or trypsin. In light-activated rod outer segments in vitro, the increase in velocity of cGMP hydrolysis can exceed 200-fold under the appropriate conditions [26]. The PDE is activated by the GTP-complexed α subunit of transducin [8,9]. Although initial experiments suggested that activation was paralleled by a transposition of the inhibitory subunit into the cytoplasm [68], more recent studies [48,49] have instead indicated that the γ subunit of the PDE remains attached to the membrane after activation, perhaps indirectly bound through transducin [69]. Most evidence suggests that Tα activates the PDE by stripping its γ subunits away from the α and β catalytic subunits [59,68–71] rather than remaining bound to $\alpha\beta$ or $\alpha\beta\gamma_2$ complex [43]. In any case, each transducin apparently activates only a single PDE [3]. The interaction between Tα and PDE is relatively weak; half-maximal activation of PDE is achieved with approximately 1 μM transducin [35,72–73]. Rod outer segment membranes, however, appear to increase the affinity of Tα for PDE [73]; furthermore, a single rhodopsin may activate all of the transducins on a disk, raising the Tα–GTP concentration locally to higher than 300 μM [3]. The relatively low affinity of the interaction between the Tα and PDE may instead allow for rapid dissociation of transducin from the PDE after the complexed GTP has been hydrolyzed, which would increase the speed at which the PDE is turned off.

It is not known whether the soluble rod PDE or the cone PDE is activated in vivo in response to light. Nonetheless, both are readily activated by purified rod transducin [35,50]. The soluble rod PDE has approximately the same sensitivity to transducin as the membrane-associated rod PDE [50]. Surprisingly, the cone PDE is much more sensitive to transducin activation than is either rod isoenzyme, requiring only one-fiftieth as much transducin to achieve 50% activation (Figure 7.3) [35]. Because the cone transducin [33] has not been purified, it is unclear whether the high affinity of rod transducin for cone PDE is representative of the affinity of the interaction between cone transducin and cone PDE. Nonetheless, this high-affinity interaction is interesting because one method of reducing the sensitivity of cones might have been to reduce the transducin–PDE affinity; instead, the affinity is dramatically higher.

The photoreceptor PDE isoenzymes can be activated by a number of nonphysiological treatments, including limited proteolysis by trypsin [45] and treatment with polycations such as histones [54] and protamine [45]. The ratio of the maximal catalytic activity of the trypsin-activated rod PDE to its Michaelis constant, k_{cat}/K_m approaches the diffusion-controlled limit [2]. Trypsin activates the photoreceptor PDEs by proteolyzing their inhibitory subunits [47]; rod PDE γ is probably sensitive to trypsin because of the large number of basic residues [63]. Limited trypsin proteolysis also removes 0.5–1 kDa from the rod α and β subunits and 4 kDa from the cone PDE, which may change the catalytic

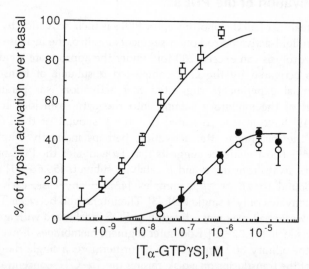

Figure 7.3 Transducin activation of bovine photoreceptor PDEs. Bovine photoreceptor PDEs were purified by ROS-1a chromatography [50]. Rod transducin was purified by ion exchange chromatography of a GTPγS extract of PDE-depleted ROS [35]. PDE (0.5 nM) and the indicated concentrations of transducin were mixed and incubated at 30°C for 10 min; PDE activity was then assayed by the addition of 1 mM cGMP for 1 min [35,50]. □ cone PDE; ● membrane-associated rod PDE; ○, soluble rod PDE. The y-axis values were calculated by:

$$\% \text{ trypsin activation over basal} = \frac{(\text{activity with transducin}) - (\text{basal activity})}{(\text{activity with trypsin}) - (\text{basal activity})} \times 100$$

properties of the enzymes [56]. The mechanism of activation by polycations is not known.

7.5.2   Kinetics of cGMP hydrolysis

Trypsin-activated rod PDE hydrolyzes cGMP with K_m of 20–150 μM [25,35,43,46], and a V_{max} of 5000–7500 s^{-1} [34,46,48]. When assayed under the same conditions, the membrane-associated rod PDE, the soluble rod PDE, and the cone PDE each hydrolyze cGMP with K_m values of about 20 μM [35,50]. The V_{max} of both of the rod isoenzymes is about twice that of the cone isoenzyme (Table 7.1).

These kinetic properties may be misleading for several reasons. First, because the free concentration of cGMP (< 6 μM) is less than the K_m and is much lower than the total amount of PDE (30 μM), the PDE always operates under conditions much different than those used for determination of K_m and V_{max} [24]. In addition, the K_m for cGMP hydrolysis actually may rise in response to light. Although initial studies of light-activated PDE in rod outer segment suspensions revealed K_m values of 50–100 μM [25], subsequent reports [24,43,74] instead suggested the rod PDE increases its K_m for cGMP by at least 10-fold when activated by light. Because the free concentration of cGMP is well below the K_m, the hydrolysis of cGMP can be approximated by first-order kinetics (rate constant $= V_{max}/K_m$). The consequence of the increased K_m therefore will be to increase the rate of cGMP hydrolysis only several-fold, not the 100-fold predicted if only the increase in V_{max} is considered [24]. Because this K_m shift has not been observed when purified PDE is activated by purified transducin [35,49], and disappears after washing rod outer segments with moderate-ionic-strength buffer [74], some other outer segment factor may cause the shift. Although it would be difficult to demonstrate experimentally, the K_m shift might occur after the first second or two of light activation and thus might not be significant for the initial hydrolysis of cGMP.

7.5.3   Characteristics of inhibitory subunits

Each of the photoreceptor PDEs has one or more types of tightly bound inhibitory subunits. Several groups initially reported that heat-stable inhibitors of the PDE were present in rod outer segments [75–78]. In bovine rod .outer segments, at least one of these inhibitors is the γ subunit of the rod PDE [42,47]. Active PDE γ subunit has been isolated by heat treatment [47,49], acid gel filtration [47], and reverse-phase chromatography [50,63]. The strength of the interaction between γ and $\alpha\beta$ has not been conclusively determined. The apparent K_d reported by one group (15 nM) [43] was probably much higher than the true K_d because the concentration of PDE (20 nM) exceeded the apparent K_d, indicating that $\alpha\beta$ was simply being titrated by added γ. The apparent K_d reported by two other groups for the interaction between γ and rod PDE (activated by trypsin) was much lower (~ 5 pM) [49,50]. Whether this affinity resembles the γ–$\alpha\beta$ affinity in non-trypsin-treated PDE is an important question that has not been addressed. The interaction between γ and the catalytic subunits is sufficiently strong, nonetheless, that it is difficult to fully activate the PDE by dilution [49].

The recent report of expression of functional γ in bacteria [79] indicates that questions about the interaction of γ with $\alpha\beta$ and about the importance of different regions within γ can soon be addressed. Indeed, Brown and Stryer have shown that the C-terminus of γ, and not the N-terminus, is important for inhibition of $\alpha\beta$. More sophisticated delineation of important regions of the molecule should follow the expression of site-specific mutagens of γ.

The γ_{sol} subunit of the soluble rod PDE appears to be identical to the γ subunit because both have identical molecular weights, reverse-phase retention time, protease sensitivities, and IC_{50}s for inhibition of trypsin-activated rod PDE (Figure 7.4) [50]. Proof that γ_{sol} is γ, however, will require more detailed structural analysis. Purified cone PDE has two types of inhibitory subunits, the 11 kDa and 13 kDa subunits [35,50]. Current evidence suggests that, like γ_{sol}, the 11 kDa cone subunit is identical to rod γ. It is not yet known whether the 11 kDa subunit is derived from cone outer segments, or whether it has exchanged from the rod PDE—or a pool of free inhibitory subunit—during the purification. If the association of the 11 kDa subunit with the cone PDE is an artifact of the purification, the 13 kDa subunit may be the only inhibitory subunit associated with the α' subunit in the cone outer segment. This subunit is a less potent inhibitor of trypsin-activated rod or cone PDE than any of the other inhibitory subunits (Figure 7.4). It is possible that a lower affinity of the 13 kDa subunit for α' allows lower concentrations of transducin to activate the cone PDE.

7.5.4 Characteristics of 15 kDa subunits

The δ subunit of the soluble rod PDE and the 15 kDa subunit of the cone PDE appear to be identical; they co-elute when chromatographed on reverse-phase columns and they co-migrate on SDS–polyacrylamide gels [50]. These subunits can be isolated by reverse-phase chromatography [50], but cannot be readily transferred to neutral solutions after purification. Acetonitrile and trifluoroacetic acid (50% and 0.1%), or formic acid (70%), keep the δ subunit in solution; the polypeptide apparently is not stable, however, in neutral buffers. When it is associated with the native cone PDE, the 15 kDa subunit is apparently resistant to several proteases that extensively degrade the α', 11 kDa, and 13 kDa subunits [56]. The function of this subunit is not yet known.

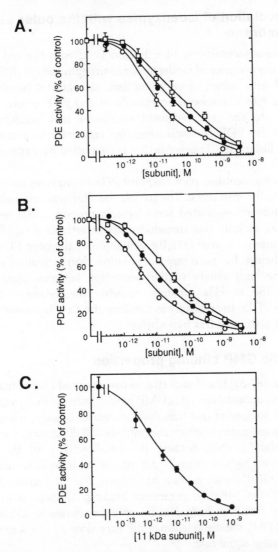

Figure 7.4 Inhibition of trypsin-activated rod and cone PDE activity by purified inhibitory subunits. Cone (A) and rod (B, C) PDEs were purified by ROS-1a chromatography [50] and were activated by trypsin. Inhibitory subunits were isolated from purified PDEs by reverse-phase chromatography. Various concentrations of inhibitory subunits were added to 1 pM trypsin-activated PDEs and activity was measured with 6 μM cGMP. (A) Inhibition of trypsin-activated cone PDE by: rod γ subunit (O; IC_{50} = 23 pM), cone 11 kDa subunit ($\bullet$; 52 pM), and cone 13 kDa subunit ($\square$; 91 pM). (B) Inhibition of trypsin-activated membrane-associated rod PDE by the same subunits (IC_{50}s: γ, 6 pM; cone 11 kDa, 19 pM; cone 13 kDa, 48 pM). (C) Inhibition of trypsin-activated membrane-associated rod PDE by γ_{sol} subunit (IC_{50} = 11 pM)

7.5.5 Association of isoenzymes with the outer segment membrane

The membrane-associated rod PDE is associated with the rod outer segment membrane in the presence of moderate-ionic-strength buffers (100–150 mM), but can be eluted with buffers of very low ionic strength (< 10 mM). Because the affinity of the interaction between transducin and PDE is low, confinement of both proteins to the outer segment membrane may accelerate the rate of activation of the PDE by increasing the encounter rate and the yield of productive collisions [3]. This prediction is supported by experimental observations [3].

The other two identified photoreceptor PDE isoenzymes do not appear to be associated with the membrane. The soluble rod PDE was originally distinguished from the membrane-associated form by this property. The solubility of this isoenzyme may indicate that transducin cannot activate it rapidly in vivo. The cone PDE is soluble as well [35]; the activation of the cone PDE by transducin could, nevertheless, be more rapid in solution than activation of the rod PDE because of the high affinity of the interaction between these proteins. It is possible that the 15 kDa subunits modulate the membrane-binding of the photoreceptor PDEs; the difficulty in handling the 15 kDa subunits, however, has prevented the testing of this hypothesis.

7.5.6 Cyclic GMP binding properties

Yamazaki et al. [80,81] first demonstrated that frog rod PDE contains two classes of high-affinity, noncatalytic [^{3}H]cGMP binding sites with K_d values of 160 and 830 nM. They suggested that transducin decreases binding, while a heat-treated rod outer segment extract (which also inhibited PDE activity) stimulates binding to the noncatalytic sites. Because the basal activity of the PDE in their preparations was high, however, the effects of transducin and inhibitor on noncatalytic cGMP binding may have been due to increased or decreased hydrolysis of [^{3}H]cGMP used to measure binding. Although more recent studies have suggested that transducin does increase the off-rate of cGMP from the frog binding site [24,82], the dissociation is still very slow ($t_{1/2} = 4$ min) and may not be physiologically significant.

Stoichiometric [^{3}H]cGMP binding to bovine rod PDE has been difficult to demonstrate, apparently because the binding sites on the rod PDE are of very high affinity, with an extraordinarily slow off-rate. The $t_{1/2}$ for the high-affinity site exceeds 4 h at 37°C [83]. At 4°C, essentially no cGMP dissociates, so that PDE isolated at this temperature retains nearly two moles of cGMP bound per mole of oligomer [83]. The soluble rod PDE also binds little [^{3}H]cGMP because of tightly bound endogenous cGMP (P. Gillespie, unpublished observation).

The cone PDE also has two moles of noncatalytic cGMP binding sites per mole of oligomer [35]. The affinity of these binding sites ($K_d = 11$ nM) is substantially lower than that of the bovine rod PDE's high-affinity site. The cone binding site also has unusually low on- and off-rates. The $t_{1/2}$ for dissociation of [^{3}H]cGMP from this site at 37°C is 15 min, considerably faster than that for the rod site. While the diffusion-set limit for an on-rate is about 10^9 M^{-1} s^{-1}, the on-rate for the cGMP–cone PDE interaction is some five orders of magnitude smaller. This low on-rate may indicate that a slow conformational change is required for binding. The binding sites of both the rod [80] and cone [35] PDEs are distinct from the catalytic sites.

The noncatalytic binding sites probably bind a majority of the cGMP in rod outer segments. The concentration of cGMP in dark-adapted bovine rods is not known; in dark-adapted frog rod outer segments, however, cGMP is present at about 60 μM (when measured with 1 mM external Ca^{2+}) [84]. Furthermore, in several species, the concentration of free cGMP has been estimated to be less than 6 μM [11,27,85]. Since the PDE concentration is approximately 30 μM and two sites are present per molecule, most of the cGMP of the rod thus can be bound by the PDE.

Whether the noncatalytic binding site has a function other than serving as a sink for cGMP is not clear. No functional changes in the properties of the rod or cone PDEs have ever been associated with occupancy of this site; it is difficult to examine the effects of occupancy on cGMP hydrolysis, however, since the affinity of cGMP for the noncatalytic binding site is so much greater than the affinity for the catalytic site. Preliminary experiments with the cone PDE suggest that cGMP has minimal effects on the hydrolysis of low concentrations of cAMP, and that the presence or absence of bound cGMP does not significantly affect the concentration dependence of activation by rod transducin [55].

7.5.7 Inhibition by drugs

Several studies have used IBMX (3-isobutyl-2-methylxanthine) a nonselective PDE inhibitor, to inhibit photoreceptor PDE [42,86–89]. IBMX increases the dark current and inhibits its suppression by light; increased amounts of light, on the other hand, can overcome this effect [86,84]. IBMX inhibits the bovine rod PDE with a K_i of approximately 10 μM [42,90]. In the intact retina, the concentration of IBMX that reduces cGMP hydrolysis by 50% is about 70 μM [84]; this value is not related to the K_i, however, since the supply of cGMP can be regulated to adjust for the decreased PDE activity. Two selective inhibitors of the cGMP-selective, cGMP-binding PDE [91,92] (dipyridamole [91,93,94] and M&B 22,948 [95,96]), also inhibit trypsin-activated rod and cone PDEs with K_i values of 125 to 400 nM [89]. Interestingly, these drugs have biphasic effects on unactivated PDEs, stimulating cGMP hydrolysis at low drug concentrations and inhibiting only at much higher concentrations. This biphasic response is apparent only at high (> 1 mM) substrate concentrations. These drugs may prove useful for determining properties of the activation of the PDE.

7.6 CONCLUSIONS AND UNRESOLVED QUESTIONS

7.6.1 Photoreceptor PDEs are part of a cGMP-specific PDE family

The photoreceptor PDEs share a substantial number of properties with the cGMP-specific, cGMP-binding PDE (cG-BPDE) that has been most thoroughly characterized in lung and platelets (see Chapter 5). All enzymes of this family, termed the cGMP-specific PDEs (family V; see Chapter 1), exhibit a 50-fold or greater selectivity for cGMP over cAMP [64]. Other similarities include noncatalytic cGMP binding sites of relatively high affinity and unknown function, similar catalytic subunit size and association into dimers, micromolar K_m values for cGMP, and submicromolar K_i values for inhibition by dipyridamole and M&B 22,948. Several photoreceptor-specific monoclonal antibodies do not, however, recognize cG-BPDE [102]; furthermore, the specific activity of cG-BPDE is much lower than that of the photoreceptor enzymes and cG-BPDE appears to be regulated by a number of factors that do not affect the photoreceptor PDEs. Nonetheless, the similarities between these classes of enzymes seem sufficiently striking to classify them together in a single family and to suggest that substantial sequence homology and functional significance may underly them.

7.6.2 PDEs in the rod

Most workers agree that the principal function of the rod PDE is to rapidly reduce the steady-state concentration of cGMP in response to light. For this purpose, the enzyme appears to be magnificently suited. Its catalytic efficiency is nearly as high as is possible, while its γ subunit keeps this activity in check in the resting state. Several important questions remain, nevertheless, about the molecular mechanisms involved in PDE activation. First, while the consensus is that transducin strips the PDE's γ subunit from the $\alpha\beta$ complex, how this occurs and what happens to the complex of transducin and the PDE γ subunit have not been clearly elucidated. Second, the K_m increase reported by several groups appears to counteract the increase in V_{max} of the enzyme. Whether the shift is a consequence strictly of transducin activation of the native PDE or whether it requires another outer segment factor is unknown. The function of the K_m shift is also not known.

Although transducin and PDE γ subunit regulation of the catalytic moiety of the PDE is reasonably well understood, other factors may regulate functions of the PDE. For example, there is ample evidence that Ca^{2+} increases the light sensitivity of native PDE in outer segments [74,97,98]. Purified rod PDE is not sensitive, however, to Ca^{2+} concentration. Furthermore, a careful study of the effects of Ca^{2+} on a reconstituted transduction system and on freshly prepared outer segments indicates that the sensitivity to Ca^{2+} falls well outside the range of the ion normally present in outer segments [98]. These authors also addressed the concern that in a reconstituted or dilute native outer segment preparation, a soluble Ca^{2+}-sensitive regulatory molecule may be diluted to a concentration too low to interact with its target molecule. If such a molecule is present, Barkdoll et al. calculate that the K_d for interaction of such a regulator with its target must be above 5 μM. These results thus cast doubt on the significance of the Ca^{2+} modulation of PDE activity.

The noncatalytic binding sites on the PDE may provide another point of regulation that is not at all understood. It is rare for an enzyme to have noncatalytic binding sites of very high affinity for its substrate. Perhaps the only other well-characterized enzymes with these properties are the F_1-ATPases from mitochondria, chloroplasts, and bacteria, which contain six ADP binding sites on subunits of the $\alpha_3 \beta_3 \gamma \delta \varepsilon$ complexes. In these enzymes, three tightly bound ADPs do not exchange with exogenous adenine nucleotides and are thought to be noncatalytic; no functional role for these ADP binding sites has been determined [99–101]. Two possible roles for the rod PDE binding site, which are not mutually exclusive, can be envisioned. First, occupancy of the binding site could alter the catalytic properties of the PDE (V_{max} or K_m). Unless the affinity of the cGMP–PDE interaction is altered, however, it is unlikely that cGMP concentrations will fall sufficiently (subnanomolar in rods) to allow for substantial dissociation of nucleotide. Second, because the amount of cGMP bound to the PDE is much larger than the amount of free cGMP, the PDE-bound nucleotide could be used to replenish cellular levels depleted after light activation. Once again, however, for this hypothesis to hold, some factor must accelerate cGMP release under the appropriate conditions. Two possible conditions are interaction with transducin or alteration in Ca^{2+} levels. Transducin stimulation of cGMP dissociation would be appropriate if the time-course of release of cGMP is substantially slower than the time-course of PDE activation, which provides a lag before the replenishment of cGMP. Since its concentration drops substantially within several hundred milliseconds after photon absorption, Ca^{2+} regulation of cGMP release would also provide a plausible way of replenishing free cGMP after light stimulation. Alternatively, cGMP affinity might be regulated by another nucleotide that rises in free concentration during or after light stimulation (e.g. GDP or ADP). It will be important to focus on the regulation of cGMP affinity for this binding site.

Given the prevailing view that the PDE needs to be associated with the membrane for rapid activation [3], it is surprising that a soluble form of the enzyme constitutes a substantial fraction of the outer segment activity. A plausible hypothesis is that the δ subunit regulates the solubility of the rod PDE, and that the amount or affinity of the δ subunit is controlled. The unusual physical properties of the δ subunit have, however, made this hypothesis unusually difficult to test. If δ does regulate PDE solubility, however, the distinction between soluble and membrane-associated PDEs developed in this chapter may be an artificial one. Instead, the significance of the δ subunit as a regulator of a single type of rod PDE should be addressed.

Although the main function of the rod PDE appears to be to rapidly reduce the cGMP concentration after a flash of light, the regulation of its activity, cGMP binding properties, and solubility probably play important roles in shaping the rod photoresponse.

7.6.3   PDEs in the cone

The cone PDE has increased transducin sensitivity, reduced affinity of inhibitory subunits, decreased affinity of cGMP for noncatalytic cGMP sites, greater sensitivity to a variety of proteases [56], and increased stimulation by dipyrida-mole and M&B 22,948, as compared to the rod PDE. Whether any of these properties of the cone PDE are responsible for the difference in sensitivity (or kinetics or adaptation) between rods and cones is not clear. Cones would be less sensitive to light if fewer PDEs were available for activation or if the cone PDE had a low specific activity. The concentration of cone PDE in the cone outer segment is, however, close to that of the rod PDE, given the abundance of the cone PDE and the estimated number and volume of bovine cones. Furthermore, the maximal catalytic activity of the cone PDE is similar to that of the rod PDE. Cones might also be less responsive to light if the cone transducin had a very low affinity for cone PDE. Because the cone PDE is much more sensitive to stimulation by rod transducin than is the rod PDE, however, it is unlikely that the properties of the cone PDE–transducin interaction alone determine the low sensitivity of cones. Other properties of the cone PDE may be important for shaping the light response of these photoreceptors. Since cones need more photons to reduce the dark current to the same degree as in rods, and since cGMP controls the light-sensitive conductance with a similar dose–response relation-ship in both rods and cones [36], free cGMP levels must be less responsive to light in cones. Potential differences between rods and cones that could account for lower sensitivity and that are consistent with the literature include the following: (1) the lifetime of the activated opsin species is short; (2) the cone transducin GTPase is high, perhaps stimulated by cone PDE inhibitory subunits; (3) low Ca^{2+} levels inhibit cone PDE to a more substantial degree than the rod PDE (by decreasing V_{max}, increasing K_m, or increasing solubility); or (4) the cone guanylate cyclase is stimulated very rapidly to match the PDEs activity. Further research should establish the source of the sensitivity difference and will show whether the cone PDEs properties are essential for this difference or for any of the other physiological differences between rods and cones.

ACKNOWLEDGEMENTS

I would like to thank Dr Joe Beavo for criticism and guidance while I was conducting my dissertation research, as well as for comments on this chapter.

Drs A. J. Hudspeth and J. Howard provided extensive critical and constructive comments on this manuscript, for which I am grateful; Dr Rabi Prusti and Ms Susan Gillespie also corrected various drafts.

REFERENCES

1. Pugh, E. N., and Cobbs, W. H. (1987) *Vision Res.*, **26**, 1613–1643.
2. Stryer, L. (1986) *Annu. Rev. Neurosci.*, **9**, 87–119.
3. Liebman, P. A., Parker, K. R., and Dratz, E. A. (1987) *Annu. Rev. Physiol.*, **49**, 765–791.
4. Hurley, J. B. (1987) *Annu. Rev. Physiol.*, **49**, 793–812.
5. Baylor, D. A., Lamb, T. D., and Yau, K.-W. (1979) *J. Physiol.*, **288**, 613–634.
6. Hecht, S., Shlaer, S., and Pirene, M. H. (1942) *J. Gen. Physiol.*, **25**, 819–830.
7. Wald, G. (1968) *Science*, **162**, 230–238.
8. Fung, B. K.-K., and Stryer, L. (1980) *Proc. Natl Acad. Sci. USA*, **77**, 2500–2504.
9. Fung, B. K.-K., Hurley, J. B., and Stryer, L. (1981) *Proc. Natl Acad. Sci. USA*, **78**, 152–156.
10. Fesenko, E. E., Kolesnikov, S. S., and Lyubarsky, A. L. (1985) *Nature*, **313**, 310–313.
11. Yau, K.-W., and Nakatani, K. (1985) *Nature*, **317**, 252–255.
12. Matthews, G. (1987) *Proc. Natl Acad. Sci. USA*, **84**, 299–302.
13. Hagins, W. A., Penn, R. D., and Yoshikami, S. (1970) *Biophys. J.*, **10**, 380–412.
14. Zuckermann, R. (1973) *J. Physiol.*, **235**, 333.
15. Hodgkin, A. L., McNaughton, P. A., Nunn, B. J., and Yau, K.-W. (1984) *J. Physiol.*, **350**, 649–680.
16. McNaughton, P. A., Cervetto, L., and Nunn, B. J. (1986) *Nature*, **322**, 261–263.
17. Matthews, H. R., Murphy, R. L. W., Fain, G. L., and Lamb, T. D. (1988) *Nature*, **334**, 67–69.
18. Nakatani, K., and Yau, K.-W. (1988) *Nature*, **334**, 69–71.
19. Lolley, R. N., and Racz, E. (1982) *Vision Res.*, **22**, 1481–1486.
20. Pepe, I. M., Panfoli, I., and Cugnoli, C. (1986) *FEBS Lett.*, **203**, 73–76.
21. Koch, K.-W., and Stryer, L. (1988) *Nature*, **334**, 64–66.
22. Goldberg, N. D., Ames, A., Gander, J. E., and Walseth, T. M. (1983) *J. Biol. Chem.*, **258**, 9213–9219.
23 Ames, A., Walseth, T. F., Heyman, R. A., Barad, M., Graeff, R. M., and Goldberg, N. D. (1986) *J. Biol. Chem.*, **261**, 13034–13042.
24. Dawis, S. M., Graeff, R. M., Heyman, R. A., Walseth, T. F., and Goldberg, N. D. (1988) *J. Biol. Chem.*, **263**, 8771–8785.
25. Yee, R., and Liebman, P. A. (1978) *J. Biol. Chem.*, **253**, 8902–8909.
26. Hodgkin, A. L., and Nunn, B. J. (1988) *J. Physiol.*, **403**, 439–471.
27. Baylor, D. A. (1987) *Invest. Ophthalmol. Vis. Sci.*, **28**, 34–49.
28. Normann, R. A., and Werblin, F. S. (1974) *J. Gen. Physiol.*, **63**, 37–61.
29. Brown, P. K., and Wald, G. (1963) *Nature*, **200**, 37–43.
30. Brown, P. K., and Wald, G. (1964) *Science*, **144**, 45–51.
31. Nunn, B. J., Schnapf, J. L., and Baylor, D. A. (1984) *Nature*, **309**, 264–266.
32. Nathans, J., Thomas, D., and Hogness, D. S. (1986) *Science*, **232**, 193–232.

33. Lerea, C. L., Somers, D. E., Hurley, J. B., Klock, I. B., and Bunt-Milam, A. H. (1986) *Science*, **234**, 77–80.
34. Hurwitz, R. L., Bunt-Milam, A. H., Change, M. L., and Beavo, J. A. (1985) *J. Biol. Chem.*, **260**, 568–573.
35. Gillespie, P. G., and Beavo, J. A. (1988) *J. Biol. Chem.*, **263**, 8133–8141.
36. Cobbs, W. H., Barkdoll, A. E., and Pugh, E. N. (1985) *Nature*, **317**, 64–66.
37. Haynes, L., and Yau, K.-W. (1985) *Nature*, **317**, 61–64.
38. Pannbacker, R. G., Fleischman, D. E., and Reed, D. W. (1972) *Science*, **175**, 757–758.
39. Keirns, J. J., Miki, N., Bitensky, M. W., and Kierns, M. (1975) *Biochemistry*, **14**, 2760–2766.
40. Miki, N., Kierns, J. J., Marcus, F. R., Freeman, J., and Bitensky, M. W. (1973) *Proc. Natl Acad. Sci. USA*, **70**, 3820.
41. Goridis, C., and Weller, M. (1976) *Adv. Biochem. Psychopharmacol.*, **15**, 391–412.
42. Baehr, W., Devlin, M. J., and Applebury, M. L. (1979) *J. Biol. Chem.*, **254**, 11669–11677.
43. Sitaramayya, A., Harkness, J., Parkes, J. H., Gonzalez-Olivia, C., and Liebman, P. A. (1986) *Biochemistry*, **25**, 651–656.
44. Hamm, H. E., and Bownds, M. D. (1986) *Biochemistry*, **25**, 4512–4523.
45. Miki, N., Baraban, J. M., Keirns, J. J., Boyce, J. J., and Bitensky, M. W. (1975) *J. Biol. Chem.*, **250**, 6320–6327.
46. Kohnken, R. E., Eadie, D. M., Revzin, A., and McConnell, D. G. (1981) *J. Biol. Chem.*, **256**, 12502–12509.
47. Hurley, J. B., and Stryer, L. (1982) *J. Biol. Chem.*, **257**, 11094–11099.
48. Fung, B. K.-K. (1985) In *Molecular Mechanisms of Transmembrane Signalling* (P. Cohen and M. D. Houslay, eds), Elsevier Science, Biomedical Division, New York, pp. 183–214.
49. Wensel. T. G., and Stryer, L. (1986) *Proteins: Structure, Function, and Genetics*, **1**, 90–99.
50. Gillespie, P. G., Prusti, R. K., Apel, E. A., and Beavo, J. A. (1989) *J. Biol. Chem.*, **264**, 12187–12193.
51. Schnetkamp, P. P. M., Klompmakers, A. A., and Daemen, F. J. M. (1979) *Biochim. Biophys. Acta*, **552**, 379–389.
52. Sitaramayya, A., and Liebman, P. A. (1983) *J. Biol. Chem.*, **258**, 12106–12109.
53. Uhl, R., Desel, H., Ryba, N., and Wagner, R. (1987) *J. Biochem. Biophys. Methods*, **14**, 127–138.
54. Hurwitz, R. L., Bunt-Milam, A. H., and Beavo, J. A. (1984) *J. Biol. Chem.*, **259**, 8612–8618.
55. Booth, D. P., Hurwitz, R. L., & Lolley, R. N. (1986) *Invest. Ophthalmol. Visual Sci.*, **27**, ARVO Suppl., 217.
56. Gillespie, P. G. (1988) PhD dissertation, University of Washington.
57. Orlov, N. Y., Kalinin, E. V., Orlova, T. G., and Freidin, A. A. (1988) *Biochim. Biophys. Acta*, **954**, 325–335.
58. Swanson, R. J., and Applebury, M. L. (1983) *J. Biol. Chem.*, **258**, 10599–10605.
59. Deterre, P., Bigay, J., Forquet, F., Robert, M., and Chabre, M. (1988) *Proc. Natl Acad. Sci. USA*, **85**, 2424–2428.
60. Baehr, W., Morita, E., Swanson, R., and Applebury, M. (1982) *J. Biol. Chem.*, **257**, 6452–6460.

61. Roof, D. J., Korenbrot, J. I., and Heuser, J. E. (1982) *J. Cell Biol.*, **95**, 501–509.
62. Ovchinnikov, Y. A., Gubanov, V. V., Khramtsov, N. V., Ischenko, K. A., Zagranichny, V. E., Muradov, K. G., Shuvaeva, T. M., and Lipkin, V. M. (1987) *FEBS Lett.*, **223**, 169–173.
63. Ovchinnikov, Y. A., Lipkin, V. M., Kumarev, V. P., Gubanov, V. V., Khramtsov, N. V., Akhmedov, N. B., Zagranichny, V. E., and Muradov, K. G. (1986) *FEBS Lett.*, **204**, 288–292.
64. Beavo, J. A. (1988) *Adv. Second Messenger Phosphoprotein Res.*, **22**, 1–38.
65. Charbonneau, H., Beier, N., Walsh, K. A., and Beavo, J. A. (1986) *Proc. Natl Acad. Sci. USA*, **83**, 9308–9312.
66. Saas, P., Field, J., Nikawa, J., Toda, T., and Wigler, M. (1983) *Proc. Natl Acad. Sci. USA*, **83**, 9303–9307.
67. Chen, C.-N., Denome, S., and Davis, R. L. (1986) *Proc. Natl Acad. Sci. USA*, **83**, 9313–9317.
68. Yamazaki, A., Stein, P. J., Chernoff, N., and Bitensky, M. W. (1983) *J. Biol. Chem.*, **258**, 8188–8194.
69. Deterre, P., Bigay, J., Pfister, C., Kuhn, H., and Chabre, M. (1986) *Proteins: Structure, Function, and Genetics*, **1**, 199–193.
70. Fung, B. K.-K., and Griswold-Prenner, I. (1989) *Biochemistry*, **28**, 3133–3137.
71. Bitensky, M. W., Whalen, M. M., and Torney, D. C. (1988) *Cold Spring Harbor Symp. Quan. Biol.*, **53**, 303–311.
72. Tyminski, P. N., and O'Brien, D. F. (1984) *Biochemistry*, **23**, 3986–3993.
73. Fung, B. K.-K., and Nash, C. H. (1983) *J. Biol. Chem.*, **258**, 10503–10510.
74. Robinson, P. R., Kawamara, S., Abramson, B., and Bownds, M. D. (1980) *J. Gen. Physiol.*, **76**, 631–645.
75. Dumler, I. L., and Etingof, R. N. (1976) *Biochim. Biophys. Acta*, **429**, 474–484.
76. Liu, Y. P., and Wong, V. G. (1979) *Biochim. Biophys. Acta*, **583**, 273–278.
77. Hurley, J. B. (1980) *Biochem. Biophys. Res. Commun.*, **92**, 505–510.
78. Hurley, J. B., Barry, B., and Ebrey, T. G. (1981) *Biochim. Biophys. Acta*, **675**, 359–365.
79. Brown, R. L., and Stryer, L. (1989) *Proc. Natl Acad. Sci. USA*, **86**, 4922–4926.
80. Yamazaki, A., Sen, I., Bitensky, M. W., Casnellie, J. E., and Greengard, P. (1980) *J. Biol. Chem.*, **255**, 11619–11624.
81. Yamazaki, A., Bartucca, F., Ting, A., and Bitensky, M. W. (1982) *Proc. Natl Acad. Sci. USA*, **79**, 3702–3706.
82. Yuen, P. S. T., Walseth, T. F., Pante, S. S., Sundby, S. R., Graeff, R. N., and Goldberg, N. D. (1987) *Fed. Proc.*, **46**, 2249.
83. Gillespie, P. G., and Beavo, J. A. (1989) *Proc. Natl Acad. Sci. USA*, **86**, 4311–4315.
84. Cote, R. H., Bierbaum, M. S., Nicol, G. D., and Bownds, M. D. (1984) *J. Biol. Chem.*, **259**, 9635–9641.
85. Cobbs, W. H., and Pugh, E. N. (1985) *Nature*, **313**, 585–587.
86. Capovilla, M., Cervetto, L., and Torre, V. (1983) *J. Physiol.*, **343**, 277–294.
87. Cervetto, L., and McNaughton, P. A. (1986) *J. Physiol.*, **370**, 91–109.
88. Ames. A., and Barad, M. (1988) *J. Physiol.*, **406**, 163–179.
89. Cerveto, L., Menini, A., Rispoli, G., and Torre, V. (1988) *J. Physiol.*, **406**, 181–198.
90. Gillespie, P. G., and Beavo, J. A. (1989) *Mol. Pharmacol.*, **36**, 773–781.

91. Coquil, J. E., Franks, D. J., Wells, J. N., Dupuis, M., and Hamet, P. (1980) *Biochim. Biophys. Acta,* **631**, 148–165.
92. Francis, S. H., Lincoln, T. M., and Corbin J. D. (1980) *J. Biol. Chem.,* **255**, 620–626.
93. McElroy, F. A., and Philip, R. B. (1975) *Life Sci.,* **17**, 1479–1493.
94. FitzGerald, G. A. (1987) *N. Engl. J. Med.,* **316**, 1247–1256.
95. Bergstrand, H., Kristoffersson, J., Lundquist, B., and Schurmann, A. (1977) *Mol. Pharmacol.,* **13**, 38–43.
96. Weishaar, R. E., Burrows, S. D., Koylarz, D. C., Quade, M. M., and Evans, D. B. (1986) *Biochem. Pharmacol.,* **35**, 787–800.
97. Torre, V., Matthews, H. R., and Lamb, T. D. (1986) *Proc. Natl Acad. Sci. USA,* **83**, 7109–7113.
98. Barkdoll, A. E., Pugh, E. N., and Sitaramayya, A. (1989) *J. Gen. Physiol.,* **93**, 1091–1108.
99. Boyer, P. D. (1987) *Biochemistry,* **26**, 8503–8507.
100. Senior, A. E. (1988) *Physiol. Rev.,* **68**, 177–231.
101. Wise, J. G., and Senior, A. E. (1985) *Biochemistry,* **24**, 6949–6954.
102. Hurwitz, R. L., Hansen, R. S., Harrison, S. A., Martins, T. J., Mumby, M. C., and Beavo, J. A. (1984) *Adv. Cyclic Nucleotide Protein Phosphorylation Res.,* **16**, 89–106.
103. Charbonneau, H., Prusti, R. K., LeTrong, H., Sonnenburg, W. K., Mullaney, P. J., Walsh, K. A., and Beavo, J. A. (1990) *Proc. Natl Acad. Sci. USA,* **87**, 288–292.
104. Li, T., Volpp, K., and Applebury, M. L. (1990) *Proc. Natl Acad. Sci. USA,* **87**, 293–297.

8

CYCLIC NUCLEOTIDE PHOSPHODIESTERASES IN LIVER
A Review of their Characterisation, Regulation by Insulin and Glucagon and their Role in Controlling Intracellular Cyclic AMP Concentrations

Miles D. Houslay and Elaine Kilgour

Molecular Pharmacology Group, Department of Biochemistry, University of Glasgow, Glasgow G12 8QQ, UK

8.1 INTRODUCTION

Cyclic nucleotide phosphodiesterases (PDEs) are present in most mammalian tissues. Generally the preferred substrates are cAMP and cGMP, although a few exceptions are known, e.g. a non-specific intestinal PDE [1,2].

Initial studies on PDEs focused on the enzyme activity in the whole tissue [3,4]. It is now apparent that this is an oversimplification of a very complex system, as multiple forms of PDE are expressed in many mammalian tissues [5,6]. In order to understand the role of these PDE isoenzymes in regulating cyclic nucleotide concentrations within the cell, it is necessary to establish their numbers, properties, subcellular locations and hormonal regulation.

Recently, polyclonal and monoclonal antisera have been raised to pure and even partially pure preparations of PDEs [7–11]. These have proved to be

Cyclic Nucleotide Phosphodiesterases: Structure, Regulation and Drug Action
Edited by J. Beavo and M. D. Houslay © 1990 John Wiley and Sons Ltd

extremely useful tools for separating out different forms of PDE and classifying them according to their immunological crossreactivity. Ultimately it will be neccessary to clone and sequence the genes for the PDE isoenzymes to establish unequivocally the extent of homology between these enzymes.

In order to understand the range of isoenzymes and their function it is necessary to identify and characterise all of the PDE forms in a particular cell type. This may allow us to appreciate the role of individual PDEs in controlling cell metabolism, growth and differentiation by observing changes in their activity, expression and regulation. To date, this has not been done. However, a serious attempt to achieve this is being tried in our laboratory where we ([12] for review) have concentrated on isolating and characterising the PDE isoenzymes present in rat liver. This chapter will review our present knowledge of the multiple forms of PDE in rat liver and will only deal with the enzymes in other tissues and species in the context of drawing comparisons with those of the liver.

8.2 MULTIPLE FORMS OF PDE IN RAT LIVER

Distinct species of PDE have been identified in rat liver plasma membranes [13], endoplasmic reticulum [14,15], mitochondria [16] and cytosol fractions [13,17] with little cAMP–PDE activity being found in either the Golgi or lysosomal fractions [14,16].

Initial separation and characterisation of PDEs has most frequently been attempted by using chromatographic techniques. However, investigators in our laboratory have also developed a rapid Percoll gradient fractionation technique which enables the separation of PDEs according to their subcellular location [18].

It has, however, proved to be extremely difficult to achieve complete separation of different PDE isoenzymes due to their high multiplicity, paucity of proteinaceous material, degrees of similarity and, in many instances, acute lability and sensitivity to proteolysis. Nevertheless, in order to understand the role of specific isoenzymes it is highly desirable to obtain homogeneous preparations, and to this end we have purified to apparent homogeneity four distinct PDEs from rat liver [9,19,20].

The different forms of PDE that have so far been isolated and characterised in particulate and soluble fractions of rat liver are discussed in this chapter. We should point out that many investigators have alluded to cAMP-PDEs as existing in two categories, namely 'low-affinity' and 'high-affinity'. This is a gross oversimplification of reality, where even one enzyme can display anomalous kinetics indicative of 'high'- and 'low'-affinity components [19]. Thus, in order to attribute specific changes in activity or function to a particular PDE, one needs to 'isolate' its activity. This can be achieved by physical separation or by using a specific inhibitor.

Cyclic nucleotide PDEs exhibit a range of K_m values from submicromolar values to hundreds of micromolar. Those in the range of 0.1–100 μM are most predominant and likely to be of importance. The few noted as exhibiting extremely high K_m values (mM range) towards cAMP, e.g. in mitochondria [16], have low activities (V_{max}), are unlikely to be functionally important and may, indeed, utilise another substrate in situ.

Hormones appear to regulate specific 'high-affinity' PDEs and it is these enzymes that this chapter will concentrate upon.

8.2.1 Particulate PDEs

8.2.1.1 'DENSE-VESICLE' PDE

When subcellular fractionation of rat hepatocyte homogenates was carried out using Percoll gradients, a species of hormone-sensitive PDE migrated to the high-density shoulder of the microsomal fraction [18]. The exact intracellular location of this enzyme, which is the only PDE in rat liver known to be activated by both insulin and glucagon, remains uncertain and it is referred to as the 'dense-vesicle' PDE.

The 'dense-vesicle' enzyme is an integral membrane protein, associated with membranes through hydrophobic interactions, and can be solubilised using detergents. However, like the aminopeptidase enzymes, which line brush border membranes, this PDE can be solubilised using mild proteolysis. Indeed, provided that membrane preparations are contaminated with lysosomes, then exposure of such an appropriate membrane fraction to hypo-osmotic buffers [21] releases an active soluble enzyme. Solubilisation by this method can be blocked by certain protease inhibitors [15,22,23], indicating that hypotonic shock releases or activates an endogenous sulphydryl protease that 'clips' the 'dense-vesicle' enzyme and releases it from its membrane environment. Although three species of PDE are solubilised by treatment of hepatic membranes with hypotonic buffer, the major proportion of the activity is attributable to the 'dense-vesicle' enzyme [9,24].

The ability to solubilise the 'dense-vesicle' PDE selectively, by this 'hypotonic shock' procedure, has facilitated the purification of the enzyme to apparent homogeneity [9]. Two methods have been used to achieve this, one based upon Cibacron Blue (Affi-gel Blue) affinity matrix and the other using a cGMP affinity matrix [9].

This enzyme is a high-affinity cAMP-specific PDE which exhibits kinetics indicative of apparent negative co-operativity. It can, however, also poorly hydrolyse cGMP whilst exhibiting linear Michaelis–Menten kinetics. Whilst it exhibits a very low K_m value for cGMP, the V_{max} is markedly reduced compared with that seen for cAMP as substrate (see Table 8.1). Indeed, the very low K_m value for cGMP, which, due to the low V_{max}, presumably reflects a high affinity

Table 8.1 Properties of rat liver particulate PDEs

	'Dense-vesicle' PDE	Peripheral plasma membrane PDE	Cyclic GMP-stimulated PDE
M_r	112000	52000	135000
Subunit M_r	57000	52000	67000
cAMP hydrolysis			
H	0.43	0.62	1.5
K_m1 (μM)	0.3	0.7	34.0
K_m2 (μM)	28.0	39.0	—
$V_{max}1$	114 munit/mg	1.2 unit/mg	4.0 unit/mg
$V_{max}2$	633 munit/mg	9.2 unit/mg	—
cGMP hydrolysis			
K_m (μM)	10.0	120.0	31.0
V_{max}	4.0 munit/mg	0.4 unit/mg	2.0 unit/mg
IC_{50} for inhibition of cAMP hydrolysis (μM)			
cGMP	2.0	180.0	NA
Amrinone	71.0	1000.0	1000.0
Milrinone	1.0	7.3	1000.0
ICI 63197	100.0	8.0	1000.0
Ro 20-1724	50.0	7.2	210.0
SQ 20009	4.6	1000.0	91.0
Dipyridamole	450.0	29.0	22.0
Carbazeran	4.5	9.0	43.0
Buquineran	14.5	99.0	790.0
ICI 118233	3.8	no inhibition	no inhibition

H, Hill coefficient; NA, not applicable; IC_{50}, concentration of the drug which elicited 50% inhibition of cAMP hydrolysis assayed at 1 μm cAMP, apart from the value for ICI 118233, which was obtained at 0.1 μM cAMP. One unit is 1 μmol/min. Taken from refs 9, 10, 19 and 20.

of this enzyme for cGMP, appears to be unique amongst the so-called cAMP-specific PDEs. This allows the 'dense-vesicle' enzyme to be easily identified, as, in the presence of low concentrations of cGMP the hydrolysis of cAMP by the enzyme can be potently and competitively inhibited (see Table 8.1) by cGMP [9,10].

Analysis of the purified 'dense-vesicle' PDE by SDS/PAGE revealed two bands (57 kDa and 51 kDa) when the purification of the protease-solubilised enzyme was carried out in the absence of protease inhibitors. However, only one band (57 kDa) was evident when protease inhibitors were included in the purification procedure at a stage immediately after (proteolytic) solubilisation.

Two-dimensional mapping of iodinated tryptic peptides of the purified 57 kDa and 51 kDa species revealed close homology between the two [9], indicating that the 51 kDa band was a degradative product of the 57 kDa species. Both sucrose density gradient analysis and gel exclusion chromatography yielded molecular mass values of about 110 kDa, indicating that the enzyme may exist as a dimer in situ. Certainly the 57 kDa and 51 kDa components seen on SDS-PAGE reflect subunits of this enzyme, as characterisation of the pure enzyme upon native polyacrylamide gels yielded a single band which co-migrated with activity that could be eluted from the gel.

Polyclonal antisera have been raised by Pyne et al. [9,10] to the purified 'hypotonic-shock'-solubilised 'dense-vesicle' PDE. The antisera immunoprecipitated a single band of 57 kDa from 'hypotonic-shock' solubilised material which contained all the PDE activity [9]. Pyne et al. [9] used these antisera to identify the 'native' membrane-bound enzyme. This enzyme was released from membranes by detergent solubilisation in the presence of protease inhibitors and then taken for analysis by Western blotting with the specific antiserum for the 'dense-vesicle' enzyme. Alternatively, Western-blot analysis of intact hepatocytes placed directly into boiling SDS-sample buffer in order to solubilise the native enzyme was performed. Immunoblot analysis of this detergent-solubilised protein revealed a major band at 62 kDa and a minor species at 63 kDa. Comparison of the 62 kDa species with the purified 'hypotonic-shock'-released enzyme, by tryptic peptide mapping, showed them to be very similar [9]. It is probable that the 57 kDa and 51 kDa species both derived from the native 62 kDa subunit by limited proteolysis. It was suggested by Pyne et al. [9] that the 63 kDa protein was a precursor containing a signal peptide and that the 57 kDa species could represent the main globular mass of the enzyme, which would be anchored to the lipid bilayer by a small hydrophobic peptide of about 5 kDa, yielding a 'native', mature subunit in the membrane of 62 kDa (Figure 8.1).

A number of agents which are known to exert positive inotropic effects in the heart, such as milrinone, amrinone, carbazeran and buquineran [25,26], have been shown to be potent and selective inhibitors of the 'dense-vesicle' PDE. In particular, the compound ICI 118233, which is a potent inotropic agent, and a highly selective inhibitor of the 'dense-vesicle' PDE in rat liver, has proved to be an extremely useful tool for studying this enzyme [10,27]. Therefore it is possible that the cardiac 'fraction III' PDE, which is purported to be the site of action of such drugs [28–31], is either related to or an equivalent species to the 'dense-vesicle' enzyme. Indeed, an antiserum to the hepatic 'dense-vesicle' PDE was found to crossreact with a 63 kDa species in rat heart as well as in white adipose tissue and kidney [10] and also in bovine heart (unpublished). Interestingly, the amounts of enzyme, expressed per mg of homogenate protein, varied between these tissues (Figure 8.2), with both heart and white adipose tissue (0.75 μg/mg protein and 0.48 μg/mg protein respectively) containing markedly higher levels than either kidney (0.14 μg/mg protein) or liver (0.15 μg/mg protein). It is probable that differential expression of this PDE has a functional significance.

Boyes and Loten [11] reported the purification from rat liver of an insulin-

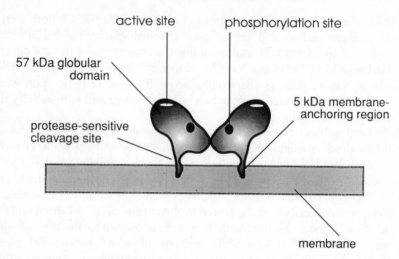

Figure 8.1   Schematic representation of the 'dense-vesicle' PDE. Based upon the biochemical analysis of this enzyme [9], the dimeric 'native', membrane-bound enzyme is shown with its large, globular domain anchored to the bilayer by a hydrophobic tail. The putative active and phosphorylation (A-kinase) sites are shown

sensitive, high-affinity cAMP-PDE. This enzyme hydrolysed both cAMP (K_m 0.24 μM, V_{max} 129 munit/mg) and cGMP (K_m 0.17 μM, V_{max} 46 munit/mg) and, as with the 'dense-vesicle' enzyme, the hydrolysis of cAMP was potently inhibited by cGMP. However, the enzyme purified by Boyes and Loten [11] appeared to differ from the 'dense-vesicle' PDE [9] in that it displayed a subunit molecular mass of 74 kDa, as compared to 62 kDa for the 'dense-vesicle' PDE and appeared to hydrolyse cAMP with linear kinetics, whilst the 'dense-vesicle' enzyme exhibited negative co-operativity. However, the range of substrate concentrations used by Boyes and Loten [11] in their analysis was somewhat restricted, and may not have allowed for the detection of negative co-operativity. Absolute comparison of these two enzymes awaits further details of the properties of the PDE purified by Boyes and Loten [11].

High-affinity cAMP-PDEs with similar properties to those of the 'dense-vesicle' PDE have been purified to apparent homogeneity from bovine platelets [32] and rat adipocytes [33]. Analysis of these purified enzymes by SDS-PAGE yielded apparent molecular masses of 61 kDa and 64 kDa respectively. Both enzymes hydrolysed cAMP and cGMP and were inhibited by agents which exert positive inotropic effects on the heart. As with the 'dense-vesicle' PDE [9] cGMP potently inhibited cAMP hydrolysis achieved by both the platelet PDE [32] and the adipocyte enzyme [33].

Beavo and co-workers reported the purification of a cAMP-PDE from bovine heart [7]. This enzyme shared a number of properties with the rat liver 'dense-vesicle' PDE, namely the sensitivity to inhibition by cardiotonic drugs and the ability of cGMP to inhibit potently cAMP hydrolysis. SDS-PAGE analysis revealed that preparations of the PDE purified from bovine heart contained a predominant 80 kDa polypeptide which was slowly degraded to a 60 kDa and a 67 kDa species. However, Beavo and co-workers subsequently showed, by immunoblot analysis of fresh tissue homogenates, that the native subunit molecular mass value for both the bovine cardiac PDE [7] and a similar bovine platelet PDE [8] was actually 110 kDa and that these enzymes were extremely susceptible to proteolysis to lower molecular mass forms. The platelet enzyme had previously been purified as a 61 kDa species by Grant and Colman [32]. Although it is impossible to be absolutely certain that the 'dense-vesicle' PDE purified by Pyne et al [9] was not derived from a 110 kDa or similar large precursor, immunoblot analysis of fresh rat liver homogenates revealed a 62 kDa species with no larger molecular mass bands being detectable [9] as did Western blot analyses of hepatocytes treated with SDS-PAGE sample buffer (Kilgour, Pyne and Houslay, unpublished) [10]. It is possible that a species difference accounts for the different molecular masses of the bovine and rat enzymes.

8.2.1.2 PERIPHERAL PLASMA MEMBRANE PDE

Rat liver plasma membranes express two forms of cAMP-PDE activity. These are an intrinsic membrane enzyme, which requires detergent to solubilise it, and a peripheral enzyme which is solubilised under conditions of high ionic strength [13].

The peripheral PDE is associated with the cytosolic surface of the plasma membrane through ionic interactions. A single species of integral membrane protein binds the PDE and serves to localise the enzyme to the plasma membrane [34]. The identity of the integral protein is as yet unknown; however, irradiation inactivation experiments indicate that either it is or has a 90 kDa component and it has been suggested that this might be provided by the insulin receptor β subunit [35,36].

The peripheral PDE has been purified to apparent homogeneity and character-ised [19]. Analysis of the purified PDE by SDS-PAGE, sucrose density gradient sedimentation and gel filtration all indicated a molecular mass of about 52 kDa, indicating that the enzyme is monomeric. The purified enzyme exhibited non-linear kinetics and hydrolysed cAMP with a high affinity (see Table 8.1).

Cyclic GMP was an extremely poor substrate for the peripheral PDE and exhibited a high K_m, so its cAMP hydrolysis capabilities are not inhibited by low concentrations of cGMP. 5'-AMP is also a weak inhibitor of this enzyme, and analysis of its product inhibition patterns suggested that a mnemonical type of mechanism might explain the enzyme's unusual kinetics [19,37]. Certainly over the range of cAMP concentrations encountered in vivo (0.3–5 μM), the inhibitory effects of cGMP and 5'-AMP will be negligible.

Polyclonal antisera have been prepared to the purified peripheral PDE and used to immunoblot the amount of enzyme present in different tissues quantitatively [10]. In marked contrast to the 'dense-vesicle' PDE, the plasma membrane PDE was expressed approximately equally in rat liver, kidney, heart and white adipose tissue, on a mg PDE/mg protein basis (Figure 8.2).

The plasma membrane PDE can be inhibited by low concentrations of the agent Ro 20-1724 and also by dipyridamole (see Table 8.1). Indeed, the ability of high-affinity PDEs to be inhibited by Ro 20-1724 is becoming used increasingly to identify a particular class of (related?) isoenzyme(s). Thus, a Ro 20-1724-sensitive PDE has also been identified in mammalian brain, where it is a major contributor to overall PDE activity and also appears to be the binding site for rolipram, an antidepressant agent [38]. It is thought that Ro 20-1724-sensitive PDE isoenzymes could have an important role in the central nervous system [6]. As yet, however, insufficient information is available concerning the brain isoenzyme to allow any direct comparison to be made with the hepatic peripheral plasma membrane PDE.

It has been suggested [39] that the 'dense-vesicle' and peripheral plasma membrane PDEs in rat liver might be related to each other. However, these two isoenzymes have different subcellular locations and are attached to cellular membranes in different ways [18], and although insulin activates both enzymes it does so by distinct mechanisms [18,40]. The enzymes also differ in their molecular sizes, kinetics, and susceptibility to inhibition by selective inhibitors as well as, crucially, in the fact that cGMP potently inhibits cAMP hydrolysis by the 'dense-vesicle' enzyme only (see Table 8.1). In addition there is a lack of immunological crossreactivity between the two PDEs and little homology is evident between

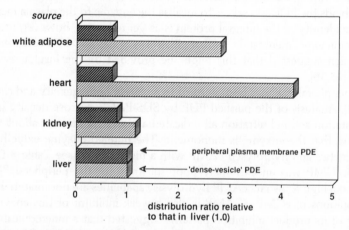

Figure 8.2 Distribution of the 'dense-vesicle' and peripheral plasma membrane PDEs from various rat tissues. Quantification of enzymes was done by immunoblotting using specific antisera. The distribution is given relative to liver [1]. (Data adapted from ref. 10.)

their iodinated (^{125}I) tryptic peptide maps (less than 20%; see [9]). All of this strongly suggests that each of these PDEs is a distinct enzyme species.

Iodinated tryptic peptide maps of this enzyme have also been compared with the rat brain Ca^{2+}-calmodulin-activated PDE with which it shares 31–43% identity as regards co-migration of peptides. In contrast, it only shares some 7–12% identity with the bovine retinal rod outer segment cGMP-specific PDE [41,42].

8.2.1.3   MEMBRANE-ASSOCIATED cGMP-ACTIVATED PDE

Beavo and co-workers were the first to demonstrate that rat liver contains cGMP-activated PDE activity in both the cytosolic and membrane fractions [43]. Such PDE activity is approximately equally distributed between those fractions. Cyclic GMP-activated PDEs have been purified to apparent homogeneity from both the cytosolic and particulate fractions of rat liver [20]. These enzymes are identified by virtue of the fact that low concentrations of cGMP can increase the hydrolysis of [20]. For further discussion on the soluble form of cGMP-activated PDE see later in this chapter, as this section will concentrate on the particulate enzyme. The properties of both enzymes are given in Table 8.2.

Purified hepatic plasma membranes treated with high-ionic-strength solutions, to strip off peripheral proteins, still exhibited cGMP-activated PDE activity and therefore the particulate form of this enzyme is believed to be an integral protein associated with the hepatocyte plasma membranes [20]. This was confirmed by the fact that it was necessary to use detergents to solubilise the particulate cGMP-activated PDE from rat liver membranes [20], showing that it interacted through hydrophobic interactions with the membrane bilayer. Indeed, ionic detergents solubilised the enzyme in a form that was activated by cGMP, whereas non-ionic detergent abolished the ability of cGMP to stimulate cAMP hydrolysis [20].

The detergent-solubilised cGMP-activated PDE purified to apparent homogeneity from rat liver membranes displayed a subunit molecular mass of 66–67 kDa and a native molecular mass of about 135 kDa, indicating that the enzyme forms a dimer [20]. This was confirmed by irradiation inactivation studies on this enzyme [44].

Although the purified particulate cGMP-activated PDE displayed similar affinities for both cGMP and cAMP, the V_{max} for cAMP hydrolysis was approximately twice that for cGMP (see Table 8.2). Furthermore, hydrolysis of cGMP obeyed linear kinetics while cAMP hydrolysis displayed apparent positive co-operativity [20]. Now, low concentrations of cGMP in the range of 0.1 μM induced up to six-fold increases in the rate of cAMP hydrolysis [20] and this was accompanied by marked changes in the kinetics of hydrolysis of cAMP. Indeed, positive co-operative kinetics were abolished and linear kinetics obeyed, with a decrease in the apparent K_m for cAMP occurring [20].

Table 8.2 Properties of the cytosolic and particulate cGMP-activated PDEs in rat liver

	Cytosolic	Particulate
M_r	66000	66000
Subunit M_r	135000	135000
K_m for cAMP (μM)	40	34
In presence of cGMP	20	20
V_{max} for cAMP (unit/mg protein)	4.8	4.0
In presence of cGMP	4.2	3.8
K_m for cGMP (μM)	25	31
V_{max} for cGMP (unit/mg protein)	1.6	2.1
IC_{50} for IBMX (μM)	44	50
In presence of cGMP	19	13
IC_{50} for arachidonic acid (mM)	1.6	0.18
In presence of cGMP	0.6	0.078
IC_{50} for (cis-16)-palmitoleic acid (mM)	1.0	NI
In presence of cGMP	0.45	1.6

IC_{50} is the concentration of inhibitor yielding 50% inhibition. IC_{50} values are for inhibition of cAMP hydrolysis, the stimulating concentration of cGMP used being 2 μM. The cAMP substrate concentrations used were 3 μM for IBMX experiments and 0.5 μM for fatty acid experiments. NI, not inhibited at concentrations up to 5 mM. Taken from ref. 20.

Studies with various inhibitors and cGMP analogues indicated that there are at least two different binding sites for cGMP on cGMP-activated PDEs [45–49]. It is probable that one site is a regulatory site which can modify the activity of the cAMP catalytic site. However, it is not yet clear whether or not both cGMP and cAMP are hydrolysed at the same or distinct catalytic sites.

IBMX (3-isobutyl-1-methylxanthine) is well established as an inhibitor of cAMP PDE activity. However, although high concentrations of IBMX inhibited both the soluble and particulate cGMP-activated PDEs from rat liver, low concentrations of this agent actually stimulated these enzymes [20]. Furthermore, this stimulatory effect was completely abolished by the presence of stimulatory concentrations of cGMP but, in contrast, the inhibitory effect of IBMX was enhanced by cGMP [20]. It has previously been reported that IBMX stimulated PDE activity in crude preparations from rat liver [45,46] and activated soluble cGMP-stimulated PDE purified from bovine liver [49]. It is presumed that IBMX interacts with both the substrate and regulatory sites of the cGMP-activated PDEs.

Cyclic GMP-activated PDE activity has been reported in various tissues, including thymic lymphocytes [50], human platelets [51], rat adipose tissue [52] and bovine adrenal medulla and heart [48]. However, apart from the hepatic enzymes, the only source of purified cGMP-activated PDE has been the soluble fraction of bovine tissues [48,53]. There are distinct differences between these bovine enzymes and the rat cGMP-activated PDEs which are discussed further in a later section on the soluble cGMP-activated PDE.

8.2.1.4   MEMBRANE ASSOCIATED PDEs WHICH HAVE NOT BEEN FULLY CHARACTERISED

Cyclic nucleotide PDE activity has been observed but not fully characterised in both the endoplasmic reticulum (ER) [14,15] and mitochondrial [16] fractions of rat liver.

The endoplasmic reticulum (ER) contributes at least 20% of the cyclic nucleotide PDE activity in a rat liver homogenate [14]. Subfractionation of a microsomal fraction yielded three rough ER fractions together with a smooth ER fraction all of which appear to contain distinct species of PDE activity [14]. These enzymes all hydrolysed both cAMP and cGMP and all expressed a higher K_m for cGMP than for cAMP [14]. To date no further characterisation or purification of these enzymes has been carried out.

Two distinct cAMP-PDE activities are associated with liver mitochondria. One is associated with the inner membrane and contributes approximately 80% of the total mitochondrial PDE activity and the other, which accounts for the remaining 20% of PDE activity, is associated with the outer membrane [16]. These enzymes hydrolyse both cAMP and cGMP but with very high K_m values indeed (mM range). On this basis, and because they express low hydrolytic rates to cyclic nucleotides, it is highly likely that neither cAMP nor cGMP are the primary substrates for these enzymes in situ.

8.2.2   Soluble PDEs

Soluble PDEs account for 40% of the total cAMP-PDE activity in rat hepatocytes [18]. However, until recently very little information has been available about these enzymes. Russell et al. [54] reported that DEAE–cellulose fractionation of a high-speed supernatant prepared from rat liver revealed the presence of three major peaks of cyclic nucleotide PDE activity, but no further characterisation has been pursued. Recently, we [17] have confirmed this report, but using the higher resolving power of Mono-Q ion exchange chromatography, have further fractionated supernatant from rat liver and isolated hepatocytes into five distinct peaks of PDE activity which appear to reflect single, distinct enzyme forms (PDE-MQ-I- to V inclusive; see Figure 8.3). A comparsion of some of the properties of these soluble PDEs is given in Table 8.3.

The first peak to elute from Mono-Q ion exchange chromatography (PDE-MQ-I) hydrolysed cAMP predominantly and contributed a substantial propor-tion (see Figure 8.1) of the total cAMP-PDE activity in the hepatocyte supernatant. This enzyme exhibited linear kinetics for the hydrolysis of both cAMP and cGMP, with K_m values of 25 μM and 237 μM respectively. Absolute

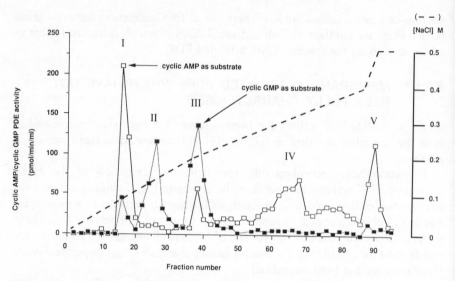

Figure 8.3   Mono-Q ion exchange profile of rat liver soluble PDE activity. A high-speed supernatant of rat liver was chromatographed on a Mono-Q column and fractions were assayed for cyclic nucleotide PDE activity. 1 μM cAMP substrate is represented by open squares whilst closed squares depict a substrate of 1 μM cGMP. The 0–0.5 M NaCl gradient is shown as a broken line. The letters indicate the occurrence of peaks PDE-MQ-I to V. (Adapted from ref. 17)

Table 8.3   Properties of the PDE activities separated by Mono-Q ion exchange chromatography of the soluble fraction from rat hepatocytes

	I	II	III	IV	V
M_r	33 150	237 500	NA	NA	NA
K_m cAMP (μM)	25	NA	38	0.6	0.6
K_m cGMP (μM)	237	5	36	NA	NA
IC$_{50}$ (μM)					
IBMX	NI[b]	6.7	323	5.2	9.8
Milrinone	NI[b]	61.0	NI[b]	7.3	20.0
Ro 20-1724	NI[b]	303.0	NI[b]	13.0	1.5
ICI 118233	NI[a]	NI[a]	NI[a]	NI[a]	NI[a]
Zaprinast	NI[a]	4.5	NI[a]	300.0	NI[a]
Rolipram	NI[a]	NI[a]	NI[a]	0.5	0.1

IC$_{50}$ is the concentration of drug required for 50% inhibition of enzyme activity at 1 μM cAMP. ND, not done. NI, not inhibited at concentrations up to either 100 μM[a] or 500 μM[b]. NA, not applicable. Taken from ref. 17.

V_{max} values could not be determined, as the PDE preparations of Lavan et al. [17] were impure, but the V_{max} ratio for cGMP/cAMP hydrolysis indicated that the activity had a slightly higher V_{max} value for cGMP compared to cAMP.

Determination of the molecular mass of PDE-MQ-I, by gel filtration chromatography, yielded a value of 33 kDa. PDE-MQ-I was very unusual in that it was insensitive to inhibition by any one of the PDE inhibitors tested, including IBMX (see Table 8.3). This is a highly unusual feature, as IBMX is generally regarded as a nonspecific inhibitor of PDE activity, and PDE-MQ-I may be unique in its insensitivity to this compound. This enzyme is also rather unusual in that, unlike most other PDEs studied [9,55], it is unable to bind to Affi-gel Blue, even in the presence of Mg^{2+}.

PDE-MQ-II accounted for a substantial proportion of the cGMP PDE activity in hepatocyte supernatant (see Figure 8.3). This enzyme displayed linear kinetics with a K_m for cGMP hydrolysis of 5 μM. Ca^{2+}–calmodulin (CaM) caused a two-fold stimulation of the cGMP PDE activity of peak II and a greater than four-fold stimulation of the cAMP activity. However, even following Ca^{2+}–CaM stimulation, PDE-MQ-II still only represented a small fraction (less than 5%) of the total cAMP activity of rat liver supernatant. Gel filtration chromatography indicated that this PDE had a molecular mass of 237.5 kDa. Of the inhibitors tested, PDE-MQ-II was most susceptible to inhibition by IBMX and least affected by Ro 20-1724 (see Table 8.3).

PDE-MQ-III hydrolysed both cAMP and cGMP, with the cAMP activity being stimulated some six-fold by 2 μM cGMP. This enzyme contributed the majority of soluble cGMP-hydrolysing PDE activity in liver and also a substantial proportion of cAMP-PDE activity. PDE-MQ-III was inhibited by high concentrations of IBMX (IC_{50} approx 323 μM; see Table 8.3) but not by either milrinone or Ro 20-1724 and is the same enzyme as the cGMP-activated PDE purified from rat liver soluble fraction by Pyne et al [20].

PDE-MQ-IV and PDE-MQ-V both specifically hydrolysed cAMP. However, even though cGMP did not appear to act as a substrate, it produced only a small (10–17%) inhibition of cAMP-PDE activity of both peaks assayed at 1 μM substrate concentration. Both peaks IV and V were, however, potently inhibited by both IBMX and by milrinone (see Table 8.3). In contrast, Ro 20-1724 was a more potent inhibitor of PDE-MQ-IV than PDE-MQ-V (see Table 8.3).

It appears, therefore, that there are at least five distinct soluble PDE activities in rat liver. Four of these PDEs hydrolyse cAMP, two with a high affinity. Reeves et al. [31] have identified four PDEs present in a soluble fraction of rat cardiac tissue, two of which expressed a high affinity towards cAMP. Therefore it seems likely that multiple cAMP-PDEs may also exist in other tissues.

Two high-affinity membrane-bound cAMP-PDEs, namely the 'dense-vesicle' PDE and the peripheral plasma membrane PDE, have been isolated and characterised in rat liver (see Sections 8.2.1.1 and 8.2.1.2). The 'dense-vesicle' PDE can be readily solubilised by endogenous proteases and therefore it is extremely important when studying soluble PDEs to use conditions which minimise protease activity. On the basis of susceptibility to inhibition by cGMP, neither PDE-MQ-IV nor PDE-MQ-V resembled the 'dense-vesicle' cAMP-PDE, which is potently inhibited by cGMP [9]. Furthermore, the compound ICI 118233, which is a potent inhibitor of the dense-vesicle PDE in rat liver, failed to inhibit any of the soluble PDE fractions and an antiserum raised to the 'dense-vesicle' PDE failed to crossreact with any of the soluble fractions [17]. It is, however, possible that the Ro 20-1724-sensitive PDE-MQ-IV may be either related to the peripheral plasma membrane enzyme or indeed may be the peripheral plasma membrane PDE which has been displaced from its membrane during the purification procedure.

Although the techniques used by Lavan et al. [17] resulted in the apparent separation of five PDE activities, it cannot be stated with certainty at this stage whether these represent functionally and structurally distinct enzymes. Cross-contamination between adjacent peaks and generation of proteolytic fragments with altered properties can occur and ultimately it will be necessary to purify and compare the properties and amino acid sequences of all five PDEs in order to establish unequivocally the range of soluble PDEs present in rat liver cytosol. However, release of the 'dense-vesicle' PDE and loss of activation of the so-called 'PDE-II' activity (PDE-MQ-II) by Ca^{2+}–CaM are highly sensitive indicators of proteolysis. As neither of these events occurred, it is extremely unlikely that these apparently highly distinct forms were generated by proteolysis.

8.2.2.1 SOLUBLE cGMP-ACTIVATED PDE

Cyclic GMP-activated PDEs have been purified from both the cytosolic and particulate fractions of rat liver [20]. The distribution of these enzymes was unaffected by the presence of protease inhibitors, indicating that the cytosolic enzyme was not derived by proteolysis of the membrane-bound PDE [20]. The particulate cGMP-activated PDE was discussed in Section 8.2.1.3, and its properties were compared with those of the soluble enzyme in Table 8.2.

Analysis of the purified cytosolic cGMP-activated PDE and SDS-PAGE revealed a single band of molecular mass 66 kDa. The enzyme appeared to form a dimer, as both gel filtration and sucrose density gradient analysis identify the native enzyme as a 135 kDa species.

The kinetic properties of the cytosolic PDE resembled those displayed by the particulate enzyme [20] (see Table 8.2). Both enzymes exhibited apparent positive co-operativity for the hydrolysis of cAMP, whilst cGMP hydrolysis obeyed linear kinetics [20]. Furthermore, the kinetic parameters for the hydrolysis of both cAMP and cGMP were the same for both enzymes (see Table 8.2). Thus cGMP increased cAMP hydrolysis by both the particulate and cytosolic PDEs with K_a values of 0.23 μM and 0.28 μM respectively. In the presence of cGMP the kinetics of cAMP hydrolysis by both enzymes altered markedly, with the activity now obeying linear kinetics [20] and a decrease occurring in the K_m value for cAMP (see Table 8.2). Interestingly, cGMP induced a smaller increase in cAMP hydrolysis by the cytosolic PDE compared with that seen for the particulate enzyme. In the presence of cGMP the maximal increase in the rate of cAMP hydrolysis by the cytosolic enzyme was between two- and three-fold, whereas the increase was between five- and six-fold for the particulate PDE [20]. Furthermore, membrane attachment of the particulate enzyme appeared to enhance the stimulatory effect of cGMP, which was reduced upon solubilisation [20].

Although the similarities in molecular size (see Table 8.2), thermostabilities [20] and kinetics [20] (see Table 8.2) imply that the soluble and particulate cGMP-activated PDEs are closely related, they are nevertheless distinct species. Thus, although the presence of stimulatory concentrations of cGMP considerably enhanced the inhibitory effect of IBMX on both the cytosolic and particulate enzymes, the decrease in IC_{50} for inhibition by IBMX was greater for the particulate enzyme than for the cytosolic species (see Table 8.2). Furthermore, IC_{50} values for inhibition of the particulate PDE and cytosolic forms of this PDE by fatty acids were markedly different (see Table 8.2). This implies that these forms of PDE have different hydrophobic domains that can alter catalytic function towards cAMP. These hydrophobic domains might be further exploited to develop selective inhibitors of these two enzymes.

Tryptic-peptide maps of the purified cytosolic and particulate enzymes revealed close homology between these two forms but at least two peptides appeared to be specific to the particulate PDE and one peptide was apparently specific to the cytosolic enzyme [20]. It would be of interest to obtain the amino acid sequence and sequence of the gene encoding for these two forms of PDE to ascertain exactly what differences account for one enzyme being directed to the plasma membranes and the other to the cytosol. Perhaps alternative splicing could account for the differences observed.

Cyclic GMP-activated PDEs have also been purified from the soluble fractions of bovine heart, bovine adrenals [48] and liver [53]. The bovine enzymes all appeared to be dimers with subunit molecular mass of 102–107 kDa, which differs markedly from the values obtained for both the particulate and cytosolic PDEs from rat liver (Table 8.2). Yamamoto et al. [56] showed that low concentrations of uncertain unsaturated fatty acids, such as cis-palmitoleic acid, could stimulate cAMP hydrolysis by the bovine liver enzyme. However, low concentrations of these fatty acids elicited only marginal stimulations of both the cGMP-activated PDEs in rat liver and, in fact, higher concentrations of these fatty acids actually inhibited the enzymes [20]. Although the K_m values for cAMP hydrolysis by both the rat hepatic PDEs (see Table 8.3) and the enzymes purified from bovine tissues (30–33 μM) [48,53] were very similar, the V_{max} values exhibited by the rat enzymes (see Table 8.2) were noticeably lower than those for bovine PDEs (120–170 unit/mg). It is therefore evident that with regard to molecular size, inhibitor sensitivity and kinetics, species differences exist between the bovine and rat cGMP-activated PDEs.

8.3 HORMONAL CONTROL OF cAMP-PDE ACTIVITY

Many hormones exert their actions on target cells by elevating the intracellular concentration of cAMP. The only known mechanism for the degradation of cAMP is by the action of PDEs; these enzymes therefore provide important sites for hormonal regulation.

Exposure of isolated hepatocytes to either insulin or glucagon results in an increase in particulate high-affinity cAMP-PDE activity [18,21]. In contrast, no hormonal effects have been detected on cytosolic PDE activity [21]. Cyclic AMP mediates many of the intracellular effects of glucagon, and the stimulation of PDE activity by this hormone probably serves to diminish and terminate the hormonal signal through accelerated degradation of cAMP.

Insulin antagonises the effects of glucagon on the liver by decreasing cAMP levels [21,57,58]. This is achieved by the activation of PDE activity [18,21] as well as by the inhibition of adenylate cyclase [59].

Percoll gradient analysis revealed that insulin activates two cAMP PDE isoenzymes in rat hepatocytes, namely the 'dense-vesicle' PDE and the peripheral plasma membrane enzyme. In contrast, glucagon activated only the 'dense-vesicle' PDE [21]. Inhibitor studies, however, have indicated that insulin also activates at least one other cAMP-PDE in rat hepatocytes [60], although the identity of this enzyme(s) is not known.

The fact that hormones regulate the activity of specific PDE isoenzymes may indicate that either cAMP is compartmentalised within the cell and that each species of PDE controls a different pool of the cyclic nucleotide or that the particular kinetics of these enzymes provide for distinct effects on cAMP metabolism. Our data on simulating hepatocyte cAMP metabolism using a computer model (see below) would favour the latter interpretation.

8.3.1 Hormonal control of the 'dense-vesicle' PDE

The 'dense-vesicle' enzyme can be activated by insulin as well as by elevation in the intracellular concentration of cAMP, as is, for example, achieved by exposing hepatocytes to glucagon [18,21]. The maximal activation of this enzyme elicited by insulin (30% activation) was much lower than the maximal 70% activation in the prescence of glucagon [18]. This implies that those hormones activate the enzyme by distinct routes.

Indeed there appears to be a complex interplay between insulin and glucagon in controlling the activity of the 'dense-vesicle' PDE. When cells were given insulin and glucagon together, the 'dense-vesicle' PDE activity was the same as that observed with insulin alone but, in sharp contrast with this, these hormones had a synergistic effect if cells were pretreated with glucagon prior to exposure to insulin [18]. We have proposed that the molecular root of such a synergistic effect is at the membrane receptor level, as treatment of hepatocytes with IBMX, dibutyryl-cAMP and cholera toxin, all of which increase intracellular cAMP levels and as a result activate the 'dense-vesicle' PDE, failed to mimic glucagon in eliciting a synergistic increase in PDE activity on subsequent exposure of cells to insulin [18]. However, one should note that glucagon can also exert a small stimulation of inositol phospholipid metabolism in hepatocytes [61]. Whilst this alone does not activate the 'dense-vesicle' enzyme, e.g. neither angiotensin nor vasopressin activate this enzyme, a small stimulation of this pathway in concert with an elevation of intracellular cAMP may provide the basis for the synergistic effect with insulin.

In both hepatocytes [18,62] and adipocytes [63] a high-affinity cAMP-PDE can be activated by agents, such as glucagon and β-adrenergic agonists, which increase intracellular cAMP concentrations. The effects of cAMP on metabolic pathways are mediated by cAMP-dependent protein kinase (A-kinase) which phosphorylates and hence modifies the activity of various enzymes ([64] for review). The increase in PDE activity which occurs when cAMP concentrations are raised has been correlated with the stimulation of A-kinase in both adipocytes and hepatocytes [65]. Percoll gradient analysis revealed that glucagon activated only the 'dense-vesicle' PDE in rat hepatocytes [18]. However, attempts to phosphorylate and activate the 'dense-vesicle' PDE purified from rat liver, with A-kinase have been unsuccessful (N. G. Anderson, E. Kilgour and M. D. Houslay, unpublished work). With the thought that the membrane environment may be important for the activation process, we attempted to phosphorylate and activate the 'dense-vesicle' PDE by treatment of a crude particulate fraction, isolated from rat hepatocytes, with A-kinase [27,66]. The enzyme was identified by assaying the membrane fraction in the presence of either the compound ICI 118233 or cGMP, both of which specifically inhibit cAMP hydrolysis by the 'dense-vesicle' PDE in rat liver [10]. By treating membranes with A-kinase we were able to elicit a two- to three-fold activation of the 'dense-vesicle' enzyme but this only occurred if the membrane fraction had been preincubated for a short time with Mg^{2+} prior to challenge with A-kinase (Table 8.4). However, if protein phosphatase inhibitors were included with Mg^{2+} during the preincubation, then A-kinase failed to activate the enzyme (Table 8.4). It therefore appeared that the action of Mg^{2+} during the preincubation was to stimulate endogenous protein phosphatases [27,66].

Immunoprecipitation of the 'dense-vesicle' PDE revealed that, providing membranes had been pretreated with Mg^{2+}, a time-dependent incorporation of label into the enzyme occurred during subsequent incubation with A-kinase and $[\gamma\text{-}^{32}P]ATP$. No increase in labelling of the enzyme occurred if the membranes had not been pretreated with Mg^{2+}, which is in accord with our failure to detect any activation of the enzyme under such conditions. The rate of incorporation of

Table 8.4 Effect of preincubation conditions on the stimulation of 'dense-vesicle' PDE activity by A-kinase

Experiment	Additions to preincubation	A-kinase	pmol per min per mg	% Stimulation
1	—	+	1.4 ± 0.1	20.1 ± 12.0
	—	—	1.2 ± 0.1	
2	MgCl$_2$ (5 mM)	+	2.2 ± 0.2^a	102.0 ± 13.6
	MgCl$_2$ (5 mM)	—	1.0 ± 0.04	
3	MgCl$_2$ + NaF + βGP	+	1.0 ± 0.03	-16.0 ± 8.0
	MgCl$_2$ + NaF + βGP	—	1.2 ± 0.1	

A crude particulate fraction prepared from rat hepatocytes was suspended in 10 mM Tris-HCl, pH 7.4, containing 0.25 M sucrose, and was preincubated at 30°C with the additions indicated. After 10 min all samples were adjusted to contain 5 mM MgCl$_2$, 50 mM NaF and 10 mM β-glycerophosphate (βGP) and the incubations were continued for a further 10 min with 0.1 mM ATP and in the presence or absence of A-kinase as indicated. Results are mean $\pm$ SEM for three observations. ^{a} Value is significantly different from the appropriate value in the absence of A-kinase, $P < 0.01$. Taken from ref. 27.

label showed good correlation with the increase in activity, and both reached maximal levels after incubation with A-kinase for ~ 10 min (Figure 8.4). We therefore proposed that phosphorylation of the 'dense-vesicle' enzyme by A-kinase accounted for its activation. We cannot, however, be certain that A-kinase itself phosphorylates the PDE directly, as it is conceivable that A-kinase could have phosphorylated another kinase, associated with the membrane fraction, which then acted upon the 'dense-vesicle' enzyme.

It appears then that when the particulate fraction is isolated from rat hepatocytes the 'dense-vesicle' enzyme is in a phosphorylated form which it is necessary to dephosphorylate to enable A-kinase to activate this PDE in isolated membranes. Indeed we have immunoprecipitated a phosphorylated form of this enzyme, from ^{32}P-labelled hepatocytes which could be dephosphorylated by treatment with MgCl$_2$ [27,66]. It is interesting that this dephosphorylation reaction had no effect whatsoever on PDE activity. In contrast, the A-kinase-mediated phosphorylation of the enzyme activated it. This strongly implies that there are at least two regulatory sites on the 'dense-vesicle' enzyme which can be phosphorylated, one being a site for phosphorylation by A-kinase, which enhances enzyme activity, and the other being a 'silent' site which prevents phosphorylation and activation by A-kinase yet has no effect upon enzyme activity. Peptide mapping should confirm whether or not the 'dense-vesicle' PDE can be phosphorylated on more than one site. We have suggested that the 'silent' phosphorylation of this enzyme might be elicited by 5'-AMP-stimulated protein kinase. This would be analogous to results found using acetyl CoA carboxylase [67], where phosphorylation by 5'-AMP kinase [68] results as a consequence of cells becoming anoxic upon breakage, leading to an increase in 5'-AMP and the

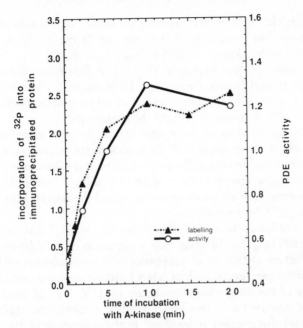

Figure 8.4 Time-course for the activation and phosphorylation of the 'dense-vesicle' PDE by A-kinase. A crude particulate fraction from rat hepatocytes was preincubated for 10 min at 30°C with 5 mM $MgCl_2$ before being incubated with A-kinase and $[\gamma\text{-}^{32}P]ATP$ for the times indicated. Activity and the time-dependent increase in labelling of the 'dense-vesicle' PDE are shown. (Adapted from ref. 27)

activation of this kinase. The resultant effect is the phosphorylation of acetyl CoA carboxylase on a site which prevents protein kinase-A from phosphorylating and activating this enzyme. Indeed, such an effect has also been seen with HMG-CoA reductase [68]. Certainly, we can observe the incorporation of phosphate (^{32}P) into this enzyme when it is isolated from broken hepatocytes [27]. Our more recent studies have been aimed at reducing this effect in order to identify a glucagon-elicited phosphorylation in intact cells. This we can now identify (E. K. Y. Tang and M. D. Houslay unpublished data). However, to date we have been unable to phosphorylate the solubilized, purified 'dense-vesicle' enzyme using a pure preparation of 5'-AMP kinase.

Recently, MacPhee et al. [69] reported the activation and phosphorylation by A-kinase of a high-affinity cAMP-PDE from human platelets and we have observed that A-kinase elicits a two-fold activation of a high-affinity cAMP-PDE associated with the ER fraction in rat adipocytes [70]. It is tempting in this respect to draw parallels with the guanine nucleotide binding protein G_i-2, which is phosphorylated on its α subunit in the basal state when immunoprecipitated from hepatocytes [71,72]. It is possible that the high basal phosphorylation of G_i-2 and the 'dense-vesicle' PDE in rat hepatocytes occurs as a consequence of the homogenisation and processing of the cells due to the activation of a kinase similar to the 5'-AMP kinase isolated by Carling et al. [68].

The activation of the adipocyte PDE by A-kinase correlated well with increased phosphorylation of the enzyme. Intriguingly, the adipocyte PDE was not phosphorylated in the basal state and hence no pretreatment with Mg^{2+} was required to enable A-kinase to activate the enzyme [70].

The molecular mechanism of insulin action remains unknown, although various events have been implicated. These include the activation of the intrinsic insulin receptor tyrosine kinase, the internalisation and subsequent processing of the insulin–receptor complex, the release of a soluble low molecular weight mediator substance(s) from plasma membranes and the activation of the putative G-protein G_{ins} ([73,74] for reviews).

The ability of insulin to activate the 'dense-vesicle' PDE was abolished when membrane trafficking was blocked. Hence depletion of ATP levels by treating cells with fructose [40,62] or dinitrophenol [40], preincubation of hepatocytes with lysosomotropic agents, such as NH_4Cl, chloroquine and methylamine, and the lowering of the incubation temperature to $22°C$ [62], all blocked insulin's activation of the enzyme. This suggests that a membrane translocation or membrane processing event is involved in the activation of the enzyme by insulin. It is possible that the internalisation and subsequent processing of the insulin–receptor complex triggers an event which results in activation of the 'dense-vesicle' PDE. However, Benelli et al. [75] reported that the activation by insulin of a high-affinity cAMP-PDE, located in Golgi fractions, preceded the internalisation of the insulin–receptor complex. Cyclic GMP inhibited cAMP hydrolysis by this enzyme and hence it is possible that it was the 'dense-vesicle' PDE. These workers, therefore, concluded that internalisation of insulin receptor was not necessary for the activation of this PDE.

Alternatively, it is possible that it is the trafficking of the 'dense-vesicle' enzyme itself that is responsible for its activation by insulin. In this light it is tempting to draw comparisons with the stimulation of glucose transport by insulin. In adipocytes insulin causes the recruitment of glucose transporters from an intracellular pool to the plasma membrane, where the transporters are subsequently activated by a second, as yet undefined, event [76,77]. This translocation process is also blocked upon ATP depletion and by lowering the incubation temperature [78]. Indeed, insulin could initiate the translocation of the 'dense-vesicle' PDE to the 'dense-vesicle' location where it is subsequently activated. Such activation could involve a change in the phosphorylation state of the enzyme.

The ability of insulin to decrease cAMP levels, raised by glucagon, was apparently unaffected by pretreatment of hepatocytes with either NH_4Cl or fructose, both of which block insulin's activation of the 'dense-vesicle' enzyme. It was therefore concluded that the 'dense-vesicle' PDE did not play a significant role in the ability of insulin to cause gross depletions in the intracellular concentration of cAMP in hepatocytes [60]. However, it is possible that the 'dense-vesicle' enzyme controls cAMP concentrations in a specific intracellular pool and that any changes in this pool are obscured when the total intracellular concentration of cAMP is measured. In adipocytes, a particulate PDE which is similar to the 'dense-vesicle' enzyme is the major PDE that is regulated by insulin [33,70]. In this system this PDE certainly appears to have a key role in the antilipolytic action of insulin [79,80].

In both hepatocytes [81] and adipocytes [82], pertussis toxin treatment of intact cells blocked the activation of the 'dense-vesicle' PDE by insulin. It is well established that pertussis toxin inhibits the guanine nucleotide regulatory protein G_i [83]. It is possible, therefore, that a 'G_i-like' guanine nucleotide regulatory protein may be involved in the activation of the 'dense-vesicle' PDE by insulin. This may be related to the possible involvement of a translocation step in the insulin-mediated activation of this enzyme, as G-proteins are believed to play an important role in secretory membrane-trafficking responses.

8.3.2 Hormonal control of the peripheral plasma membrane PDE

Activation of the peripheral plasma membrane PDE by insulin has been demonstrated in intact hepatocytes [18] as well as in isolated plasma membranes [84,85]. Lysosomotropic agents had little effect on this activation, yet completely blocked activation of the 'dense-vesicle' PDE [40], indicating that insulin activates the plasma membrane PDE and 'dense-vesicle' PDE by distinct routes.

Although cAMP levels elevated by either glucagon or by treatment of cells with cAMP analogues activated the peripheral plasma membrane PDE, the increased cAMP levels potentiated the stimulatory effect of insulin on this enzyme in intact hepatocytes [18]. Furthermore, insulin was found to cause a two-fold activation of cAMP-PDE activity in plasma membranes isolated from rat liver, but only when cAMP and Mg^{2+}-ATP were also present [84]. The presence of all three ligands was essential for the activation process to ensue and no such activation occurred if any one of the ligands was not present. The activation exhibited a K_a for cAMP of 1.6 μM [84], indicating that at basal intracellular cAMP levels of 0.3–0.5 μM [86,87] little activation of the enzyme would be elicited by insulin. However, after exposure to glucagon, cAMP levels rise to 2.0–4 μM [18,58,88], which would allow activation by insulin. This could explain the proposal [88] that it is necessary to first elevate cellular cAMP levels before they can be depressed by insulin.

Purification of the activated peripheral PDE from insulin-treated membranes showed that the phosphorylation of this enzyme was alkali-labile, indicating that it had occurred on serine residues [84]. We have proposed that rather than insulin activating a membrane-bound cAMP-dependent protein kinase, either cAMP may have to be bound to this enzyme for phosphorylation and activation to occur by an unidentified kinase, or the binding of insulin to its receptor results in a conformational change in the PDE such that it becomes a substrate for a membrane-bound cAMP-dependent protein kinase [5].

It was somewhat suprising to find that treatment of intact hepatocytes with cholera toxin also resulted in the activation of the peripheral plasma membrane PDE [18]. This is because it is well established that cholera toxin activates the guanine nucleotide regulatory protein G_s, resulting in elevated cAMP levels due to the activation of adenylate cyclase [83]. Clearly, the activation of the peripheral PDE by cholera toxin could not be due to the activation of G_s, as neither glucagon, which also activates G_s, nor elevated cAMP levels activated the enzyme [18]. The addition of insulin to cholera toxin-treated cells elicited no further increase in activation of the peripheral PDE [18], suggesting that insulin and cholera toxin achieve their effects by the same route. This led to the proposal that a unique guanine nucleotide binding protein which, like G_s, was activated by cholera toxin, could be involved in the activation of the peripheral PDE by insulin.

The possible involvement of a unique guanine nucleotide binding protein in the activation of the peripheral PDE is supported by the observation that GTP and its non-hydrolysable analogues activated the enzyme in isolated plasma membranes [85].

Certainly, when a Percoll gradient fractionation procedure was used to isolate plasma membranes, either insulin (10 nM) alone (11% activation over control) or GTP (0.1 nM) alone (26% activation over control) produced only small stimulations of the enzyme, yet activation (71% over control) was much greater in the presence of both insulin and GTP [89]. Therefore the activation of the peripheral PDE by insulin appears to be GTP-dependent. It is possible that the putative guanine nucleotide binding protein G_{ins}, which is thought to mediate at least some of the metabolic effects of insulin, is involved in the activation of the peripheral PDE [5]. An analogy can be drawn with the activation of a cGMP-PDE in retinal rods which is catalysed by rhodopsin through a distinct guanine nucleotide binding protein transducin [90] and is a substrate for both cholera toxin and pertussis toxin [91].

Although treatment of intact hepatocytes with glucagon had no effect on the activity of the peripheral PDE *per se*, prior treatment of these cells with glucagon totally abolished the ability of insulin to activate the enzyme [18]. This effect was not due to any increase in the intracellular concentration of cAMP, as pretreatment with 8-bromo-cAMP or dibutyryl-cAMP actually augmented insulin's stimulatory effect, whilst these compounds did not have any effect when added alone [18]. This densensitization to insulin's action was not mediated by cAMP [18] and appeared to have much similarity to the desensitization of adenylate cyclase that can be elicited by glucagon, and has been shown to be a cAMP-independent process [59,92].

Adenosine and its non-hydrolysable analogue N^6-(phenylisopropyl)-adenosine (PIA) potently prevented glucagon from blocking insulin's activation of the peripheral PDE [93]. Thus, following pretreatment of hepatocytes with PIA, the ability of insulin to decrease intracellular cAMP levels, raised by glucagon, was increased markedly [60]. Therefore the peripheral plasma membrane PDE is capable of exerting a potent effect in decreasing intracellular cAMP levels. However, glucagon and adenosine modulate the ability of insulin to activate this PDE and the role that this enzyme plays in reducing cAMP levels in vivo will clearly depend on the relative amounts of these modulators that the cell is exposed to.

A specific antiserum (PM1) has been used to show that this PDE is indeed phosphorylated in intact hepatocytes challenged with insulin [72]. However, phosphoamino acid analysis showed that this enzyme was in fact phosphorylated upon alkali-stable tyrosine residues. This contrasted with our previous membrane studies which indicated that the incorporated phosphate was alkali-labile [37]. Such experiments indicate that the mechanism of activation and phosphorylation of this PDE in intact cells appears to differ from that seen using isolated membranes.

Further evidence that this PDE was phosphorylated on tyrosine residues came from studies using two other antisera. One of these was raised against phosphotyrosine and was capable of immunoprecipitating species of molecular mass 180 kDa, 95 kDa, 52 kDa and 39 kDa. The 52 kDa species immunoprecipitated by this antisera was recognised by the antiserum PM1. Indeed, it was also possible to show that PDE activity could be bound to an affinity matrix of the anti-phosphotyrosine antiserum and that the amount of activity absorbed was increased in insulin-treated cells. Kwok and Yip [94] prepared an antiserum against a ~50 kDa protein which they had found to be phosphorylated on tyrosine residues upon challenge of hepatocytes with insulin. The material immunoprecipitated by this antiserum was also recognised upon Western blotting by the antiserum PM1 [72].

As discussed above, prior challenge of hepatocytes with glucagon blocked the ability of insulin to activate this PDE. Under such conditions of prior challenge with glucagon it was shown that this PDE was not phosphorylated, offering a molecular explanation for the failure of insulin to activate it. This was observed using the antiserum PM1 to immunoprecipitate this PDE and was also seen for the 52 kDa species immunoprecipitated with the anti-phosphotyrosine antiserum.

The peripheral plasma membrane PDE may thus provide a specific target for activation by the insulin receptor tyrosyl kinase.

8.4   THE CONTROL OF HEPATOCYTE INTRACELLULAR cAMP CONCENTRATIONS

8.4.1   After challenge with glucagon

Challenge of intact hepatocytes with glucagon over the physiological range of concentrations leads to a rapid, albeit transient, elevation in the intracellular concentrations of cAMP [18]. The initial rise is due to the marked stimulation of adenylate cyclase activity which is achieved through coupling of the glucagon receptor to the stimulatory G-protein, G_s (Figure 8.5). However, the production of cAMP is in itself transient due to the rapid desensitization of adenylate cyclase that follows [59]. This takes the form of an 'uncoupling' of the ability of the glucagon receptor to couple to G_s. Such an event is not due to receptor loss but probably reflects the C-kinase-mediated phosphorylation of the glucagon receptor itself rather than any effect on either G_s or G_i. Indeed, under such circumstances the function of G_i is actually inhibited [95].

Certainly, desensitization is mimicked by tumour-promoting phorbol esters and by diacylglycerols which can activate C-kinase, as well as by hormones such as angiotensin and vasopressin which act on hepatocytes to stimulate inositol phospholipid metabolism [92,96]. Indeed, glucagon itself can elicit a small stimulation of inositol phospholipid metabolism [61,97–99] which may account for the cAMP-independent ability to elevate intracellular Ca^{2+} [100]. Glucagon also elevates the intracellular concentrations of diacyclglycerol [101], which, by activating protein kinase-C, may mediate the desensitization of glucagon-stimulated adenylate cyclase.

The rapid return of cAMP concentrations to their basal levels is mediated by PDE activity. It probably reflects the functioning of a 'pool' of moderate 'affinity' PDE, perhaps provided by the considerable amount of cGMP-stimulated cyclic

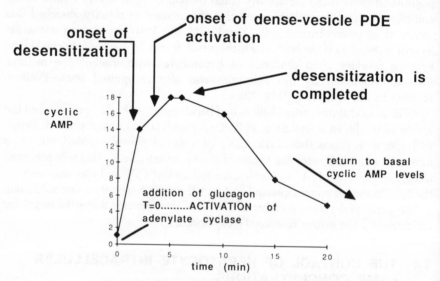

Figure 8.5 Glucagon causes a transient rise in hepatocyte intracellular cAMP concentration. The effect of treating hepatocytes with 10 nM glucagon on the intracellular [cAMP] is shown. The times at which desensitization and activation of the 'dense-vesicle' occur are indicated

nucleotide PDE together with the heightened activity of the 'dense-vesicle' enzyme which will have been phosphorylated and activated by the glucagon-mediated elevation of cAMP. Indeed, support for the protein kinase-A-mediated activation of a cAMP-PDE, in intact hepatocytes challenged with glucagon, has come from an elegant series of experiments performed by Corbin and co-workers [102]. They treated hepatocytes with a cAMP analogue called 8-p-chlorophenyl-thio cAMP (8PCPT-cAMP) at concentrations low enough not to activate protein kinase-A but high enough to be metabolised by cAMP-PDE activity. They then showed that when hepatocytes were stimulated by glucagon, the degradation of this analogue was actually enhanced in the intact hepatocytes. This led them to suggest that when glucagon increased intracellular cAMP concentrations, it led to the protein kinase-A-mediated activation of a cAMP-PDE which could elicit a significant enhancement of the metabolism of cAMP. As discussed above, we have identified this enzyme as the 'dense-vesicle' cAMP-PDE.

Our laboratory now has sufficient kinetic data on all hepatocyte PDEs and also on the activation and desensitization of adenylate cyclase to begin to make an initial attempt at simulating hepatocyte cAMP metabolism. The scheme and relevant constants are shown in Figure 8.6. In these simulations we explore the effect of causing a range of activations of the 'dense-vesicle' cAMP-PDE upon cAMP accumulation and the activity of this enzyme in glucagon-challenged hepatocytes (Figure 8.7). As can be seen, the activation of this enzyme was increased from a conservative 0.5-fold to the maximal three-fold that has been achieved in vitro [27,66]. Such simulation studies indicated that the activity and the glucagon-mediated activation of this enzyme can have a marked effect on the accumulation of cAMP (Figure 8.4). Indeed, the data of Gettys et al. [102] indicated that kinase-mediated PDE activation might lead to a ~60% reduction in the magnitude of cAMP accumulation. This compares with a simulated decrease in intracellular cAMP concentrations of at least 30% upon the A-kinase-mediated activation of the dense-vesicle enzyme by some three-fold (Figure 8.4). Furthermore, we noted that the glucagon-stimulated peak accumulation of cAMP in hepatocytes treated with ICI 118233, to block selectively the activity of the 'dense-vesicle' PDE, was comparable to that found in our simulation when the activity of this enzyme was set to zero (Figure 8.7a). Furthermore, experimental data for the peak increase of cAMP after challenge of intact hepatocytes with 10 nM glucagon approximated that seen in our model when activation of the 'dense-vesicle' enzyme was set at three-fold. Thus taking into account the generalisations and simplifications, we see that the simulation data appear to be in good agreement with experimental measurements of intracellular cAMP metabolism and the experimental determination of the A-kinase-mediated activation of the 'dense-vesicle' enzyme. This suggests that this enzyme is an important enzyme in regulating intracellular cAMP concentrations when cells are treated with ligands that elicit the activation of protein kinase-A. Interestingly, activation of this enzyme primarily modulates the peak height (maximal) of the transient increase

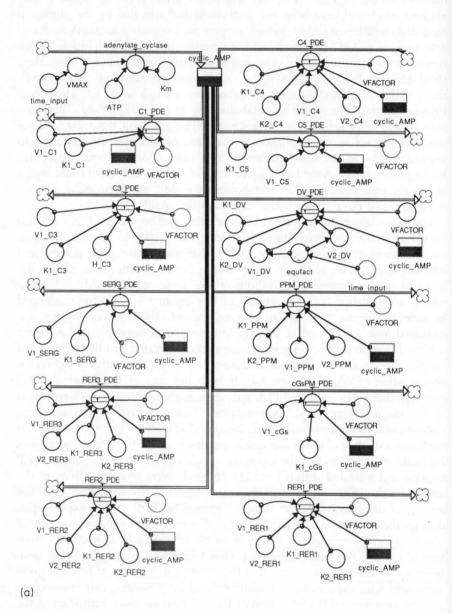

(a)

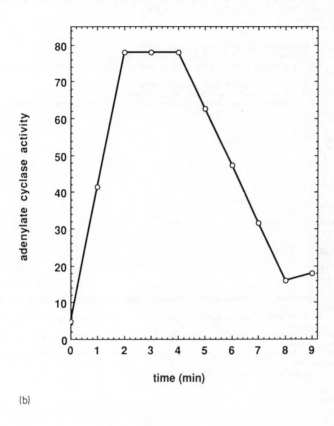

(b)

Figure 8.6   Scheme for the simulation of hepatocyte intracellular cAMP concentrations. The STELLA software (v.2.2 High Performance Systems Inc., Lyme, NH 3768, USA) program running on an Apple Macintosh II was used to simulate hepatocyte cAMP metabolism. In (a) the interrelationships of the various enzymes are shown for synthesis (adenylate cyclase) and degradation (PDEs). Computation was by Euler's method, with similar results being obtained using both 2nd- and 4th-order Runge–Kutta. The V_{max} of adenylate cyclase was controlled (b) to indicate a stimulatory challenge with 10 nM glucagon at $T = 0$ min. The desensitization of the enzyme was programmed to follow a time-course that has been shown experimentally [92]. The data used for the kinetic constants has been obtained experimentally. Thus the relative velocities of each of the enzymes assessed at 1 μM cAMP has been determined and this, together with a knowledge of K_m values, can be used to compute the relative V_{max} values for all of the PDE species. The VFACTOR is used to relate the entire PDE capacity of the cell to the entire adenylate cyclase activity and was set at 0.095 in this study. The term 'equfact' was

(*continued*)

☐ cyclic_AMP = cyclic_AMP + dt * (adenylate_cyclase - DV_PDE - PPM_PDE - cGsPM_PDE - SERG_PDE - RER3_PDE
RER2_PDE - RER1_PDE - C1_PDE - C3_PDE - C5_PDE - C4_PDE)
INIT(cyclic_AMP) = 18

○ adenylate_cyclase = (VMAX*ATP)/(Km+ATP)
○ ATP = 4000
○ C1_PDE = (VFACTOR*cyclic_AMP*V1_C1)/(K1_C1+cyclic_AMP)
○ C3_PDE = (V1_C3*VFACTOR*cyclic_AMP^H_C3)/(K1_C3+cyclic_AMP^H_C3)
○ C4_PDE = ((V1_C4*cyclic_AMP*VFACTOR)/(K1_C4+cyclic_AMP)+(V2_C4*cyclic_AMP*VFACTOR)/(K2_C4+
cyclic_AMP))
○ C5_PDE = (V1_C5*cyclic_AMP*VFACTOR)/(K1_C5+cyclic_AMP)
○ cGsPM_PDE = (V1_cGs*cyclic_AMP*VFACTOR)/(K1_cGs+cyclic_AMP)
○ DV_PDE = ((V1_DV*VFACTOR*cyclic_AMP)/(K1_DV+cyclic_AMP)+(V2_DV*VFACTOR*cyclic_AMP)/(K2_DV+
cyclic_AMP))
○ H_C3 = 1.6
○ K1_C1 = 25
○ K1_C3 = 38
○ K1_C4 = .7
○ K1_C5 = 0.6
○ K1_cGs = 34
○ K1_DV = 0.3
○ K1_PPM = 0.7
○ K1_RER1 = 0.8
○ K1_RER2 = 3.8
○ K1_RER3 = 2.6
○ K1_SERG = 44
○ K2_C4 = 4.5
○ K2_DV = 29
○ K2_PPM = 39
○ K2_RER1 = 63
○ K2_RER2 = 200
○ K2_RER3 = 56
○ Km = 100
○ PPM_PDE = ((VFACTOR*V1_PPM*cyclic_AMP)/(K1_PPM+cyclic_AMP)+(VFACTOR*V2_PPM*cyclic_AMP)/(K2_PPM+
cyclic_AMP))
○ RER1_PDE = ((V1_RER1*cyclic_AMP*VFACTOR)/(K1_RER1+cyclic_AMP)+(V2_RER1*cyclic_AMP*VFACTOR)/(
K2_RER1+cyclic_AMP))
○ RER2_PDE = ((V1_RER2*cyclic_AMP*VFACTOR)/(K1_RER2+cyclic_AMP)+(V2_RER2*cyclic_AMP*VFACTOR)/(
K2_RER2+cyclic_AMP))
○ RER3_PDE = ((V1_RER3*cyclic_AMP*VFACTOR)/(K1_RER3+cyclic_AMP)+(V2_RER3*cyclic_AMP*VFACTOR)/(
K2_RER3+cyclic_AMP))
○ SERG_PDE = (V1_SERG*VFACTOR*cyclic_AMP)/(K1_SERG+cyclic_AMP)
○ time_input = TIME
○ V1_C1 = 260
○ V1_C3 = 390
○ V1_C4 = 10
○ V1_C5 = 20
○ V1_cGs = 175
○ V1_DV = 30*equfact
○ V1_PPM = 50
○ V1_RER1 = 1
○ V1_RER2 = 20
○ V1_RER3 = 4
○ V1_SERG = 3
○ V2_C4 = 20
○ V2_DV = 190*equfact
○ V2_PPM = 400
○ V2_RER1 = 11
○ V2_RER2 = 400
○ V2_RER3 = 24
○ VFACTOR = .095
⊘ equfact = graph(time_input)
(0.0,1.00),(2.00,1.20),(4.00,1.59),(6.00,2.31),(8.00,2.70),(10.00,3.00),(12.00,3.00),(14.00,3.00),(16.00,3.00),
(18.00,3.00),(20.00,3.00)
⊘ VMAX = graph(time_input)
(0.0,80.00),(2.00,80.00),(4.00,5.00),(6.00,5.00),(8.00,80.00),(10.00,80.00),(12.00,48.40),(14.00,16.40),
(16.00,20.40),(18.00,28.80),(20.00,35.60)

(c)

in hepatocyte cAMP concentration. The properties of this computer model and its use in evaluating the contribution of specific PDEs to hepatocyte cAMP metabolism is elaborated on elsewhere [103].

8.4.2 After challenge with insulin

Insulin antagonises the metabolic effects of glucagon and β-adrenergic agents in both liver and fat by both inhibiting adenylate cyclase and activating specific PDE isoenzymes, resulting in a decrease in cellular cAMP levels [5] (Figure 8.8).

Intriguingly, in hepatocytes the relative contributions of adenylate cyclase and the different PDE isoenzymes to this effect of insulin are determined by the extracellular glucagon concentration [59]. Thus, pretreatment of hepatocytes with glucagon completely obliterates the ability of insulin to inhibit adenylate cyclase [59] and to activate the peripheral plasma membrane PDE [18].

This indicates that different actions of insulin will be seen depending upon whether glucagon was added before, during or after challenge with insulin. It appeared from Percoll gradient analysis that insulin activated only the 'dense-vesicle' PDE and the peripheral plasma membrane PDE in rat hepatocytes [18]. Therefore, when the ability of insulin to activate the peripheral PDE [18] and to inhibit adenylate cyclase [59] was blocked by pretreatment with glucagon, and the activation of the 'dense-vesicle' enzyme was blocked by pretreatment of hepatocytes with fructose or lysosomotropic agents [40], it was expected that insulin would have no effect on intracellular cAMP levels. Surprisingly, under such conditions insulin was still able to lower intracellular cAMP concentrations that had been raised by glucagon [60]. This implies that there is at least a third PDE

Figure 8.6 (*contd.*)
used to control the activation of the 'dense-vesicle' enzyme both as regards its magnitude and its time-dependence. These were altered along experimentally determined lines [18,27,66]. The equations defining the simulation are shown in (c). The enzymes referred to are the soluble cytosol species PDE-MQ-I (C1), PDE-MQ-III (C3), PDE-MQ-IV (C4), PDE-MQ-V (C5), 'dense-vesicle' PDE (DV), peripheral plasma membrane PDE (PPM), plasma membrane cGMP-stimulated cyclic nucleotide PDE (cGsPM), smooth endoplasmic reticulum/Golgi PDE (SERG) and the three rough endoplasmic reticulum PDEs (RER1, RER2, RER3). The K_m (K1) and V_{max} (V1) values for these enzymes are given in part (c); in some instances, these include either a Hill coefficient (H) or upper (V2, K2) and lower (V1, K1) limiting values for these constants. The values for 'equfact' are altered in the various simulations to reflect the degree of activation of the 'dense-vesicle' enzyme as shown in the legend to Figure 8.7. At 1 μM cAMP the relative velocities for these enzymes are DV = 15; PPM = 25; cGsPM = 5; SERG = 3; R3 = 2; R2 = 9; R1 = 1; C1 = 10; C3 = 10; C4 = 10 and C5 = 10. We have ignored the minimal contribution of both the mitochondrial enzymes [16] and the cGMP-specific PDE-MQ-II [17]. (Data taken from refs 9, 14, 17, 18, 20, and 27)

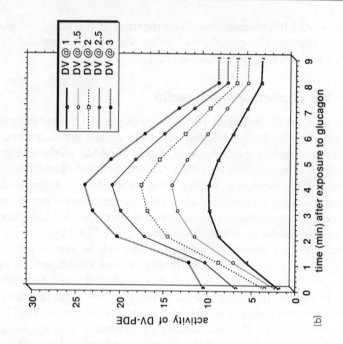

(b)

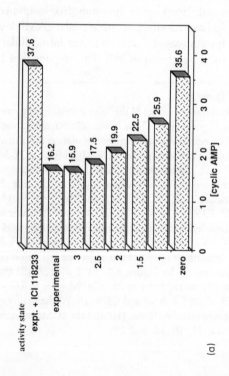

(a)

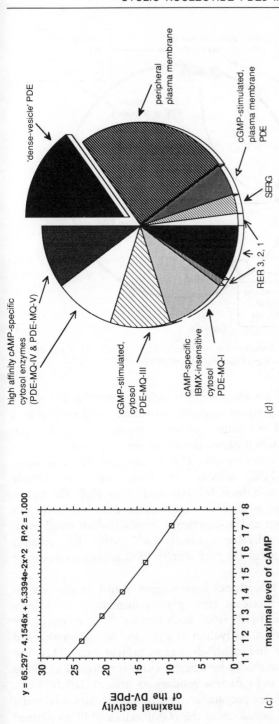

Figure 8.7 Assessment of the effect of activation of the 'dense-vesicle' PDE on the glucagon-stimulated rise in intracellular cAMP concentration. These figures are the results of simulations of hepatocyte cAMP metabolism using the scheme shown in Figure 8.6. The variable altered here was the V_{max} of the 'dense-vesicle' PDE, through the 'equfact' term (see legend to Figure 8.6), in order to assess its effect on intracellular cAMP metabolism. (a) Change in peak intracellular cAMP concentrations occurring 6 min after the addition of glucagon when activation (V_{max} increases) of the 'dense-vesicle' enzyme was set with no enzyme activation (1), 50% (1.5), two-fold (2), 2.5-fold (2.5) and three-fold (3). Simulation data are also given when the activity of the DV-PDE was set at zero, i.e. equivalent to the fully inhibited enzyme. This is compared with experimental data for hepatocytes treated with ICI 118233 (100 μM) to selectively and fully inhibit the DV-PDE (expt. + ICI 118233) some 6 min after challenge with 10 nM glucagon, and also for hepatocytes challenged only with 10 nM glucagon. (b) Change in the activity of the 'dense-vesicle' enzyme over a similar time-course with the indicated changes in V_{max} as above. (c) Relationship between the maximal observed activity of the 'dense-vesicle' PDE and the maximal level of intracellular cAMP accumulation. (d) Relative activities of hepatocyte PDE isoforms

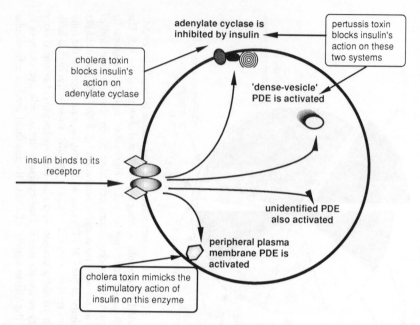

Figure 8.8 Insulin and cAMP metabolism in hepatocytes. A scheme showing the points of interaction with adenylate cyclase and cAMP-PDEs

which is activated by insulin in a manner which is rapidly reversed and therefore was not observed by Percoll gradient analysis. Activation could be due to a soluble substance which is diluted during gradient analysis or to an inherent instability in the modification which causes the activation.

The identity of this third insulin-sensitive PDE is uncertain. The two other major species of cAMP-PDE activity in liver are the soluble Ca^{2+}–CaM-activated enzyme and the cGMP-activated cAMP-PDE. The former enzyme is an unlikely candidate, as insulin antagonises the actions of hormones such as vasopressin, angiotensin and α_1-adrenergic agonists which result in an increase in the presumptive activator, cytosolic Ca^{2+}. Also, this enzyme contributes a very small fraction of the high-affinity PDE activity in liver (see above).

Reports from a number of laboratories have suggested that insulin acts on target cells to potentiate the release from plasma membranes of soluble 'mediators' within the target cell [104–106]. Such species mimic many of the metabolic effects of insulin such as activation of pyruvate dehydrogenase and inhibition of adenylate cyclase and cAMP-dependent protein kinase [12]. The identity of such species still remains to be defined. However, present evidence suggests that phosphoinositol and galactose residues are present [107,108].

Interestingly, an insulin mediator preparation from rat hepatocytes achieved a reversible and dose-dependent activation of the cGMP-activated PDEs purified

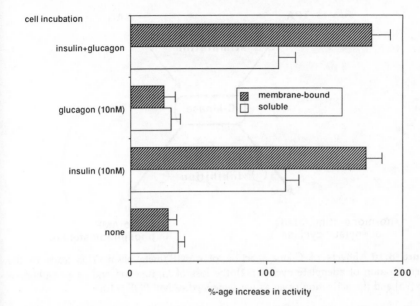

Figure 8.9   Activation of cGMP-stimulated cAMP-PDE by an insulin 'mediator' preparation. A soluble 'mediator' preparation from insulin- but not glucagon-treated cells activates both the membrane-bound and soluble forms of this enzyme from liver. (Adapted from ref. 109)

from both the cytosolic and particulate fractions of rat liver [109]. In contrast neither was such 'mediator' activity found in glucagon-treated cells, nor did glucagon pretreatment block 'mediator' action [109] (Figure 8.9). Thus such a direct interaction between the soluble mediators and the PDE would be expected to be diluted out during Percoll gradient analysis. In contrast, the mediator preparation had no effect on the activity of apparently homogeneous preparations of the 'dense-vesicle' PDE and the peripheral plasma membrane PDE [109].

It cannot be stated with certainty at this stage that the insulin mediator(s) activate the cGMP activated PDEs in situ. This will depend on the accessibility of these enzymes to such mediators within the cell. Also, the possibility that the mediator(s) activate the 'dense-vesicle' PDE or peripheral PDE cannot be dismissed, as the membrane environment of these enzymes in situ could be essential to their activation. For example, the mediator(s) could activate the PDEs via a cascade mechansim involving an intermediate kinase or phosphatase which would be lost during the purification of the 'dense-vesicle' and peripheral plasma membrane PDEs. Certainly, in adipocytes insulin mediator preparations have been reported to activate high-affinity cAMP-PDE activity in a crude membrane fraction [106,108]. However, we would suggest that this may be a membrane-bound cGMP-activated cyclic nucleotide akin to that described by us in liver [20,109].

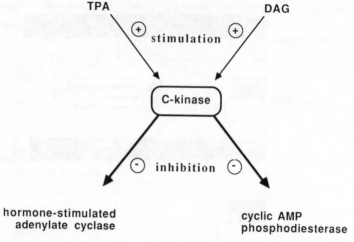

Figure 8.10 Effects of C-kinase action of cAMP metabolism. This leads to the desensitization of adenylate cyclase [92], the loss of G_i function and its phosphorylation [95] and the activation of an as yet undefined cAMP-PDE [110]

8.4.3 Other regulators

Treatment of intact hepatocytes with the tumour promoting phorbol ester TPA has been shown to markedly affect cAMP metabolism (Figure 8.10). It apparently exerts inhibitory effects on both adenylate cyclase and on cAMP-PDE activity [110]. The inhibitory effects on PDE activity are not seen upon its addition to hepatocyte homogenates or to pure enzymes and presumably reflect the C-kinase-mediated phosphorylation of a target enzyme. This species, however, has not been identified, although tentative evidence suggests that it is a soluble PDE. Also, it is not known whether such a target PDE is also affected by hormones which stimulate inositol phospholipid metabolism.

8.5 CONCLUSIONS

Multiple forms of cAMP-PDEs have been noted in many cells by many investigators. However, the functional significance of individual species is far from clear. Studies on hepatocytes done in this laboratory have tried to focus on a single cell type in order to comprehend the variety and function of the various PDE-isoenzymes (Figure 8.11). The properties and regulation of key enzymes have now been identified. This has led to the proposal that the phosphorylation of the 'dense-vesicle' enzyme by cAMP-dependent protein kinase plays an important role in attenuating the elevation of intracellular cAMP concentrations by glucagon.

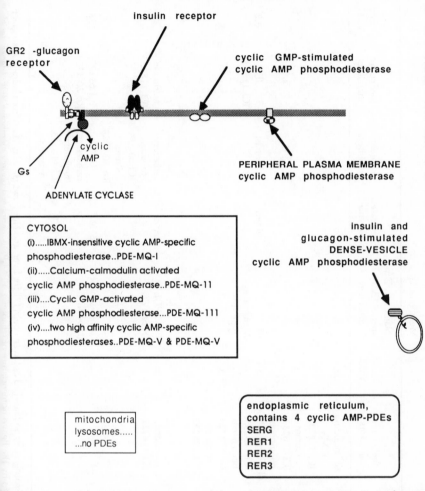

Figure 8.11   Schematic representation of the various PDE isoenzymes in rat hepato-cytes. Indicates the subcellular distribution of cAMP-metabolising enzymes in hepato-cytes

We have recently completed [111] an analysis of the PDE activities of rat kidney. Interestingly, unlike liver, where 60% of the cAMP-PDE activity is membrane-bound, in kidney less than 10% of the activity is supplied by membrane-bound enzymes. The kidney expresses soluble PDE species with similar properties and elution characteristics to the cytosol enzymes from rat hepatocytes (MQ-II, MQ-III, MQ-IV and MQ–V), except that MQ-I appears not to be expressed. This shows clearly that tissue-selective expression of PDE isoforms occurs and that the relative importance of these enzymes to cyclic nucleotide metabolism will vary between tissues. Thus inhibitors may not only

Table 8.5 Inhibitor sensitivities and substrate preferences of major PDE species of the rat

	Cytosol				Membrane				
MQ-identity (Beavo identifier)	Substrate preference	Potent inhibitors	Ineffective inhibitors	V.cA/V.cG	Name of PDE species (Beavo identifier)	Substrate preference	Potent inhibitors	Ineffective inhibitors	V.cA/V.cG
I (type IV)	cAMP	none	all, including IBMX, ICI 118233						
II (type I)	cGMP	zaprinast, IBMX	rolipram, ICI 118233, buquineran						
III Cytosol (type IIC)	either	none/zaprinast	all except IBMX, ICI 118233, buquineran	3	III Membrane (type IIB)	either	zaprinast	all, ICI 118233	3.2
IV (type IV)	cAMP	carbazeran, rolipram, ICI 63197, milrinone, buquineran	Ro 20-1724, amrinone, ICI 118233	8	PPM-PDE (type IV)	cAMP	carbazeran, rolipram, ICI 63197, milrinone, Ro 20-1724	amrinone, ICI 118233	23
V (type IV) (?PDE-4)	cAMP	carbazeran, rolipram, ICI 63197, milrinone, buquineran, Ro 20-1724	amrinone, ICI 118233	16	'dense-vesicle' (type III)	cAMP	cGMP, carbazeran, buquineran, ICI 118233, milrinone	ICI 63197, Ro 20-1724	158

Cytosol species were resolved by Mono-Q chromatography and labelled as Lavan et al. [17], for liver, and Hoey and Houslay [111] from kidney. Also given are the classes these enzymes can be attributed to by the Beavo nomenclature of Chapter 1. PPM-PDE, peripheral plasma membrane PDE; 'dense-vesicle'

ɔe selective for certain isoforms but may also display tissue-specific effects on cyclic nucleotide metabolism.

It will be of much interest to correlate these biochemically identified isoforms with the products of the PDE genes that have recently been identified. We list in Table 8.5 the basic properties of the enzymes that we have identified and characterised from rat tissues. It is tempting to conclude that PDE-MQ-V may be coded for by the so-called PDE-4 gene (Chapter 10) and it may even be that this species is the membrane-released form of the peripheral plasma membrane PDE. The conjunction of biochemistry, immunological and molecular biological techniques should soon resolve the controversy of PDE isoforms. They, together with computer simulation studies, may also lead to the resolution of the functioning of specific species in particular cells.

ACKNOWLEDGEMENTS

We thank the MRC, AFRC, British Heart Foundation and California Metabolic Research Foundation for financial support.

REFERENCES

1. Kelly, S. J., Dardinger, D. E., and Butler, L. G. (1975) *Biochemistry*, **14**, 4983–4988.
2. Kelly, S. J., and Butler, L. G. (1977) *Biochem. J.*, **16**, 1102–1104.
3. Drummond, G. I., and Perrot-Yee, S. (1961) *J. Biol. Chem.*, **236**, 1126–1129.
4. Butcher, R. W., and Sutherland, E. W. (1962) *J. Biol. Chem.*, **237**, 1244–1250.
5. Houslay, M. D. (1986) *Biochem. Soc. Trans.*, **14**, 183–193.
6. Beavo, J. A. (1988) *Adv. Second Messenger Phosphoprotein Res.*, **20**, 98–122.
7. Harrison, S. A., Reifsnyder, D. H., Gallis, B., Cadd, G. G., and Beavo, J. A. (1986) *Mol. Pharmacol.*, **29**, 506–514.
8. MacPhee, C. H., Harrison, S. A., and Beavo, J. A. (1986) *Proc. Natl Acad. Sci. USA*, **83**, 6660–6663.
9. Pyne, N. J., Cooper, M. E., and Houslay, M. D. (1987). *Biochem. J.* **242**, 33–42.
10. Pyne, N. J., Anderson, N., Lavan, B. E., Milligan, G., Nimmo, H. G., and Houslay, M. D. (1987) *Biochem. J.*, **248**, 897–901.
11. Boyes, S., and Loten, E. G. (1988) *Eur. J. Biochem.*, **174**, 303–309.
12. Houslay, M. D., Wakelam, M. J. O., and Pyne, N. J. (1986) *TIBS*, **11**, 393–394.
13. Marchmont, R. J., and Houslay, M. D. (1980) *Biochem. J.*, **187**, 381–392.
14. Cercek, B., Wilson, S. R., and Houslay, M. D. (1983) *Biochem. J.*, **213**, 89–97.
15. Wilson, S. R., and Houslay, M. D. (1983) *Biochem. J.*, **213**, 99–105.
16. Cercek, B., and Houslay, M. D. (1982) *Biochem. J.*, **207**, 123–132.
17. Lavan, B., Lakey, T. and Houslay, M. D. (1990) *Biochem. Pharmacol.* **38**, 4123–4136.
18. Heyworth, C. M., Wallace, A. V. and Houslay, M. D. (1983) *Biochem. J.*, **214**, 99–110.
19. Marchmont, R. J., Ayad, S. R., and Houslay, M. D. (1981) *Biochem. J.*, **195**, 645–652.
20. Pyne, N. J., Cooper, M. E., and Houslay, M. D. (1986) *Biochem. J.*, **234**, 325–334.

21. Loten, E. G., Assimacopoulos-Jeannet, F. D., Exton, J. H., and Park, C. R. (1978) *J. Biol. Chem.*, **253**, 746–757.
22. Loten, E. G., Francis, S. H., and Corbin, J. D. (1980) *J. Biol. Chem.*, **225**, 7838–7844.
23. Makino, H., Kanatsuka, A., Osegawa, M., and Kumagai, A. (1982) *Biochim. Biophys. Acta*, **704**, 31–36.
24. Anderson, N. G., and Houslay, M. D. (1989) *Biochem. Soc. Trans*, **17**, 217–218.
25. Bristol, J. A., and Evans, D. B. (1983) *Med. Res. Rev.*, **3**, 259–287.
26. Farah, A. E., Alousi, A. A., and Schwarz, R. P. (1984) *Annu. Rev. Toxicol.*, **84**, 275–328.
27. Kilgour, E., Anderson, N. G., and Houslay, M. D. (1989) *Biochem. J.*, **260**, 27–36.
28. Endoh, M., Yamashita, S., and Taira, N. (1982) *J. Pharmacol. Exp. Ther.*, **221**, 775–783.
29. Kariya, T., Willie, L. J., and Dage, R. C. (1982) *J. Cardiovasc. Pharmacol.*, **4**, 509–514.
30. Weishaar, R. E., Cain, M. H., and Bristol, J. A. (1985) *J. Med. Chem.*, **28**, 537–545.
31. Reeves, M. L., Leigh, B. K., and England, P. J. (1987) *Biochem. J.*, **241**, 535–541.
32. Grant, P. G., and Colman, R. W. (1984) *Biochem. J.*, **23**, 1801–1807.
33. Degerman, E., Belfrage, P., Newman, A. H., Rice, K. C., and Manganiello, V. C. (1987) *J. Biol. Chem.*, **262**, 5797–5807.
34. Houslay, M. D., and Marchmont, R. J. (1981) *Biochem. J.*, **198**, 703–706.
35. Houslay, M. D., Wallace, A. V., Wilson, S. R., Marchmont, R. J., and Heyworth, C. M. (1983) In *Hormone and Cell Regulation* (Dumont, J. N., Denton, R. M. and Nunez, J., eds), Vol. VII, Elsevier Biomedical Press, Amersterdam, pp. 105–120.
36. Houslay, M. D., Wallace, A. V., Marchmont, R. J., Martin, B. R., and Heyworth, C. M. (1983) *Adv. Cyclic Nucleotide Res.*, **16**, 159–176.
37. Marchmont, R. J., and Houslay, M. D. (1981) *Biochem. J.*, **195**, 653–660.
38. Schwabe, U., Miyake, M., Ohga, V., and Daly, J. W. (1976) *Mol. Pharmacol.*, **12**, 900–910.
39. Major, G. N., Loten, E. G., and Sneyd, J. G. T. (1983) *Int. J. Biochem.*, **15**, 217–223.
40. Wilson, S. R., Wallace, A. V., and Houslay, M. D. (1983) *Biochem. J,* **216**, 245–248.
41. Takemoto, D. J., Hansen, J., Takemoto, L. J., and Houslay, M. D. (1982) *J. Biol. Chem.*, **257**, 14597–14599.
42. Takemoto, D. J., Hansen, J., Takemoto, L. J., Houslay, M. D., and Marchmont, R. J. (1984) *Adv. Cyclic Nucleotide Protein Phosphorylation Res.*, **16**, 55–64.
43. Beavo, J. A., Hardman, J. G., and Sutherland, E. H. (1971) *J. Biol. Chem.*, **246**, 3841–3846.
44. Houslay, M. D., Wallace, A. V., Marchmont, R. J., Martin, B. R., and Heyworth, C. M. (1984) *Adv. Cyclic Nucleotide Res.*, **16**, 159–176.
45. Erneux, C., Couchie, D., Dumont, J. E., Baraniak, J., Stec, W. J., Garcia Abbad, E., Petridis, G., and Jastorff, B. (1981) *Eur. J. Biochem.*, **115**, 503–510.
46. Erneux, C., Miot, F., Boeynaems, J., and Dumont, J. E. (1982) *FEBS Lett.*, **142**, 251–254.
47. Erneux, C., Miot, F., van Haastert, P. J., and Jastorff, B. (1985) *J. Cyclic. Nucleotide Protein Phosphorylation Res.*, **10**, 463–470.
48. Martins, T. J., Mumby, M. C., and Beavo, J. A. (1982) *J. Biol. Chem.*, **257**, 1973–1979.
49. Yamamoto, T., Yamamoto, S., Osbourne, J. C., Manganiello, V. C., and Vaughan, M. (1983) *J. Biol. Chem.*, **258**, 14173–14177.

50. Franks, D. J., and MacManus, J. P. (1971) *Biochem. Biophys. Res. Commun.*, **42**, 844–849.
51. Hidaka, H., and Asano, T. (1976) *Biochim. Biophys. Acta.*, **429**, 485–497.
52. Klotz, U., and Stock, K. (1972) *Naunny-Schmiedebergs Arch. Pharmacol.*, **274**, 54–62.
53. Yamamoto, T., Manganiello, V. C., and Vaughan, M. (1983) *J. Biol. Chem.*, **258**, 12526–12533.
54. Russell, T. R., Terasaki, W. L., and Appleman, M. M. (1973) *J. Biol. Chem.*, **248**, 1334–1340.
55. Morrill, M. E., Thompson, S. J., and Stellwagen, E. (1979) *J. Biol. Chem.*, **254**, 4371–4374.
56. Yamamoto, T., Yamamoto, S., Manganiello, V. C., and Vaughan, M. (1983) *Arch. Biochem. Biophys.*, **229**, 81–89.
57. Pilkis, S. J., Claus, T. H., Johnson, R. A., and Park, C. R. (1975) *J. Biol. Chem.*, **250**, 6328–6336.
58. Blackmore, P. F., Assimacopoulos-Jeannet, F., Chan, T. M., and Exton, J. H. (1979) *J. Biol. Chem.*, **254**, 2828–2834.
59. Heyworth, C. M., and Houslay, M. D. (1983) *Biochem. J.*, **214**, 547–552.
60. Heyworth, C. M., Wallace, A. V., Wilson, S. R., and Houslay, M. D. (1984) *Biochem. J.*, **222**, 183–187.
61. Wakelam, M. J. O., Murphy, G. J., Hruby, V. J., and Houslay, M. D. (1986) *Nature (Lond.)*, **323**, 68–71.
62. Boyes, S., Allan, E. H., and Loten, E. G. (1981) *Biochim. Biophys. Acta*, **672**, 21–28.
63. Manganiello, V. C. (1987) *J. Moll. Cel. Cardiol.*, **19**, 1037–1040.
64. Cohen, P. (1985) *Eur. J. Biochem.*, **151**, 439–448.
65. Gettys, T. W., Blackmore, P. F., Redman, J. B., Beebe, S. J., and Corbin, J. D. (1987) *J. Biol. Chem.*, **262**, 333–339.
66. Kilgour, E., Anderson, N. G., and Houslay, M. D. (1989) *Biochem. Soc. Trans*, **16**, 1025–1026.
67. Munday, M. R., Carling, D., and Hardy, D. G. (1988) *FEBS Lett.* (in press).
68. Carling, D., Zammit, V. A., and Hardie, D. G. (1987) *FEBS Lett.*, **223**, 217–222.
69. MacPhee, C. H., Reifsnyder, D. H., Moore, T. A., Lerea, K. M., and Beavo, J. A. (1988) *J. Biol. Chem.*, **263**, 10353–10358.
70. Anderson, N. G., Kilgour, E., and Houslay, M. D. (1989) *Biochem. J.*, **262**, 867–872.
71. Rothenberg, P. L., and Kahn, C. R. (1988) *J. Biol. Chem.*, **263**, 15546–15552.
72. Pyne, M. J., Cushley, W., Nimmo, H. G., and Houslay, M. D. (1989) *Biochem. J.*, **261**, 897–904.
73. Denton, R. M. (1986) *Adv. Cyclic Nucleotide Res.*, **20**, 293-327.
74. Houslay, M. D., and Siddle, K. (1989) *Med. Bull.*, **45**, 264–284.
75. Benelli, C., Desbuquois, S., and De Galle, B. (1986) *Eur. J. Biochem.*, **156**, 211–220.
76. Joost, H. G., Weber, T. M., and Cushman, S. W. (1988) *Biochem. J.*, **249**, 155–161.
77. Calderhead, D. M., and Liehard, G. E. (1988) *J. Biol. Chem.*, **263**, 12171–12174.
78. Siegel, J., and Olefsky, J. M. (1980) *Biochemistry*, **19**, 2183–2190.
79. Beebe, S. J., Redman, J. B., Blackmore, P. F., and Corbin, J. D. (1985) *J. Biol. Chem.*, **260**, 15781–15788.
80. Manganiello, V. C., and Elks, M. L. (1986) In *Mechanism of Insulin Action* (Belfrage, P., Donner, J. and Stalfors, P., eds), Elsevier Scientific Publishing Co. Inc., New York, pp. 147–166.

81. Heyworth, C. M., Grey, A. M., Wilson, S. R., Hanski, E., and Houslay, M. D. (1986) *Biochem. J.*, **235**, 145–149.

82. Elks, M. L., Watkins, P. A., Manganiello, V. C., Moss, J., Hewlett, E. and Vaughan, M. (1983) *Biochem. Biophys. Res. Commun.*, **116**, 593–596.

83. Gall, C. J., and Pearson, C. K. (1986) *TIBS*, **11**, 171–175.

84. Marchmont, R. J., and Houslay, M. D. (1980) *Nature (Lond.)*, **286**, 904–906.

85. Heyworth, C. M., Rawal, S., and Houslay, M. D. (1983) *FEBS Lett.*, **154**, 87–91.

86. Exton, J. H., Harper, S. C., Tucker, A. L., Flagg, T. L., and Park, C. R. (1973) *Biochim. Biophys. Acta*, **329**, 41–57.

87. Smith, S. A., Elliott, K. R. F., and Pogson, C. I. (1978) *Biochem. J.*, **176**, 817–825.

88. Jefferson, L. S., Exton, J. H., Butcher, R. W., Sutherland, E. W., and Park, C. W. (1968) *J. Biol. Chem.*, **243**, 1031–1038.

89. Heyworth, C. M., Whetton, A. D., Wong, S., Martin, B. R., and Houslay, M. D. (1985) *Biochem. J.*, **228**, 593–603.

90. Bitensky, M. W., Wheeler, M. A., Rasenick, M. M., Yamazaki, A., Stein, P. J., Halliday, K. R., and Wheeler, G. L. (1982) *Proc. Natl Acad. Sci. USA*, **79**, 3408–3412.

91. Gilman, A. G. (1984) *Cell*, **36**, 577–579.

92. Murphy, G. J., Hruby, V. J., Trivedi, D., Wakelam, M. J. O., and Houslay, M. D. (1987) *Biochem. J.*, **243**, 39–46.

93. Wallace, A. V., Heyworth, C. M., and Houslay, M. D. (1984) *Biochem. J.*, **222**, 117–182.

94. Kwok, Y. C., and Yip, C. C. (1987) *Biochem. J.*, **248**, 27–33.

95. Murphy, G. J., Gawler, D. J., Milligan, G., Wakelam, M. J. O., Pyne, N. J., and Houslay, M. D. (1989) *Biochem. J.*, **259**, 191–197.

96. Murphy, G., and Houslay, M. D. (1988) *Biochem. J.*, **249**, 543–547.

97. Charest, R., Prpic, V., Exton, J. H., and Blackmore, P. F. (1985) *Biochem. J.*, **227**, 79–98.

98. Blackmore, P. F., and Exton, J. H. (1986) *J. Biol. Chem.*, **261**, 11056–11058.

99. Whipps, D. E., Armstrong, A. E., Pryor, H. J., and Hapestrap, A. P. (1987) *Biochem. J.*, **241**, 835–845.

100. Mine, T., Kojima, I., and Ogata, E. (1988) *Biochim. Biophys. Acta.*, **970**, 166–171.

101. Bocckino, S. B., Blackmore, P. F., and Exton, J. H. (1985) *J. Biol. Chem.*, **260**, 14201–14207.

102. Gettys, T. W., Blackmore, P. F., and Corbin, J. D. (1988) *Am. J. Physiol.*, **254**, E449–E453.

103. Houslay, M. D. (1990) *Cellular Signalling*.

104. Larner, J., Galasko, G., Cheng, K., DePaoli-Roach, A. A., Huang, L., Daggy, P., and Kellog, J. (1979) *Science*, **206**, 1408–1410.

105. Kiechle, F. L., Jarett, L., Kotagal, N., and Popp, D. A. (1981) *J. Biol. Chem.*, **256**, 2945–2951.

106. Saltiel, A., Siegel, M. I., Jacobs, S., and Cuatrecasas, P. (1982) *Proc. Natl Acad. Sci. USA*, **79**, 3513–3517.

107. Mato, J. M. (1989) *Cellular Signalling*, **1**, 143–146.

108. Saltiel, A. R., and Cuatrecasas, P. (1986) *Proc. Natl Acad. Sci. USA*, **83**, 5793–5797.

109. Pyne, N. J., and Houslay, M. D. (1988) *Biochem. Biophys. Res. Commun.*, **156**, 290–296.

110. Irvine, F., Pyne, N. J., and Houslay, M. D. (1986) *FEBS Lett.*, **208**, 455–459.

111. Hoey, M., and Houslay, M. D. (1990) *Biochem. Pharmacol.* (in press).

STRUCTURAL AND FUNCTIONAL ANALYSES OF PHOSPHODIESTERASES USING MOLECULAR BIOLOGICAL TECHNIQUES

PART D

STRUCTURAL AND FUNCTIONAL ANALYSES OF PHOSPHODIESTERASES USING MOLECULAR BIOLOGICAL TECHNIQUES

MOLECULAR GENETICS OF THE CYCLIC NUCLEOTIDE PHOSPHODIESTERASES

Ronald L. Davis

Department of Cell Biology, Baylor College of Medicine, Houston, Texas 77030, USA

9.1 INTRODUCTION

The importance of the cyclic nucleotides as intracellular messenger molecules, conveying the information presented to a cell by hormones, neurotransmitters, light, or other signals, is well recognized. This recognition has stimulated a great deal of work towards understanding the cellular machinery dedicated to the synthesis and degradation of these small molecules. The focus of this chapter is on recent results obtained regarding the degradative enzymes through the analysis of the genes which code for the enzymes.

The cyclic nucleotide phosphodiesterases (PDEs) comprise a complex class of enzymes whose function is to hydrolyze 3′,5′-cyclic nucleotides to 5′-nucleotides. These enzymes have been analyzed extensively using standard biochemical methods. This has provided for some understanding of the complexity of the enzymes, but in a number of instances, it has produced a good deal of confusion. For example, the enzymes are quite sensitive to proteolysis, so that proteolytic fragments which have enzymatic activity can be, and have been purified. This has produced ambiguity in even basic information, such as the molecular size of a particular type of PDE. Moreover, partially purified enzyme preparations have been characterized, without the realization that the preparation contained more than one enzyme form. These, and other problems, have shown that bio-chemical characterization alone is inadequate to understand the PDEs and their regulation.

Despite these problems, Beavo [1] has been able to distill sufficient information from the literature to divide the enzyme class into several different families, based on substrate affinity, substrate specificity and selective sensitivity to cofactors or drugs. The families include: (1) cyclic nucleotide (hydrolyzes both cAMP and

Cyclic Nucleotide Phosphodiesterases: Structure, Regulation and Drug Action
Edited by J. Beavo and M. D. Houslay © 1990 John Wiley and Sons Ltd

cGMP) PDEs which require Ca^{2+} and calmodulin (CaM) for maximal activity (CaM-PDE); (2) cGMP-PDEs, which are generally specific for cGMP as substrate; (3) cGMP-stimulated cyclic nucleotide PDEs (cGS-PDE); (4) cGMP-inhibited cyclic nucleotide PDEs (cGI-PDE); and (5) the cAMP-PDEs. The last of these are specific for cAMP as substrate (see Chapter 1). Undoubtedly, this classification oversimplifies the diversity of the PDEs, since most if not all of these families appear to include several related subtypes.

There are several distinct biological mechanisms for producing a multiplicity of enzymes with related functions. Some of the differences observed in biochemical properties could, in principle, be due to post-translational modifications of the enzymes by proteolysis, phosphorylation, glycosylation, etc. Different types of PDEs could be coded for by several different genes, there existing one or more genes, for instance, for each family of PDE. In addition, differential processing of RNAs from a single gene by alternative splicing could produce enzymes of identical sequence over most of their length, but with differences in one or more regions which might confer somewhat different functions. Although the molecular information which bears on these questions is still incomplete, it is likely that all three of these mechanisms are operant.

Because of the importance of the PDEs in biological processes, several groups have approached the study of the enzymes from a molecular genetic perspective. This approach has been rewarding, having produced the sequences for several enzymes, a better understanding of the sequence relationships and evolutionary origin of different PDEs, information regarding the complex structure of certain PDE genes, a partial understanding of the regulation of some PDEs at the level of RNA abundance, and, perhaps most importantly, insight into the biological effects of altering PDE gene function in an organism by mutation or over-expression. Given these initial successes, we can expect much more information from this approach in the future. Indeed, it seems apparent from the progress with this approach to date that even the basic task of identifying all PDEs will best be accomplished by molecular genetics. Thus, the PDEs would be classed first by the gene which encodes them, and second by the different RNAs from a given PDE gene, in the event that it codes for more than one type of RNA molecule.

Not surprisingly, the eukaryotes *Saccharomyces cerevisiae*, *Drosophila melanogaster* and *Dictyostelium discoideum* have played particularly important roles in the molecular genetic analysis of PDE genes. Cyclic nucleotide PDE genes have been cloned from *Dictyostelium* and yeast and cAMP-PDE genes have been cloned from yeast and *Drosophila*. In addition the isolation of PDE genes and mutants in the cAMP system in these organisms has provided methods for isolating the mammalian homologs, based on cross-hybridization or function. In addition, one type of mammalian PDE, the retinal cGMP-PDE, has been cloned using the more traditional method of obtaining partial amino acid sequence to design oligonucleotide probes for the isolation of the appropriate genes.

9.2 *DROSOPHILA* cAMP-PDE GENE

Drosophila contains two major types of PDE, a CaM-PDE and a cAMP-PDE [2]. No molecular information beyond a partial biochemical characterization is available regarding the CaM-stimulated form. The cAMP-PDE has been focused upon much more intently, in part because mutations in the cAMP-PDE gene cause deficits in the learning and memory ability of the fly.

The gene which encodes the *Drosophila melanogaster* cAMP-PDE is housed at the dunce locus. It was the first of several genes discovered originally in a search for mutants which affect the ability of the fly to learn or to remember ([3] for review). A fly can learn several different types of information; standard learning situations usually involve coupling temporally a sensory stimulus, such as an olfactory (the presence of a volatile organic compound) or visual (a certain wavelength of light) cue, with a positive (a sucrose reward) or negative (electrical shock) reinforcement. Depending on the specific learning situation, dunce mutants are either learning mutants, memory mutants, or both. In several situations, the mutants demonstrate a partial learning ability and a rapid memory loss. In some, they learn as well as normal flies, displaying only the forgetfulness. In others, they never seem to learn the required information. Whether these experimental disparities reflect true differences in fly behavior as a function of the type of information presented or whether they represent inadequacies in certain learning situations is not certain. Nevertheless, the behavioral information as a whole is consistent with the conclusion that the dunce gene, and its product cAMP-PDE, participate in the biochemical processes serving learning and/or memory.

The dunce locus was cloned by chromosomal walking from the nearby locus, Sgs-4 [4]. Sgs-4 codes for a protein component of the glue synthesized in the larval salivary gland, which it uses to cement itself to a solid support upon pupariation. Previous cytogenetic studies had positioned dunce at chromomere 3D4, just five salivary gland chromosome bands away from Sgs-4 at 3C11 (Figure 9.1). Using a cloned probe representing sequences just to the right of the Sgs-4 gene, approximately 100 kb of genomic DNA more proximal than Sgs-4 was collected. This provided genomic probes of the dunce locus to study the encoded RNA molecules [5], to isolate and characterize cDNA clones [6], and to study the structure of the dunce gene [7].

The sequence of the isolated cDNA clones provided the information necessary to conclude unambiguously that dunce codes for cAMP-PDE. An open reading frame defined by the cDNA clones predicts a molecule of about 65 kDa (see correction of original sequence [6] in ref. 21), within the previous mass estimates for the molecule [2]. Significantly, the predicted amino acid sequence of the enzyme exhibits homology with other PDEs, and this sequence, along with the sequences for the yeast cAMP-PDE (see below), bovine CaM-PDE, and bovine cGS-PDEs, allowed the identification of a conserved domain of approximately

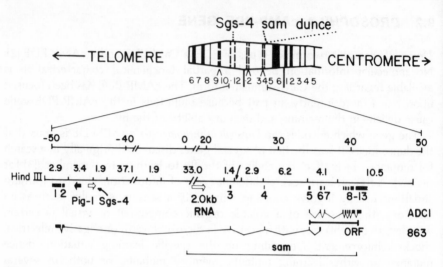

Figure 9.1 Structural organization of the *Drosophila* dunce gene. Diagram of the X-chromosome 3C6-3E5 chromosomal interval showing the Sgs-4 gene at 3C11 and the *sam* (*sperm amotile*) and dunce genes, both of which were mapped to band 3D4. An arbitrary co-ordinate system in kb, from − 50 to 50, is drawn relative to a *Hind*111 map of the 10 kb region. Dunce exons are numbered 1–13, being defined by the two overlapping cDNA clones, ADC1 and 863. The open reading frame for the cAMP-PDE is encompassed within exons 5–13. The positions and transcriptional direction of Pig-1, Sgs-4 and an anonymous gene which codes for a 2.0 kb transcript are illustrated. Genetic data [7] indicate that the *sam* complementation group resides within the bracketed region. (Reproduced by permission from *Nature*, Vol. 329, pp. 721, Copyright © 1987 Macmillan Magazines, Ltd)

275 amino acid residues within the various PDEs [8]. Interestingly, the sequence of the *Drosophila* enzyme is more similar to the conserved domain of the bovine CaM-PDE than to the yeast cAMP-PDE, even though its biochemical properties are much more similar to those of the yeast enzyme.

The analysis of the gene's structure and the encoded RNAs has revealed a remarkable complexity. At least six RNA molecules, which exist at very low abundance levels, are coded for by the locus in adult flies. These differ from one another in sequence due to the use of multiple transcriptional start sites, alternative splicing of mRNA precursors, and processes which generate different 3'-termini [5] (C.-N. Chen and R. L. Davis, unpublished). All of the heterogeneity in RNA molecules uncovered to date is outside of the open reading frame. Therefore, this heterogeneity in 5' and 3' untranslated sequences may be important for RNA stability or translational control.

The genomic sequences which code for these RNA molecules are distributed over more than 100 kb of the *Drosophila* genome [6] and a precise definition of the ends of the gene is not yet accomplished. Even with this partial understanding, however, the gene is certainly one of the largest *Drosophila* genes characterized to date and it may eventually prove to be the largest. Although it is unknown why some genes are large and others small, popular speculation is that size itself imposes a transcriptional control by limiting the rate at which a gene can be transcribed. Since preliminary work indicates that a rat homolog of the *Drosophila* dunce gene and the bovine gene for the retinal α subunit of cGMP-PDE (see below) are both very large; size itself may be a unifying feature of PDE genes reflecting a limitation on transcriptional rate.

The most remarkable feature of the *Drosophila* dunce gene is that it contains a very large intron, of 79 kb, in which several other genes reside (Figure 9.1). Although originally thought to reside to the left of the dunce locus, the Sgs-4 locus was recently found to be contained within the enormous intron. In addition, a gene named Pig-1, which is also expressed in salivary glands, resides very close to Sgs-4, within the dunce intron. Moreover, an anonymous gene which codes for a 2.0 kb RNA in adult flies resides within the same intron. The observation of a 'genes within genes' type of organization exemplified by the dunce locus is a new one with respect to the arrangement which eukaryotic genes may take. Only a few other genes are known to be arranged in this fashion and at present the biological significance of the organization remains unclear. However, it opens new questions regarding the evolution of the organization, the relationship between the transcription of dunce and its intronic genes, and the transcription of dunce and processing of its mRNA precursors.

9.3   YEAST AND *DICTYOSTELIUM* PDE GENES

Baker's yeast has two known types of PDEs, one with a high K_m for cAMP and a low substrate specificity and a second with a low K_m (0.2 μM) and specificity for cAMP as substrate [9–11]. The low-K_m form has been purified to homogeneity [9]. This enzyme has molecular mass of about 61 kDa, although if protease inhibitors are omitted during purification, the active fraction contains subfragments of 45 kDa and 17 kDa. In addition, the enzyme can apparently bind reversibly to ribosomes and other cellular particles.

The gene for the yeast, low-K_m PDE was isolated in two laboratories, using two related, functional assays. Both involved examining cells for sensitivity to nutrient starvation and/or heat shock; two phenotypes associated with conditions which elevate cAMP levels in yeast. The RAS genes in yeast, RAS1 and RAS2, are structurally and functionally homologous to the mammalian *ras* oncogenes [12] and are involved in controlling yeast adenylate cyclase activity [13]. A strain carrying the mutation *ras2*[val19] exhibits high cAMP levels, apparently due to the constitutive activation of adenylate cyclase by the mutant protein. By searching for yeast genomic sequences which could suppress the heat sensitivity conferred by the *ras2*[val19] mutation when present in high copy numbers on an extrachromosomal plasmid, Wigler and co-workers [14] isolated the structural gene for the low-K_m PDE. This gene is named PDE2. A variation in this approach was used by Wilson and Tatchell [15]. They first searched for yeast mutants which could suppress the inefficient growth of cells carrying a null $ras2$ mutation. One mutation, at a locus originally termed SRA5, suppressed the inefficient growth and exhibited a deficiency in PDE activity. Because the mutant exhibited sensitivity to nitrogen starvation in a normal RAS background and in the presence of exogenous cAMP, extrachromosomal sequences which rescued this phenotype were identified. The identified sequences were those from the PDE2 locus.

The open reading frame contained within the PDE2 gene is 526 amino acids long, predicting a product in excellent size agreement with the purified enzyme [9]. The predicted amino acid sequence is homologous to other PDEs, including the $Drosophila$ cAMP-PDE and bovine CaM- and cGMP-stimulated PDEs [8].

Interestingly, disruption of the PDE2 gene or overexpression in a wild-type background has minimal effects on phenotype [14,15]. No loss in viability of spores occurs with PDE2 disruption, although overexpression slowed cell growth in synthetic medium [14]. Thus, the PDE2 gene is not an essential function. However, as noted above, mutation of the PDE2 gene does produce a sensitivity to nitrogen starvation in the presence of exogenous cAMP. Presumably, the sensitivity is the sum of the increased cAMP levels due to the $pde2$ mutation and the added cAMP.

The second form of PDE found in yeast is quite different in biochemical properties and in sequence from the low-K_m form coded by the PDE2 gene. A highly purified preparation of the enzyme exhibits a monomeric molecular mass of 43 kDa, but sedimentation equilibrium analysis suggests that the enzyme exists as a dimer of 88 kDa [10]. Fujimoto et al. [11] have reported that the enzyme hydrolyzes cAMP with a K_m of 250 μM and cGMP with a K_m of 160 μM. In addition, the enzyme may contain two molecules of tightly bound Zn^{2+} per monomer of enzyme [10].

The PDE1 gene of yeast, which codes for this enzyme, was also isolated by searching for genomic sequences which suppress the phenotypes of the $ras2^{val19}$ mutation when carried on a high-copy-number plasmid [16]. An open reading frame of 369 amino acids, predicting a protein molecule of 42 kDa, was identified in one such plasmid. A remarkable observation regarding the enzyme's sequence is that it demonstrates absolutely no homology with any other known PDE, save one. The exception is the cAMP-PDE of the slime mold, $Dictyostelium\ discoideum$ (see below). This homology with the $Dictyostelium$ PDE, and the lack thereof with all other known PDEs, suggests that two unrelated DNA sequences have evolved into PDE genes. It is uncertain whether any additional PDE subclasses will be found, but certainly, the two subclasses which now exist offer one way of discriminating the enzymes.

The PDE1 gene has also been disrupted by insertion, and, like PDE2, this gene is not essential [16]. Moreover, strains constructed to carry disruptions in both PDE genes are also viable. However, the double mutant exhibits heat shock and starvation sensitivity, similar to $ras2^{val19}$ mutants. This is expected, since the $ras2^{val19}$ mutation and the $pde1$ $pde2$ mutation both produce yeast with elevated cAMP levels. However, even though the $pde1$ $pde2$ strain has no detectable PDE activity, cAMP levels are elevated only two-fold. Presumably, cAMP production is subject to feedback limitations when cAMP levels are elevated, the cAMP is secreted from the cells, or there exists another, undetected PDE in the cells which is responsible for keeping cAMP levels in check.

A cyclic nucleotide PDE gene of *Dictyostelium discoideum* has been isolated and characterized. This PDE is found attached to the outside surface of the cell and in the medium, released by secretion. It participates in the process of amoebae aggregation, which occurs when the cells are starved. In this process, cAMP released by signaling cells serves as a chemoattractant for other cells. The PDE serves to degrade the cAMP between cAMP signal pulses.

The extracellular PDE has been purified from the culture medium as a doublet protein with 48 kDa and 50 kDa polypeptide constituents [17]. It exhibits a high affinity for cAMP with a K_m of 8 μM, while the K_m for cGMP is 25 μM [18]. The purified enzyme was used to obtain partial amino acid sequence information for the design of oligonucleotide probes. Sequence analysis of an isolated cDNA clone predicts a protein of 452 amino acids and 51 kDa [19].

The *Dictyostelium* cyclic nucleotide PDE cDNA clone was recently used to select a genomic clone for structural analysis and for transformation experiments [20]. Transformation of the PDE gene containing the complete coding region, with 2.3 kb of 5′ flanking and 0.6 kb of 3′ flanking DNA, yielded transformant strains with 25–150 copies of the gene per cell, and an increased activity of extracellular PDE proportional to the gene copy number. This demonstrates that the DNA fragment used contains all of the necessary information for the production of extracellular PDE activity. However, the normal PDE gene produces a 1.8 kb transcript in growing and developing cells, as well as a 2.2 kb transcript shortly after starvation. The transformants showed an increase only in the abundance of the 1.8 kb transcript, indicating that the necessary information for producing the 2.2 kb transcript was not present. Interestingly, only the extracellular PDE activity, and not the membrane-bound, was elevated in the transformants. It is possible, therefore, that sequences specific to the 2.2 kb transcript are responsible for directing the coded enzyme to the membrane.

The crucial question, of course, is what effect does overexpression of the PDE have upon the normal physiology and development of the cells? Even though the levels of cAMP during signaling must be very low due to the excess PDE, the cells are still able to aggregate [20]. However, they aggregate much more quickly than control cells, and they do so without forming the streams of cells typical of normal cells. The precocious aggregation of the cells has been postulated to be the result of a greater sensitivity of cAMP receptors in transformed cells as a consequence of the lower cAMP levels, so that the cells detect and respond quickly to much lower cAMP levels than normal cells. In addition, the overexpression of the PDE prevents the further development of the cell aggregate, consistent with the hypothesis that cAMP is involved in subsequent differentiation and morphogenesis.

9.4 MAMMALIAN cAMP-PDE GENES

Recently, cDNA clones and genomic clones were obtained for mammalian cAMP-PDE genes using two different approaches. First, a probe of the *Drosophila* dunce gene was used to isolate mammalian counterparts by cross-hybridization at low stringency [21]. Second, a search for mammalian cDNA clones whose expression suppresses the phenotypes associated with $ras2^{val19}$ mutation in yeast yielded a cAMP-PDE clone [22]. The net outcome of these studies has been the isolation of several distinct cDNA clones, representing mRNAs from at least two different genes which appear to encode low-K_m cAMP-PDEs.

A fragment of a *Drosophila* cDNA clone which contains most of the region coding for the conserved domain of the cAMP-PDE was used to survey a variety of different species for homologous sequences [21]. Virtually every organism surveyed by Southern blotting contained one or more hybridizing bands, potentially representing authentic counterparts of the *Drosophila* cAMP-PDE gene. The probe was subsequently used to isolate a rat genomic clone and several cDNA clones from a rat brain cDNA library. Sequence analysis of the longest cDNA clone (RD1) reveals that it contains an open reading frame which translates into a conceptual protein product of about 68 kDa with a very high homology to the *Drosophila* cAMP-PDE (Figure 9.3). Indeed, the amino acid identity across the conserved domain is a striking 75%, demonstrating an extreme selective pressure to maintain this sequence in the last 600 million years, since the separation of the vertebrate and invertebrate Phyla.

Two additional rat brain cDNA clones selected with the *Drosophila* probe [21] appear identical to RD1 over the majority of their lengths, but show sequence variations towards their predicted N-termini. The modular nature of the sequence differences (Figure 9.2) indicates that all three clones are likely to represent RNAs from a single gene, but code for PDEs with some sequence variation. Thus, the gene identified, named ratdnc-1 because of its homology with the *Drosophila* dunce gene, may code for a family of related cAMP-PDEs.

In contrast to the single gene which codes for this class of enzyme in *Drosophila*, mammalian organisms appear to have at least two, and probably several, structural genes for the cAMP-PDEs. Genomic blotting experiments at high stringency using a portion of a genomic clone representing the ratdnc-1 gene have identified at least one other related gene [21]. In addition, cDNA clones isolated from rat testis with *Drosophila* dunce probes [23] are similar in

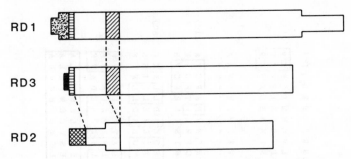

Figure 9.2 Schematic diagram to show the sequence relationships of predicted cAMP-PDEs isolated using a *Drosophila* dunce probe. Wide portions indicate the open reading frame and narrow portions of each figure represent untranslated sequences. The cDNA clones, RD2 and RD3, are partial, with truncations in the protein coding region. Regions where sequences are identical are unmarked. Different fill patterns indicate sequence divergence

sequence to RD1 but not identical, indicating that these represent RNAs from a gene other than ratdnc-1. This also applies to the rat brain cDNA clone isolated by virtue of its ability to suppress yeast $ras2^{val19}$ mutants [22] (see below). Hybridization experiments at very low stringencies would likely reveal addition-al cAMP-PDE genes.

The RNAs homologous to the ratdnc-1 gene were examined by RNA blotting experiments using a probe common to RD1, RD2 and RD3 [21]. The probe detects homologous RNAs in a variety of tissues, including brain, cerebellum, lung, testes and heart. The major hybridizing RNA is 4.4 kb in non-neuronal tissues and 4.0 kb in brain and cerebellum. Thus, there may be tissue-specific transcription and/or processing of transcripts from the ratdnc-1 gene. It may be that the RNA size differences between neuronal and non-neuronal tissue reflect protein heterogeneity beyond that suggested by structural differences in the rat brain cDNA clones (Figure 9.2).

The isolation of a cAMP-PDE cDNA clone based on functional suppression of the phenotypes caused by the $ras2^{val19}$ mutation in yeast utilized a rat brain cDNA library constructed in the yeast expression vector, pADNS [22]. This vector employs the yeast alcohol dehydrogenase promoter and terminator sequences for transcription of foreign sequences. The library was transformed into $ras2^{val19}$ yeast cells and the transformants tested for heat-shock sensitivity. One transformant, which clearly required the plasmid for resistance to heat shock, was isolated.

Sequence analysis of the insert identified an open reading frame capable of coding for a molecule of at least 562 amino acids and 62 kDa [22]. The putative amino acid sequence of this protein is highly homologous with the PDE coded for by the *Drosophila* dunce gene. Indeed, the amino acid sequence identity is 77% across the PDE conserved domain, similar to that found with the cDNA clone

```
                                                                          (25)
Dro cA   :  M F Q H Q T N P C G P T N R R R P R D Q E I

Ratdnc-1 :  H Q E P R Y P K A R R H T P A W P P T Q S R S W T G C A S T S W R P S R P I A A S P T W R R L S C K F R
Dro cA   :  F F C E T C S K P W L V G G W W D Q F K R
DPD      :  M E T L E E E L D W C L D Q L E T I Q T Y R S V S E E M A S N K F K R

                                                                          (63)
Ratdnc-1 :  M L N R E L T H L S E M S R S G N Q V S E Y I S N T F L D K Q N E V E I P S - - - - - - - - -
Dro cA   :  M L N K E L S H L S E S K S G N Q I S E Y I C S T F L D K Q Q E F D L P S L R V - E D N P E L V A A
DPD      :  M L N R E L T H L S E M S R S G N Q I V S E Y I S N T F L D K Q D V E I P S - - - - - - - - -

                                                                          (110)
Ratdnc-1 :  - P T P R Q R A F Q Q P P P S V L R Q S Q P M S Q I T G L K K - L V H T - G S L N T N V P R F C V K
Dro cA   :  N A A C G Q Q S A - C Q Y A R S P R C G P P M S Q I S G V K R P L - S H T N S F T G E R L P T F C V E
DPD      :  - P T Q K D R - I - E K K K Q Q L M T Q I S G V K K - - L M H H S S L N N T S I S R F C V N

                                                                          (160)
Ratdnc-1 :  T D Q E D L L A Q E L E N L S K W G L N I F C V S E Y A G G R S L S C I M Y T I F Q E R D L L K K F
Dro cA   :  T P R E N E L G T L L G E L D T W C L Q I F S I C E F S V N R P L T C V A Y T I F Q S R E L L T S L
DPD      :  T E N E D H L A K E L E D L N V A G Y S H N R P L T C I M Y A I F Q E R D L L K T F

                                                                          (210)
Ratdnc-1 :  H I P V D T M M M Y M L T L E D H Y H A D V L Q S T H V L L A A T P A L D A V F
Dro cA   :  M I P P K T F L N F W S D N P F H N S L H A A D V T Q S T H V L L S T P A L E C V F
DPD      :  K I S S D T F V Y H L E D H Y H S D V A Y H N S L H A A D V Q S T H V L L D A V F

                                                                          (260)
Ratdnc-1 :  T D L E I L A A L F A A A I H D V D H P G V S N Q F L I N T N S E L A L M Y N D E S V L E N H H L A
Dro cA   :  T P L E V G G A L F A A A I H D V D H P G L T N Q F L V N N S S E L A L M Y N D E S V L E N H H L A
DPD      :  T D L E I L A A T F A A A I H D V D H P G V S N Q F L I N T N S E L A L M Y N D E S V L E N H H L A
```

```
Ratdnc-1 :  V G F K L L Q E E N C D I F Q N L S K R Q R Q S L R K M V I D M V L A T D M S K H M T L L A D L K T  (310)
Dro cA   :  V A F K L L Q Q Q C C D I F Q N M Q K K Q R Q T L R K M V I D I V L S T D M S K H M S L L A D L K T
DPD      :  V G F K L L Q E Q H C D I F Q N L K K Q R Q T L R K M V I D M V L A T D M S K H M S L L A D L K T

Ratdnc-1 :  M V E T K K V T S S G V L L L D N Y S D R I Q V L R N M V H C A D L S N P T K P L E L Y R Q W T D R  (360)
Dro cA   :  M V E T K K V A C C G V L L L D N Y T D R I Q V L E N N L V H C A D L S N P T K P L P L Y K R W V A L
DPD      :  M V E T K K V T S S G V L L L D N Y T D R I Q V L R N M V H C A D L S N P T K S L E L Y R Q W T D R

Ratdnc-1 :  I M A E F F Q Q G D R E R E R G M E I S P M C D K H T A S V E K S Q V G F I D Y I V H P L W E T W A  (410)
Dro cA   :  L M E E F F L Q Q D D K E R E R G M E I D I S P M C D R H T A S T I E K S Q V G F I D Y I V H P L W E T W A
DPD      :  I M E E F F Q Q G D K E R E R G M E I S P M C D K H T A S V E K S Q V G F I D Y I V H P L W E T W A

Ratdnc-1 :  D L V H P D A Q D I L D T L E D N R D W Y H S A I R Q S P S P P L E E E P G C L G H P S L P D K F Q  (460)
Dro cA   :  S L V H P D A Q D I L D T L E D N R D Y Y Q S M I P P S P V D E N P Q E D R I R F Q V T L E
DPD      :  D L V Q P D A Q D I L D T L E D N R N W Y Q S M I P P S P P L D Q R S R D C Q G L M Q K F Q F E

Ratdnc-1 :  F E L T L E E E E E D S L E V P G L P T T E E T F L A A E D A R A Q A V D W S K V K G P S T T V V  (510)
Dro cA   :  E S D Q E N L A E L E E C D E S G E T T T T G T T C T T A A S A L R A C G C G G G G C G C H A P R T
DPD      :  L T L E E E D S E G P E K E G P Q Y F S S T K T L C V I D P E N R D S L E E T D I D I A T E D K

Ratdnc-1 :  E V A E R L K Q E T A S A Y G A P Q E S M E A V G C S F S P G T P I L P D V R T L S S S E E A P G L  (560)
Dro cA   :  C C C Q N Q P Q H G G M
DPD      :  S L I D T

Ratdnc-1 :  L C L P S T A A E V E A P R D H L A A T R A C S A C S G T S G D N S A I I S T P G R W G S C G D P A  (610)
```

Figure 9.3 Aligned amino acid sequences of the *Drosophila* dunce encoded cAMP-PDE and the products of its mammalian homologs, RD1 and DPD

identified by direct homology to the *Drosophila* dunce gene. To determine whether the dunce-related cDNA clone could compensate for the absence of endogenous yeast PDE activity, the plasmid was transformed into a strain carrying disruptions in both PDE genes. The plasmid did rescue the heat-shock sensitivity associated with the lack of yeast PDE activity and conferred substantial cAMP-PDE activity to the cells. The K_m for the new cAMP-PDE activity was estimated to be 3.5 μM. No cGMP hydrolysis was detected in the yeast extracts.

The latter method used to isolate a mammalian cAMP-PDE cDNA clone, which developed from the broader goal of identifying sequences which can suppress constitutive *ras2* phenotypes, potentially offers a general way to isolate additional mammalian cyclic nucleotide PDE clones. Any sequence which confers cAMP-PDE activity might suppress *ras2*^{val19} or *pde1 pde2* phenotypes, but it is unclear whether an associated high-affinity cGMP-PDE activity will be deleterious. In addition, the yeast strain lacking endogenous activity but carrying a mammalian cAMP-PDE gene may offer an ideal way to screen for new inhibitors of cAMP-PDE, since these should restore the mutant phenotypes to the suppressed cells. PDE mutants which are more or less sensitive to the inhibitors or have an altered capacity to hydrolyze cAMP could, in principle, also be identified using this expression system.

9.5 RETINAL cGMP-PDE GENES

Perhaps the most abundantly expressed PDE is the cGMP-PDE of the vertebrate retina. This enzyme is highly enriched in photoreceptor cells and has the unique property of being activated by light. The role of the enzyme in phototransduction ([24,25,26] for reviews) and its potential involvement in retinal dystrophies ([27,28,29] for reviews) have made it an important focus of research.

Biochemical investigations of retinal cGMP-PDEs have revealed some apparent heterogeneity. The enzyme, as purified from bovine rod outer segments [30], consists of three subunits, α (88 kDa), β (84 kDa) and γ (11 kDa). The γ subunit is a heat-stable inhibitor. A variant polypeptide form, termed α', has an apparent mass of 94 kDa and is found in cone photoreceptor cells [31,32]. It appears likely that each of the three larger polypeptides contains one or more catalytic sites [31]. An additional retinal cGMP-PDE has been characterized [33] from the interphotoreceptor cell matrix, an extracellular compartment. The specific function of this enzyme is unknown, but it could potentially be involved in processes such as cGMP-mediated intercellular communication, akin to the intercellular communication processes used by *Dictyostelium*. The cGMP-PDE from the matrix has apparent subunit sizes of 47 kDa and 45 kDa.

Because of its abundance, the rod outer segment cGMP-PDE is relatively easy to purify. Two different groups have purified the enzyme from bovine retina, obtained partial amino acid sequence information, and isolated cDNA clones using oligonucleotide probes. Ovchinnikov and co-workers first reported the isolation and sequence of cDNA clones representing mRNAs for the γ [34] and the α [35] subunits of the enzyme. The sequence of the γ subunit cDNA clone and the putative protein product revealed no features critical to this discussion and will not be considered further. The amino acid sequence for the α subunit, deduced from the sequence of overlapping cDNA clones, predicts a primary translation product of 859 amino acid residues with a mass of approximately 99 kDa, substantially larger than that predicted from SDS gel electrophoresis of the purified protein. The size discrepancy may result from anomalous migration of the protein in standard gel systems. In addition, comparison of the predicted sequence with other PDEs, including the yeast and *Drosophila* cAMP-PDEs, revealed that this polypeptide also contains the conserved domain found in the major subclass of PDEs.

A portion of a sequence claimed to be the β subunit, deduced from cDNA clones, has also been reported [35]. It, too, contains amino acid residues resembling the conserved domain of other PDEs. However, there are numerous differences within this homologous domain between the α and the β sequences, potentially indicating that these two polypeptides are products of distinct, but related, genes. The conservation of the domain in this enzyme suggests that the members of the major subclass of PDEs evolved from a common ancestor, with subsequent specialization to participate in processes ranging from visual transduction to memory.

More recently, cDNA clones representing the α subunit of both the human and the bovine rod cell cGMP-PDEs were isolated and characterized [36]. The open reading frame is of identical size (859 amino acids) between the species and has 94% identical amino acids. Transcripts detected by RNA blotting experiments were large and heterogeneous, with a broad band of 4.9–5.3 kb and representing two or more RNAs appearing in the human retinal RNA population and two relatively distinct RNAs of 4.6 and 4.0 kb appearing in bovine cells. Genome reconstruction experiments and genome blotting experiments using a variety of probes indicate that there exists but a single bovine gene with homology to the cDNA clones. In addition, a minimal size estimate for the gene obtained from the genome blotting experiments is about 140 kb. Thus, this gene, like the *Drosophila* dunce gene, is very large and complex. As mentioned previously, this complexity may be a general feature of PDE genes from higher eukaryotes.

REFERENCES

1. Beavo, J. (1987) *Adv. Second Messenger Phosphoprotein Res.*, **22**, 1–38.
2. Davis, R., and Kauvar, L. (1984) *Adv. Cyclic Nucleotide Protein Phosphorylation Res.*, **16**, 393–402.
3. Dudai, Y. (1988) *Annu. Rev. Neurosci.*, **11**, 537–563.

4. Davis, R. L., and Davidson, N. (1984) *Mol. Cell. Biol.*, **4**, 358–367.
5. Davis, R. L., and Davidson, N. (1986) *Mol. Cell. Biol.*, **6**, 1464–1470.
6. Chen, C.-N., Denome, S., and Davis, R. L. (1986) *Proc. Natl Acad. Sci. USA*, **83**, 9313–9317.
7. Chen, C.-N., Malone, T., Beckendorf, S., and Davies, R. L. (1987) *Nature*, **329**, 721–724.
8. Charbonneau, H., Beier, N., Walsh, K. A., and Beavo, J. A. (1986) *Proc. Natl Acad. Sci. USA*, **83**, 9308–9312.
9. Suoranta, K., and Londesborough, J. (1984) *J. Biol. Chem.*, **259**, 6964–6971.
10. Londesborough, J., and Suoranta, K. (1983) *J. Biol. Chem.*, **258**, 2966–2972.
11. Fujimoto, M., Ichikawa, A., and Tomita, K. (1974) *Arch. Biochem. Biophys.*, **161**, 54–63.
12. Kataoka, T., Powers, S., Cameron, S., Fasano, O., Goldfarb, M., Broach, J., and Wigler, M. (1985) *Cell*, **40**, 19–26.
13. Toda, T., Uno, I., Ishikawa, T., Powers, S., Kataoka, T., Broek, D., Cameron, S., Broach, J., Matsumoto, K., and Wigler, M. (1985) *Cell*, **40**, 27–36.
14. Sass, P., Field, J., Nikawa, J., Toda, T., and Wigler, M. (1986) *Proc. Natl Acad. Sci. USA*, **83**, 9303–9307.
15. Wilson, R. B., and Tatchell, K. (1988) *Mol. Cell. Biol.*, **8**, 505–510.
16. Nikawa, J.-I., Sass. P., and Wigler, M. (1987) *Mol. Cell. Biol.*, **7**, 3629–3636.
17. Orlow, S. J., Shapiro, R. I., Franke, J., and Kessin, R. H. (1983) *J. Biol. Chem.*, **256**, 7620–7627.
18. Shapiro, R. I., Franke, J., Luna, E. J., and Kessin, R. H. (1983) *Biochim. Biophys. Acta*, **758**, 49–57.
19. Lacombe, M.-L., Podgorski, G. J., Franke, J., and Kessin, R. H. (1986) *J. Biol. Chem.*, **261**, 16811–16817.
20. Faure, M., Podgorski, G. J., Franke, J., and Kessin, R. H. (1988) *Proc. Natl Acad. Sci. USA*, **85**, 8076–8080.
21. Davis, R. L., Takayasu, H., Eberwine, M., and Myres, J. (1989) *Proc. Natl Acad. Sci. USA*, **86**, 3604–3608.
22. Colicelli, J., Birchmeier, C., O'Neill, K., and Wigler, M. (1989) *Proc. Natl Acad. Sci. USA*, **86**, 3599–3603.
23. Conti, M., and Swinnen, J. V. (1990) In *Cyclic Nucleotide Phosphodiesterases: Structure, Regulation and Drug Action* (Beavo, J. and Houslay, M. D., eds), John Wiley & Sons, Chichester, pp. 243–266.
24. Stryer, L. (1986) *Annu. Rev. Neurosci.*, **9**, 87–119.
25. Liebman, P. A., Parker, K. R., and Dratz, E. A. (1987) *Annu. Rev. Physiol.*, **49**, 765–792.
26. Hurley, J. B. (1987) *Annu. Rev. Physiol.*, **49**, 793–812.
27. Lolley, R. N., Navon, S. E., Fung, B. K.-K., and Lee, R. H. (1987) In *Degenerative Retinal Disorders: Clinical and Laboratory Investigations*, Alan R. Liss, Inc., Town, pp. 269–287.
28. Chader, G. J., Fletcher, R. T., Barbehenn, E., Aguirre, G., and Sanyal, S. (1987) In *Degenerative Retinal Disorders: Clinical and Laboratory Investigations*, Alan R. Liss, Inc., New York, pp. 289–307.
29. Farber, D. B., Flannery, J. G., Bird, A. C., Shuster, T., and Bok, D. (1987) In *Degenerative Retinal Disorders: Clinical and Laboratory Investigations*, Alan R. Liss, Inc., New York, pp. 53–67.

30. Baehr, W., Devlin, M. J., and Applebury, M. L. (1979) *J. Biol. Chem.*, **254**, 11669–11677.
31. Hurwitz, R., Bunt-Milam, A. H., Chang, M. L., and Beavo, J. A. (1985) *J. Biol. Chem.*, **260**, 568–573.
32. Gillespie, P. G., and Beavo, J. A. (1988) *J. Biol. Chem.*, **263**, 8133–8141.
33. Barbehenn, E. K., Wiggert, B., Lee, L., Kapoor, C. L., Zonnenberg, B. A., Redmond, T. M., Passonneau, J. V., and Chader, G. J. (1985) *Biochemistry*, **24**, 1309–1316.
34. Ovchinnikov, Y. A., Lipkin, V. M., Kumarex, V. P., Gubanov, V. V., Khramtsov, N. V., Akhmedov, N. B., Zagranichny, V. E., and Muradov, K. G. (1986) *FEBS Lett.*, **204**, 288–292.
35. Ovchinnikov, Y. A., Gubanov, V. V., Khramtsov, N. V., Ischenko, K. A., Zagranichny, V. E., Muradov, K. G., Shuvaeva, T. M., and Lipkin, V. M. (1987) *FEBS Lett.*, **223**, 169–173.
36. Pittler, S. J., Baehr, W., Wasmuth, J., McConnell, D., Champagne, M., van Tuinen, P., Ledbetter, D., and Davis, R. L. (1989) *Genomics*, **6**, 272–283.

STRUCTURE AND FUNCTION OF THE ROLIPRAM-SENSITIVE, LOW-K_m CYCLIC AMP PHOSPHODIESTERASES: A FAMILY OF HIGHLY RELATED ENZYMES

Marco Conti and Johannes V. Swinnen

The Laboratories for Reproductive Biology, Departments of Pediatrics and Physiology, University of North Carolina at Chapel Hill, Chapel Hill, NC 27599, USA

10.1 INTRODUCTION

Probably the most elusive and poorly defined of all the phosphodiesterases (PDEs) are the enzymes commonly referred to as low-K_m PDEs [1]. This is a group of enzymes that share the characteristics of being insensitive to Ca^{2+} and calmodulin (CaM) and of hydrolyzing cAMP with a K_m in the micromolar or submicromolar range. Aside from this similar affinity for the substrate, a broad range of properties has been attributed to these enzymes. The uncertainty in defining these PDEs is reflected in the variety of names used for them. They have been termed peak III PDEs [2] on the basis of their elution from a DEAE–cellulose column, cAMP-PDEs and high-affinity or low-K_m PDEs on the basis of their selectivity and affinity for the substrate [2,3], type III [4], or type IV PDEs [5]. The fact that these PDEs are present only in trace amounts, that they are localized in different cellular compartments, and that they are remarkably unstable and prone to proteolysis has rendered the purification of these forms a formidable task. To date, purification data on these PDEs are still sparse.

A major advance in the characterization of these forms has been the realization that these PDEs can be separated into two subtypes based on the inhibition by

Cyclic Nucleotide Phosphodiesterases: Structure, Regulation and Drug Action
Edited by J. Beavo and M. D. Housley © 1990 John Wiley and Sons Ltd

cGMP [6] and sensitivity to different PDE inhibitors [1,2]. On the basis of these two criteria, it is now accepted that both cGMP-inhibited and cGMP-insensitive low-K_m PDEs exist in many tissues. With some remarkable exceptions, the cGMP-inhibited PDEs are usually susceptible to cardiotonic drugs [2] or cilostamide [7] inhibition, while the cGMP-insensitive enzymes are preferentially inhibited by Ro 20-1724 and rolipram [1,2].

In this chapter we will review the available information about the cGMP-insensitive, Ro 20-1724- and rolipram-inhibited PDEs. We will analyze their properties and their regulation. The reader should be warned that conflicting results are not the exception in this area and that many issues are still unsettled. It also should be pointed out that selective PDE inhibitors as a diagnostic tool to differentiate these PDEs have become available only recently, and many of the older reports have to be viewed in retrospect. In many instances we will refer to our data derived from the Sertoli cell, a system whose characteristics make it particularly useful in studying the function and regulation of these cAMP-PDEs.

10.2 MOLECULAR STRUCTURE OF THE cAMP-PDEs

10.2.1 Isolation and purification of the cAMP-PDEs

The initial reports on the purification of low-K_m PDEs that in retrospect fit the characteristics of the rolipram-sensitive, low-K_m PDE described specific activities ranging between 50 and 200 (nmol/min)/mg protein [8,9,10]. These are 10–100-fold lower than those reported for the Ca^{2+}–CaM-dependent PDE [11,12,13] or the cGMP-inhibited PDEs [14,15,16] (5–10 (μmol/min)/mg protein). Even though these preparations demonstrated the presence of a single band on SDS-PAGE [8,10], it is unclear whether the recovered band corresponds to the protein possessing PDE activity. The possibility has often been put forward that PDE loses its activity during purification either because an activating factor is removed or because proteolytic clipping reduces the catalytic activity [3]. Furthermore, the yield from the different procedures has usually been low. A possible improvement in the cAMP-PDE purification protocol is the inclusion of an affinity chromatographic step. Fougier et al. have indeed reported that rolipram can be coupled to AH–Sepharose and that this matrix can be used to purify a rat heart cAMP-PDE 102-fold with a 35% yield [17]. In our experience, the use of several fast HPLC chromatographic steps alleviates some of the above-mentioned instability problems [18]. Overexpression of recombinant cAMP-PDEs in a bacterial system might also be a promising alternative path to obtaining homogeneous cAMP-PDE preparations (see below).

It is generally agreed that, due to the low abundance of the protein, 5000–10000-fold concentration or more is required for a complete purification [1]. This should yield preparations with specific activities of approximately 1–10

(μmol/min)/mg protein. Indeed, several laboratories have recently reported specific activities in this range. Table 10.1 is a summary of the properties of the cAMP-PDEs thus far characterized. A low-K_m PDE has been purified 6000-fold to a specific activity of 1.8 (μmol/min)/mg protein from dog kidney extracts [19]. These preparations showed a major band of 82 kDA on SDS-PAGE. Houslay and collaborators have also reported the purification of a cGMP-insensitive, peripheral, membrane-bound rat liver cAMP-PDE to a specific activity of 380 (nmol/min)/mg protein [27]. This protein behaves as a monomeric structure with a molecular mass of 52 kDa. Using dibutyryl-cAMP-treated Sertoli cells that possess a starting activity of 0.5–1 (nmol/min)/mg protein, we have purified a cAMP-PDE 1000-fold to a specific activity of 1–2 (μmol/min)/mg protein [18]. Two major bands of 85 kDa and 66–67 kDa are present on SDS-PAGE and silver staining, indicating that preparations of this specific activity are, in our hands, still not homogeneous. Furthermore, although the PDE activity eluted from the first three purification steps as a single symmetrical peak [18], the final hydroxyapatite chromatography consistently yields a major peak and a shoulder (Figure 10.1). Although the two species have similar kinetic properties, it it likely that two very similar but distinct PDE species are present at this stage of purification. Antibodies generated against these preparations recognize the 85 kDa but not the 67 kDa polypeptide by Western blot analysis, and do not immunoprecipitate any PDE activity [18]. Possible explanations for this are that this antibody recognizes epitopes that become exposed after SDS-PAGE and are not accessible in the native PDE, or that the 85 kDa peptide is devoid of cAMP hydrolytic activity. The 85 kDa protein appears to be a major component of the Sertoli cell extract, and the concentration of the 85 kDa band in the fraction of both gel filtration and hydroxyapatite chromatography does not parallel the activity eluted from the column. On the other hand, the 66–67 kDa band is only a minor component of the Sertoli cell extract, and the intensity of the band closely follows the peak of activity on both gel filtration and hydroxyapatite chromatography. For these reasons, we favor the hypothesis that the 67 kDa protein corresponds to the cAMP-PDE. If this assumption is correct, we calculated that the specific activity of the Sertoli cell PDE corresponds to 5–6 (μmol/min)/mg protein. It should be pointed out that the starting material is an extract from Sertoli cell cultures activated by a 24 h cAMP analog treatment [18], and that this treatment might affect the state of activation of the enzyme.

From the above data (see also Table 10.1) it can be concluded that substantial divergence of opinion exists on the size of this PDE and on its specific activity when in pure form. The reason for these discrepancies is puzzling. It has often been proposed that limited proteolysis can generate PDEs of low molecular weight [3]. This has been a common finding also in our laboratory over the past five years. However, in light of our recent data on PDE cloning, it is proposed that marked heterogeneity in the different PDE forms is to be expected and that different tissues express different enzymes that have similar kinetic characteristics

Table 10.1 Summary of the properties of the low-K_m cAMP-PDEs derived from different tissues and different species

	Dog kidney	Dog heart	Human lung	Human lympho-cytes	Human heart	Human Leuke-mic C.	Rat brain	Rat liver	Rat glioma	Rat Ser-toli C.	Rat heart	Mouse lym-phoma	Rabbit ven-tricle
References	[8,9,19]	[20]	[10]	[21-23]	[24]	[23]	[25,26]	[27,28]	[29]	[18]	[17]	[30,31]	[32]
K_m cAMP (μM)	2.2	1.1	0.7	1.0	2.0	1.6	1.0	0.7	2.0	1.6	—	0.53	0.93
K_m cGMP (μM)	3.12	>1	—	—	50	—	—	120	—	>50	—	0.04	—
V_{max} cAMP ((μmol/min)/mg)	10	—	0.1	—	—	0.009	—	0.93	—	5-6	—	—	—
Isoelectric point	4.8	4.6	4.8	4.9	—	—	—	—	—	5.6	6.7, 4.75	—	—
Molecular mass native (kDa)	71-74	—	60	60	45	26	53	52	54	135	—	104	—
Molecular mass SDS-PAGE (kDa)	82	—	60	—	—	26	60.6	52	—	67	—	—	—
ED$_{50}$ Inhibitors (μM)													
cGMP	600	1500	1000	—	—	—	—	180	—	>50	—	>100	142
Ro 20-1724	2.2	18	—	15	3.8	—	4.0	7.2	—	2.2	—	—	—
Rolipram	—	1.0	—	—	3.0	—	1.0	—	—	0.9	1.0	—	—
Cilostamide	—	31	—	—	—	—	—	—	—	59	—	—	12
Milrinone	—	—	—	—	—	—	—	7.3	—	—	—	—	—
MIX	11	—	21	16	—	—	24	—	—	10	—	—	—
SQ20009	0.1	—	1.0	1.4	—	—	—	1000	—	—	—	—	—

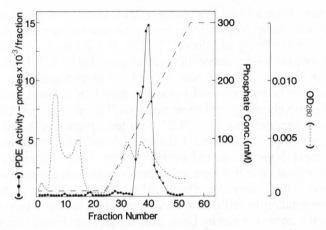

Figure 10.1   Hydroxyapatite high performance liquid chromatography (HPLC) of the partially purified Sertoli cell cAMP-PDE. Sertoli cells from 25-day-old Sprague-Dawley rats were treated for 24 h in medium containing 1 mM dibutyryl-cAMP [18]. At the end of the treatment, the cells were harvested and the soluble cAMP-PDE activity partially purified by three sequential HPLC chromatographic steps [18]. An aliquot of the pool from the third step (specific activity 300 (nmol/min)/mg protein) was loaded onto a HPA–hydroxyapatite HPLC column pre-equilibrated with a phosphate buffer containing protease inhibitors [18]. After application of the sample, the column was washed with the equilibration buffer until absorbance had returned to the baseline. The PDE activity was eluted by applying a 10–300 mM phosphate gradient. The activity of the fractions was measured with 1 μM cAMP as substrate

but that differ in terms of molecular mass and amino acid sequence. Finally, molecular structure differences due to species-specific differences in cAMP-PDEs should also be taken into consideration.

10.2.2   Molecular cloning of the cAMP-PDEs

The cloning of the mammalian PDEs is still in its infancy, but it represents a major development in the field. The characterization of the 'dunce' mutation in *Drosophila* and the ras complementation studies in yeast have provided invaluable tools for the characterization of the mammalian cAMP-PDEs. Davis and collaborators have shown that the 'dunce' gene codes for a high-affinity PDE with a region homologous to the bovine Ca^{2+}–CaM-dependent PDE and the cGMP-stimulated PDE [33]. Furthermore, Sass and collaborators have isolated a PDE cDNA from yeast that again shares regions of homology with the *Drosophila* and other mammalian PDEs [34,35]. These findings on the homology of invertebrate PDEs with sequences available from mammalian enzymes [33,34,35] have set the stage for screening mammalian libraries to isolate PDE cDNA clones. Using a *Drosophila* 'dunce' cDNA, Davis et al. have isolated (see Chapter 9) and characterized cDNAs from a rat brain library that share many similarities with the *Drosophila* 'dunce' PDE [36]. Using a similar approach, in our laboratory we have used a 'dunce' PDE cDNA provided by R. L. Davis to screen rat testicular libraries. We have now isolated and characterized four groups of clones which we believe represent transcripts from four different genes [37,38]. We have tentatively designated these genes ratPDE1 through ratPDE4 (Figure 10.2). Clones corresponding to ratPDE1 code for a protein with undetermined molecular weight [37], since no full-length clones have been isolated. Its molecular mass, however, should be higher than 47 kDa. Messenger RNAs encoding this putative PDE are expressed predominantly in the germ cells of the rat testis and in the kidney [37]. The partial nucleotide sequence available from clones corresponding to ratPDE2 shows a nucleotide sequence very similar or identical to the clones isolated by Davis from the rat brain library [36], indicating that these two groups of cDNAs are derived from the same gene [37]. On the basis of the sequencing data provided by Davis, this cDNA encodes a protein [36] of calculated molecular mass 68 kDa. We probably have [37,38] the complete coding sequence of ratPDE3 and ratPDE4. They encode proteins of 67 kDa and 64 kDa, respectively. These latter two cDNAs have been expressed both in

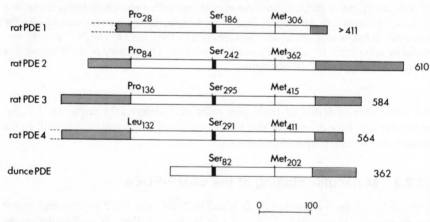

Figure 10.2 Structure of the major open reading frames of the four groups of PDE cDNA clones isolated from rat testicular libraries. The open reading frames are reported as bars. Numbers to the right of the bars indicate the number of residues present in each open reading frame. The domain highly conserved in the four rat sequences and in the *Drosophila* 'dunce' sequence [33] corresponds to the empty portion of the bars. Shaded areas correspond to the less homologous N- and C-terminus domains. The filled areas starting with a serine correspond to the region homologous to the cAMP binding domain of the RII regulatory subunit of the cAMP-dependent protein kinase [47]. The Met residue corresponds to the position of the motif Met–Ala/Glu–Glu–Phe–Phe present in all five sequences, and is used as a reference point

bacteria and eukaryotic cells, and their expression leads to large increases in PDE activity in the transfected cells. The expression in a bacterial system has been particularly useful because bacteria do not contain any high-affinity cAMP-PDE [39,40], simplifying the task of characterizing the recombinant PDE activity. This approach should also be useful in establishing if the anomalous kinetic behavior often attributed to cAMP-PDEs [3,9,25,27] is intrinsic to the catalytic site or is due to a mixture of closely related enzymes possessing slightly different K_m values. This activity expressed in *E. coli* and corresponding to ratPDE3 has been further characterized, and the data collected indicate that the recombinant enzyme has properties similar or identical to the enzyme partially purified from the Sertoli cell [18]. The recombinant ratPDE3 hydrolyzes cAMP with an affinity of 1.5–2 μM, and cAMP hydrolysis is not inhibited by cGMP concentrations ranging between 0.1 and 50 μM. That ratPDE3 codes for a member of the cAMP-PDE family is further supported by the finding that the recombinant PDE activity is inhibited by Ro 20-1724 and rolipram but only poorly by cilostamide [38]. Although the characterization of recombinant ratPDE4 is not completed, all available data indicate that this latter cDNA codes for a PDE with characteristics very similar to those of ratPDE3. In conclusion, although formal proof of the identity of these clones with the PDEs that we have isolated from the testis is still missing, all available data support this conclusion. It is also apparent that the PDEs encoded by the four groups of clones are highly related enzymes, probably possessing similar kinetic characteristics. A cDNA clone corresponding to ratPDE4 has also been isolated by Wigler and collaborators by expressing rat brain cDNAs in yeast to suppress the phenotypic effects of the RAS2 mutations [41].

The above findings demonstrate that the cAMP-PDEs form a heterogeneous family of isoenzymes. This heterogeneity, based on the presence of at least four different, but homologous, genes is probably only one level of the complexity present in this family of enzymes. Post-transcriptional modifications increase the number of possible forms present in the cells. We have retrieved two clones corresponding to ratPDE3 (ratPDE3.1 and ratPDE3.2) [38]. These two cDNAs are identical in the coding region except for an insertion of 85 bases in ratPDE3.2. The boundaries of this insert contain a consensus sequence for splicing. If this 85 base insert is spliced out in the mature RNA, as it is found in ratPDE3.1, a reading frame shift moves the initiation ATG codon further down the sequence. This results in a putative open reading frame coding for a protein with a catalytic domain identical to ratPDE3.2, but with a shorter N-terminus region (calculated molecular mass = 57 kDa). It should be pointed out that expression of the two cDNAs in bacteria leads to the appearance of PDE activities with identical substrate specificity and sensitivity to inhibitors, indicating that the intervening sequence does not alter the basal catalytic properties (Conti, M. and Swinnen, J. V., unpublished observation). This, together with the observation that multiple transcripts are present in several organs [37,38], suggests that the number of proteins corresponding to the four rat PDE genes probably exceeds four. This unsuspected heterogeneity might provide an explanation for the inconsistencies that have been experienced during the purification of these enzymes (Table 10.1). Another factor to be considered is that not all four genes are expressed in a single cell. Our preliminary data show that in the testis, PDE1 and PDE2 are expressed predominantly in germ cells, and PDE3 and PDE4 are expressed mainly in somatic cells [37]. Therefore, more than one PDE gene is expressed in one cell type and different genes are selectively expressed in different cells. The physiological significance of this multiplicity of forms with different patterns of expression is largely unknown, but it supports the concept that second messenger degradation and inactivation is regulated in a complex fashion. Efforts will have to be directed to clarify which individual member of this cAMP-PDE family is involved in each of the known cAMP-PDE regulations.

10.2.3   Structure–function relationship of the cAMP-PDEs

Although the exact structure of the cAMP-PDEs is far from being understood, the cloning and the purification data provide some preliminary information about the characteristics of these molecules. Here we will report some of these features.

A survey of the literature on the size of the native cAMP-PDE shows molecular masses ranging between 26 kDa and 150 kDa (see Table 10.1). The possibility that some of the described molecular masses are the result of partial proteolysis has been often proposed but other possibilities should also be taken into consideration. The dog kidney PDE [19] and the rat liver peripheral PDE [27] have an apparent size of 71 kDa and 52 kDa on gel filtration, similar to the size derived from SDS-PAGE, indicating that these PDEs exist as monomers. Our data on the Sertoli cell, however, point to a different conclusion, suggested also for the cAMP-PDE derived from human lymphocytes [21,42]. All our experimental data using HPLC gel filtration or sucrose density gradient centrifugation indicate that the native cAMP-PDE present in the Sertoli cell behaves as a molecule of 135–150 kDa. If these values are compared with the cloning and the SDS–PAGE data indicating that the PDE is a polypeptide of 66–67 kDa, it can be concluded that the enzyme is a multimeric structure, possibly formed of two subunits. This structure is reminiscent of the structure of the Ca^{2+}–CaM-dependent PDE, which is a dimer of two identical or dissimilar subunits of 59–62 kDa [11,12,43]. It should also be pointed out that, since we find mRNA for at least two ratPDE genes (ratPDE3 and ratPDE4) expressed in the Sertoli cell [38], it is possible that the dimeric enzyme is composed of two dissimilar subunits. Why do the rat liver or the dog kidney enzymes behave differently from the Sertoli cell and lymphocyte enzymes? It is possible that the presence of different genes coding for multiple mRNAs by alternate splicing might be at the basis of differences in cAMP-PDE oligomerization. One could speculate that the PDE domain involved in subunit interactions is present in the N- or C-terminus (see below) of not all the cAMP-PDEs or that alternate splicing night dictate whether the binding site for subunit interaction is maintained or not in the mature product.

By comparing the sequences of the four groups of PDE clones that we have isolated, the primary structure of these proteins can be subdivided into three major domains: a highly conserved domain probably corresponding to the catalytic site and the C- and N-terminus domains (Figure 10.2). The four rat PDE sequences contain a core of 270 amino acids that is highly conserved both at the level of nucleotide and amino acid sequence (Figure 10.3). It should be pointed out that the 270 amino acid region also shares substantial homology with partial sequences available for the Ca^{2+}–CaM-dependent PDE [35], the cGMP-stimulated PDE [35], a retina cGMP-PDE [44] (all three from bovine tissues and a yeast PDE [34] (Figure 10.3) but not with the *Dictyostelium discoideum* PDE [46]. In agreement with what has been previously observed [33,35], we believe that this region contains the catalytic site. This is based on the following observation. Davis et al. were the first to notice that a short region within the dunce-PDE sequence [33] is homologous to the cAMP binding site of the RII regulatory subunity of the cAMP-dependent protein kinase [47]. All our clones contain this highly conserved domain [37]. Furthermore, truncated cDNA clones lacking this putative binding site do not show PDE activity when expressed in bacteria [38]. Finally, Charbonneau et al. have reported that an antibody raised against a peptide corresponding to a similar region reacts with a 35 kDa fragment of the CaM-dependent PDE that is catalytically active [48]. That this 270 amino acid domain plays an important role in the structure of this protein is suggested by the comparison of the amino acid sequences of all published PDE sequences (Figure 10.3). It is striking that some residues, including several histidines, are conserved in all nine sequences. These conserved residues probably have an important function in the interaction of the protein with the substrate or for the folding of the protein. Of the nine PDEs compared, some have a higher affinity for cGMP than for cAMP. The presence of a homologous region in spite of the difference in substrate specificity should not be surprising. Considering the conformation of the cyclic nucleotides [49], cAMP and cGMP are very similar in the ribose and phosphate moieties, and differ only in the purine ring. It is therefore likely that the catalytic domain interacting with the phosphate is probably similar in both cAMP- and cGMP-PDEs. On the other hand, as proposed for the cyclic nucleotide binding domain of the bacterial CAP protein [50,51], and in the RII subunit of the cAMP-dependent protein kinases [52,53], the sites interacting with the purine ring might be localized in domains different from the sequence that interacts with the ribose and the phosphate group.

It is known that 5′-nucleotide PDEs cleave the phosphodiester bond through an intermediate step [54] in which a threonine is phosphorylated as a covalent intermediate in the catalytic mechanism [55]. How the phosphate moiety of the cyclic nucleotides interacts with the catalytic site of a cyclic nucleotide PDE is largely unknown [125]. In the center of a domain of seven residues conserved in

(A)

ratPDE1

ratPDE2

ratPDE3

ratPDE4

dunce PDE

CaM PDE

CGS PDE

ret PDE

sac PDE

(B)

```
                 *                                       *
         *    *   ***                  **  *     *                     **   *    *          *
         *    *   *** *          *              *   *             *    ** * ***  *        **
ratPDE1  QIPADTLLRYLLTLEGHYHSN-VAYHNSIHAADVVQSAHVLLGTPALEAVFTDLEVLAAIFACAIHDVDHPGVSNQFLINTNSELALMYNDS-
ratPDE2  H..V..MMM.M....D...AD-.....L...L.T...A....D.......I...L..A.........AL..................E-
ratPDE3  K..V...IT..M...D...AD-.....N.......T....S..........I.....S...................E-
ratPDE4  K.SS..FVT.MM...D....D-......L....A..T....S.....D.......I......A........................E-
dunce-PDE M..PK.F.NFMS...D..VKD-NPF...L......T..TN...N......G..P...GG.L..AC.......LT....V.SS.........E-
CaM-PDE  K..VSC.IAFAEA..VG.XKYKNP-.L......T.TV.YIMLGTGIMHWL.E..I..MV..A....YE.T.TT.N.H.QXR.DV.IL..R-
cGS-PDE  K.DCP..A.FC.MVKKG.-RD-IP...WM..FS.SHFCYL.YKNLE.TNYLE.M.IF.LFIS.MC..L..R.TN.S.QVASK.V..AL.SSEG
ret-PDE  H..QEA.V.FMYS.SKG.-R-RIT...WR.GFN.G.TMFS..V.GK.KRY.....A..MVT.AFC..I..R.TN.LYQMKSQNP..KLHG-.-
sac-PDE  L.ADNK...LL.F...SS..Q-VNKF..FR..I..M.ATWR.C-.YL.KDNPVQT---.LLCM.AIG...G...TN...L.C.CE..V.QNFKNV-
```

```
          *  *                                                    **
         ** *  *       **          *           *                ******                                   *
ratPDE1  LENHHLAVGFKLLQGENCDIFQNLSTKQKLSLRRMVIDMVLATDMSKHMSLLADLKTMVETKKVTSLGV---------LLLDNYSDRIQVLQ
ratPDE2  ..............E.........KR.RQ...K............T...........S..--------.........R
ratPDE3  ..............E.........TK..RQ...K.A..I............N............S..--------........R
ratPDE4  ..............E.H.........TK..RQT..K...........................S..--------.......T....R
dunce-PDE ........A.....NQG....C.MQK...RQT..K....I..S..............AGS.--------.......T.....E
CaM-PDE  .....VSAAYR.M.E.EMNVLI...KDDWRD..WL..E...S....G.FQQIKNIRNSLQ----QPE.L--------............KAKTMS
cGS-PDE  M.R...F.QAIAI.NTHG.N..DHF.R.DYQRMLDLMR.II....LAH.LRIFK..QK.A.--------.G--------YDRT.KQHHSLL.C
ret-PDE  ..R....EF.KT..RD.SLN......NRR.HEHAIH.MDIAII...LAL---YCKK-R..FQ-.I.DQSKTYETQQEWTQYMM..QTRKE.VMAM
sac-PDE  ...F.REL.QQ..SEH-WPLKLSI.K.KF----DFISEAI.....AL.SQYEDR--------------------.MHE.PMKQ.-T.I
```

```
              * *                            **       *            *
         *   *** **        **                  **    *           **    * *
ratPDE1  SL-VHCADLSNPAKPLPLYRQWTERIMAEFFQQGDRERESGLDI-SPMCDKHTAS-VEKSQV-GFIDYIAHPLWETWADLVHPDAQELLDTLEC
ratPDE2  NM-.........T...E......D...........AH....R.ME.-..................V.................DI....
ratPDE3  NM-.........T...Q.......D...E....R.......R.ME.-.........N.....-..........V.................DI.....
ratPDE4  NM-.........T.S.E.......D...E.........K...R.ME.-....................V..............Q....DI.....
dunce-PDE N.-...........T.......KR.VALL.E...L...K.....M...........R.N.T-I.........V............S....DI.....E
CaM-PDE  LI-L.A..I.H....SWXKLHR..MAL.E....L....K.A.L..PF-..L..RKSTM-.AQ..I-.....F.VE.----FSL.TDSTEKIIIP.IEE
cGS-PDE  L.-MTSC...DQT.-----------K...S..L.KAM.NRP-ME.M.REK.Y-IPEL-I-S.MEH..M.---IYKL.QDLFPKAA-ELYER
ret-PDE  MM--TAC...AIT..WEVQSKVALLVA...WE...L..TVLQQNPI..M.RNK.DELP.L.-.----FVCTFVYKEFSRF-.EEITPM..GITM
sac-PDE  ..IIKA..I...VTRT.SISAR.AYL.TL..NDCALL.TFHKAHR--.EQ.CFGD---Y.-N.DSPKEDLESIQNILVNVTD-..DI-IK.HPHI
```

ll nine PDE sequences available there is a threonine residue [37]. In addition, a serine is also a residue conserved in all PDEs in close proximity to the region of homology with RII [37] (Figure 10.5). These residues might play an important role in the enzyme interaction with the phosphate ring.

Conversely, the N-termini and the C-termini of these nine PDE sequences which were compared show little homology with each other [33–38,41,44], opening the possibility that these regions determine differences in the function of these proteins. For instance, it is possible that different regulatory domains are present in these two regions. This idea is in line with the finding that the CaM binding site is at the N-terminus of Ca^{2+}–CaM-dependent PDE [1]. It is also possible that sites for post-translational modification are present in these two regions. For instance, a possible site for phosphorylation by a cAMP-dependent protein kinase was found at the N-terminus of ratPDE1. It is, then, possible that the activity of a catalytic moiety conserved in the four groups of rat PDE clones is differentially modulated by post-translational modifications of the N- and C-termini. Another appealing possibility is that these different regions contain a signal for subcellular compartmentalization.

10.3   FUNCTION AND REGULATION OF THE cAMP-PDEs

Many reports demonstrate the role of high-affinity cAMP-PDEs in a wide variety of physiological regulatory processes. Substantial information comes from indirect studies using second-generation, selective cAMP-PDE inhibitors like rolipram and Ro 20-1724 [2]. Obviously these observations need to be confirmed

---

Figure 10.3   Homology between the deduced amino acid sequences of the four groups of the rat cDNA clones with the sequences of PDEs from different species. (A) Schematic representation of the coding region of the four groups of cDNA clones isolated from testicular libraries (ratPDE1, ratPDE2, ratPDE3, ratPDE4) [37,38], the *Drosophila* 'dunce' cAMP-PDE (dunce-PDE) [33], a CaM-sensitive PDE from bovine brain (CaM-PDE) [35], a cGMP-stimulated PDE from bovine heart (CGS-PDE) [35], the α subunit of the rhodopsin-sensitive cGMP-PDE from bovine retina (ret-PDE) [44] and a cAMP-PDE from *Saccharomyces cerevisiae* (sac-PDE) [34]. The region of homology compared in (B) is represented by black bars. Shaded bars indicate additional regions that are homologous. The other protein regions are represented by open bars and unknown regions are indicated by broken lines. The scale line is equal to 100 amino acid residues. (B) Comparison of the conserved domain of the proteins reported in (A). A dot in the sequence indicates identity of the residue with that of ratPDE1. Gaps (-) are introduced for alignment. Residues that are conserved [45] in all compared sequences are marked with an asterisk. Amino acid residues are represented by the single letters code. The following groupings were applied: A, S, T; L, I, V, M; D, E; N, Q; K, R, H; F, W, Y; C, P, G. Two asterisks indicate the identity of that residue in all the compared sequences

by more direct biochemical studies. Furthermore, on the basis of the presence of a number of highly related cAMP-PDEs, it will be necessary to define which member of the cAMP-PDE family of enzymes is involved in each of the processes described below. We will first summarize some general and theoretical aspects of the role of cAMP-PDE in intracellular cAMP regulation. We will then review data on the involvement of a cAMP-PDE regulation in biological processes like regulation of cellular responses, regulation of the entry and exit from the cell cycle, and regulation of gonadal and CNS functions.

10.3.1   Theoretical considerations

Inactivation of the second messenger cAMP occurs through degradation to 5'-AMP or extrusion from the cell. The cAMP degradation usually accounts for more than 80% of cAMP disposal [56,57]. Several techniques have been used to measure cAMP degradation in the intact cell by quantitating the fractional turnover constant (K_{ft}) or the decay constant (K_{decay}) and are reviewed in detail elsewhere [58–61]. These techniques have the obvious advantage of measuring the actual PDE activity in the cell, thus avoiding artifacts caused by cell homogenization. Unfortunately, this approach has only rarely been used to confirm that changes in PDE activity measured in cell extracts indeed occur in the intact cell in vivo.

Most cells contain both low- and high-affinity PDEs. This is reflected in complex kinetics of cAMP degradation that depend on the intracellular concentration of the cyclic nucleotide [62]. The difference in K_m between the high-affinity and the low-affinity PDEs is such that at low concentrations of cAMP (less than one μM) the low-K_m PDEs hydrolyze cAMP with velocities that are 1/4–1/2 maximal, while the high-K_m PDEs are practically operating with zero-order kinetics [62]. When intracellular cAMP increases from concentrations of 0.1 μM to 1–10 μM, the low-K_m PDEs rapidly reach the V_{max}, while the velocity of the hydrolysis by high-K_m PDE continues to increase. Under basal conditions or when hormones produce only minor increases in cAMP, the high-affinity enzymes then play a major role in cAMP disposal, while with high-cAMP concentrations the low-affinity enzymes take on the task of degrading the large amount of cAMP that accumulates under saturating hormone concentrations. Although the 10–50 μM K_m of the low-affinity PDEs would indicate that saturation of this PDE system rarely occurs in the intact cell, it has been reported that, in platelets stimulated with prostaglandins, cAMP hydrolysis reaches saturation at cAMP concentrations of approximately 10 μM [63]. On the basis of this finding, it has been proposed that the high-K_m activity is not active under those experimental conditions in the intact platelets [63]. Further experiments are obviously required to clarify this important issue.

A consequence of the presence of both low- and high-affinity PDEs is that cAMP degradation in the intact cell follows apparent negatively co-operative kinetics [62]. This negative co-operativity of cAMP degradation has been proposed to explain the often observed positive co-operativity in the responses [62] to hormones like ACTH [64], TSH [65] or gonadotropins [66]. Furthermore, the presence of low-K_m PDEs at concentrations that exceed those of high-K_m enzymes is reflected in several distinct features of the cellular cAMP response. These have been used as diagnostic tools to demonstrate the presence of active cAMP PDEs in the intact cell [67]. In a cell in which cAMP hydrolysis is predominantly catalyzed by a high-affinity cAMP-PDE, the concentration–response curve to low hormone doses is concave upward on a linear scale, indicating a positive co-operative behavior [62,67]. In cells containing high levels of low-K_m PDE, the cAMP decay does not follow first-order kinetics. In fact, the decay constant varies in relation with the intracellular cAMP concentration [62,67]. Finally, as a corollary of the above finding, the time-course of cAMP accumulation in a cell containing an active low-K_m degradation system is dependent on the hormone concentration [61,67], since the time required to reach cAMP steady-state levels is dependent on the rate of cAMP degradation.

Since these theoretical predictions have been confirmed by experimental observations [61,67], the concept that cAMP-PDEs have a major role in regulating intracellular cAMP is strengthened further. It should also be remembered that hormonal and neurotransmitter responses have usually been measured under receptor-saturating conditions. This does not reflect the in vivo conditions in which cells are exposed only to hormone concentrations that produce fractional occupancy. Thus, under more physiological conditions, changes in intracellular second messengers are minimal and in a range of concentrations in which high-affinity PDEs are playing a major role in cAMP degradation. This line of thinking is particularly true for trophic hormones, which exert their effects by minimal but sustained receptor occupancy (see Chapter 8).

A final point should be considered. It is readily understood that a change in PDE activity causes a change in cyclic nucleotide steady-state levels. However, this might not be the only necessary outcome. It is in fact possible that a change in PDE activity is reflected only in a change in turnover without a change in steady-state levels of the second messenger. An increased flux through the adenylate cyclase–PDE system might play an important role in determining the cell biological response. Although this possibility has been often laid out [63,68], the exact significance of an altered cyclic nucleotide flux is poorly understood. Goldberg and collaborators have shown that in the retina a change in cGMP turnover without changes in cGMP steady-state levels follows light activation [68]. It also has been shown that in a reconstituted system, protein kinase activation is more sensitive to cAMP when a CaM-dependent PDE is present in the reconstitution system [69]. Comparison with other signal transduction systems might help to shed some light on this phenomenon. For instance, changes in Ca^{2+} fluxes without changes in steady-state levels of the cation are thought to play an important role in cellular signaling [70].

10.3.2 Biochemical mechanisms of activation of the cAMP-PDE by hormones and neurotransmitters

More than a decade ago it was shown that several hormones and neurotransmitters increase the PDE activity of the target cell (see also Chapter 8). There are numerous reports showing that in cells which are responsive to epinephrine, such as fibroblasts [71,72], lymphoma cells [73] or C6 glioma cell lines [74–76], stimulation of intracellular cAMP leads to various degrees of activation of a soluble PDE. Stimulation of PDE has been reported also in vivo for epinephrine in the pineal gland [77]. Similar activation of PDE follows PGE1 stimulation of fibroblasts [71]. The gonadotropin FSH, whose action is mediated by cAMP, induces a large increase in PDE activity in the target cells of both the male [78–80] and the female [81] gonads. In addition, FSH injection in immature male rats produces an increase in testicular PDE activity in vivo [82]. Thus, PDE activation by hormones that increase intracellular cAMP appears to be a widespread phenomenon. Although there has been general agreement that a low-K_m PDE is activated by these hormones [71,72,74,78], until recently very little was known about the form of PDE involved in this activation or about the mechanism causing this stimulation in PDE activity. The time-course of the PDE activation varies between different systems. In some instances it has been shown that PDE activation, mostly particulate, is very rapid [83–86]. It is now clear that a mechanism by which a particulate low-K_m PDE is rapidly activated in the cell is by a cAMP-dependent phosphorylation [87–89] (Chapter 8). Thus, hormones that increase intracellular cAMP activate a cAMP-dependent protein kinase. This in turn phosphorylates a PDE, promoting its catalytic activity. This mechanism has been shown to operate for the particulate, cGMP-inhibited PDE [87–89]. Houslay and collaborators have shown that a peripheral cAMP-PDE that is not inhibited by cGMP is activated by insulin via a phosphorylation [90]. It is not clear whether this post-translational modification is the cause of activation also of other cGMP-insensitive, rolipram-inhibited cAMP-PDEs. In those systems in which this latter form of low-K_m PDE is likely to be involved, the activation has a much longer time-course. In C6 glioma cells, for instance, activation is maximal only after 3 h [74,75,91]. In the Sertoli cell, the activation of PDE requires 6–12 h to develop [78–80]. Furthermore, as shown in several systems, long-term hormonal activation of PDE is blocked if protein and RNA synthesis is inhibited [72,79,80,91]. The mechanism underlying this long-term activation of PDE has been recently clarified, when cloning of the cAMP-PDEs has provided the tools to measure PDE mRNA levels (Figure 10.4). In the Sertoli cell and in the C6 glioma cell line we have shown that hormones and cAMP increase the levels of transcripts corresponding to ratPDE3 [38] and to a much lesser extent of those corresponding to ratPDE4 (Conti and Swinnen, unpublished observation), which are low-K_m, rolipram-inhibited cAMP-PDEs. Since this induction of the mRNA parallels the increase in activity, it can be concluded that the PDE activation is dependent on increased mRNA levels and synthesis of this PDE. Preliminary data from our laboratory indicate also that activation of transcription of a gene rather than mRNA stabilization is the probable cause of the increase in mRNA steady-state levels. Whether this activation is the sole cause of increased PDE activity remains to be determined. It is, for instance, possible that post-translational modifications of this PDE, like a phosphorylation, are still necessary for the newly synthesized PDE to express its hydrolytic activity.

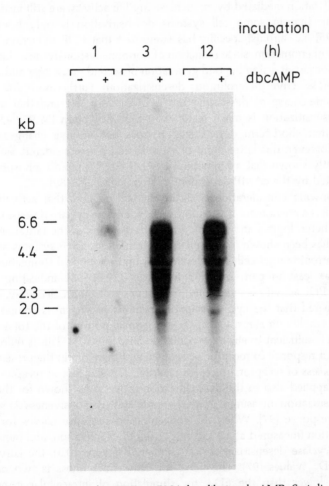

Figure 10.4 Regulation of ratPDE3 mRNA by dibutyryl-cAMP. Sertoli cell cultures were prepared as reported [38]. On the 4th day of culture, medium was removed and cells were incubated for the indicated times in the absence or presence of 0.5 mM dibutyryl-cAMP. At the end of the incubation, medium was removed and RNA extracted as previously detailed [38]. The Northern blot with 10 μg of total RNA was hybridized with a [^{32}P]-labeled ratPDE3 cDNA [38] and autoradiographed

10.3.3   Activation of the cAMP-PDE and cell desensitization

This demonstration that hormones which increase cAMP also stimulate PDE activity has supported the concept that the desensitization state that follows hormonal stimulation is due to changes in both synthesis and degradation of cAMP. However, while extensive investigations have been performed to convincingly link desensitization to an impaired cAMP synthesis [92,93], the data on desensitization mediated by an increase in PDE activity are still inconclusive. The finding that in some cell systems desensitization is only homologous [59,92,94,95], i.e. agonist-specific, has suggested that PDE induction is not a ubiquitous phenomenon, since reduction of hormone responsiveness due to PDE activation occurs at a step beyond generation of second messenger and is clearly agonist-non-selective (heterologous desensitization). Furthermore, the findings that the time-course of desensitization is often very fast and that adenylate cyclase desensitization is much faster than PDE activation [59,92,94,95] have indicated that short-term desensitization does not involve changes in PDE activity. However, this latter point of view should be reconsidered, since it has been recently shown that a high-affinity cAMP-PDE is rapidly phosphorylated and activated by the cAMP-dependent protein kinase [87–89].

The following considerations indicate, in our opinion, that activation of a cAMP-PDE is a phenomenon as important as adenylate cyclase desensitization in long-term heterologous and homologous refractoriness of the target cell.

First, it has been shown in several epinephrine-responsive systems as well as for gonadotropin target cells that desensitization to hormonal stimulation can be reversed, at least in part, by PDE inhibitors [59,96–98], indicating that an increase in PDE activity is indeed involved in desensitization. Secondly, it should be remembered that the cyclic nucleotide-responsive system is composed of a number of amplifying steps. Occupancy of a small portion of the total receptor population is sufficient to elicit a maximal reponse [99–101]. This is reflected in a high ratio of response to receptor occupancy and therefore in the presence of an apparent excess of receptors (receptor reserve). This concept of receptor reserve should be applied also to the desensitization process. As shown for the Sertoli cell, desensitization measured in terms of cAMP responsiveness is a highly amplified response [97]. When the concentration–response curves for cellular desensitization (measured as cAMP response), PDE activation, and induction of adenylate cyclase desensitization are compared (Figure 10.5), the curves have different ED_{50} values [97]. The ED_{50} for cell refractoriness is two orders of magnitude lower than the ED_{50} for stimulation of intracellular cAMP. The stimulation of the PDE activity occurs at slightly higher concentrations of FSH than refractoriness, and adenylate cyclase desensitization requires even higher doses. This finding is probably due to the fact that PDE induction is a phenomenon more amplified than adenylate cyclase desensitization. This is quite possible, because PDE induction requires activation of many steps in the cAMP-

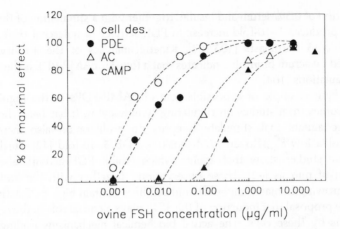

ovine FSH concentration (µg/ml)

Figure 10.5 Comparison of the FSH concentration–response curves for cell refractoriness (○), PDE stimulation (●), adenylate cyclase desensitization (△), and cAMP accumulation (▲) in the Sertoli cell. Cells were preincubated with graded FSH concentrations for 24 h, and the different FSH-dependent responses were measured as described [97]. FSH stimulation of cAMP accumulation during a 1 h period is reported for comparison. Data are reported as percentage of the maximal response [97]. (Redrawn from ref. 97.)

dependent pathway, including gene activation, while adenylate cyclase desensitization might be the result of only a phosphorylation step [93]. Therefore one can define a range of hormone concentrations at which desensitization occurs in the presence of minimal adenylate cyclase desensitization but in the presence of substantial PDE induction. This confirms that, under some conditions, induction of PDE might be the predominant cause of cellular refractoriness. It should also be mentioned that if cell responsiveness in desensitized cells is measured in the presence of IBMX (3-isobutyl-1-methylxanthine) as a PDE inhibitor, the desensitization curve is shifted to the right and overlaps the adenylate cyclase desensitization curve (Conti, unpublished observation). This indicates that one of the mechanisms that causes a decrease in cell responsiveness is removed, so that intact cell refractoriness closely matches adenylate cyclase desensitization.

10.3.4 Low-K_m PDEs and regulation of the cell cycle

Even though it has been a common finding that growth factors stimulate cell replication by mechanisms other than activation of the cAMP-dependent pathway, it has often been proposed that regulation of intracellular cAMP levels plays an important role in modulating the entry and exit from the cell cycle [102].

Several early reports studied the association of mitogenic activity and regulation of cAMP degradation and found that mitosis is usually associated with changes in PDE activity. Pledger and collaborators have reported that serum or a combination of 0.5% serum and insulin, together with a stimulation of the mitotic activity, produces 3–10-fold increase in PDE activity of quiescent or replicating BHK 21 c/13 fibroblasts [103,104]. Kinetic analyses of the activity that is stimulated by serum or insulin indicates that a low-K_m cAMP-PDE is the target of these regulations [104].

A striking example of a possible role of cAMP-PDEs in the regulation of mitosis comes from studies on circulating lymphocytes. It has been shown that mitogenic agents that stimulate lymphocyte proliferation also increase the activity of a low-K_m, Ro 20-1724-sensitive PDE 5–10-fold [42]. Furthermore, correlative studies show that agents which inhibit PDE activity also inhibit induction of mitosis [42]. These observations, together with the finding that rapidly growing human lymphoblastoid cell lines contain high PDE activity, has led to the proposal that induction of this PDE plays a crucial role in the regulation of mitosis by these cells. The actual biochemical mechanisms leading to this activation are unknown.

10.3.5   Regulation of cAMP-PDEs and gametogenesis

The regulations involved in the production of male and female gametes comprise complex hormonal interactions. The cyclic nucleotide-dependent pathway is thought to play a major role in these processes, but the biochemical mechanisms involved are poorly understood. That a PDE plays a crucial role in the regulation of gametogenesis is inferred from the findings on 'dunce' mutations of *Drosophila melanogaster* [105]. Mutations affecting the 'dunce' locus, which includes a gene coding for a low-K_m PDE, produce an altered expression of this cAMP-PDE and a phenotype characterized, among other alterations, by female fly sterility [105]. Our finding that cAMP-PDEs homologous to the 'dunce' PDE are expressed in mammalian germ cells [37] again suggests an involvement of a 'dunce'-like PDE in the regulation of gametogenesis.

One of the earliest events that occurs with the resumption of meiosis is a decrease in cAMP levels in amphibian and mammalian oocytes [106–110]. Although regulation of adenylate cyclase and cyclic nucleotide synthesis plays a role in this action [111–113], it has also been shown that in *Xenopus* oocytes activation of a PDE is probably a primary cause of this transient decrease in intracellular cAMP [114]. In support of this, it has long been known that low concentrations of PDE inhibitors block the resumption of meiosis in rat and mouse denuded oocytes [107]. The characteristics of the PDE involved in this regulation are not known. The observation that papaverine is a potent inhibitor of both oocyte PDE and of the resumption of meiosis argues in favor of a regulation of a low-K_m PDE [114], similar to the rolipram-inhibited PDE. On the other hand, since insulin and IGF-1 also induce germinal vesicle breakdown and meiotic resumption and activate the xenopus oocyte PDE, it is possible that this low-K_m PDE is a cGMP-inhibited, rolipram-insensitive PDE [114]. The characterization of this enzyme will be a crucial issue to resolve, also in light of the finding that the activated protooncogene c-*ras* has been shown to activate the oocyte PDE [115] and to induce meiotic resumption [115,116]. Thus, further clarification of the mechanisms of activation of this PDE in the oocytes might shed some light on the physiological function of the ras family of protooncogenes.

Further support of a role of a cAMP-PDE in the regulation of the gametogenic process comes from our studies in the male rat. Maturing germ cells express a complex pattern of PDEs, including several cAMP-PDEs [117]. Transcripts corresponding to ratPDE1 and ratPDE2 are, in the rat testis, present predominantly in the germ line [37]. Furthermore, ratPDE1 appears not to be a PDE present in all tissues, because transcripts corresponding to this enzyme are, in the several organs tested, expressed only in germ cells of the testis, and transcripts of different size are present in the kidney. More interestingly, transcripts corresponding to this gene are expressed mainly during the meiotic prophase of spermatogenesis [37]. This selective regulation of a gene corresponding to a cAMP-PDE suggests that PDE regulation and the consequent changes in cAMP turnover are finely regulated during gametogenesis. It should also be re-emphasized that ratPDE1 is highly homologous to the *Drosophila* 'dunce' PDE, and mutations affecting this enzyme disrupt fly fertility [105].

Regulation of a cAMP-PDE has also been reported in the somatic cell involved in the regulation of gamete production. A cAMP-PDE is regulated by FSH both in the male Sertoli cell [80] and in the female granulosa cell [81]. That this regulation is a physiological phenomenon is shown by studies on the follicular development in the ovary [118]. It has been reported that a high affinity cAMP-PDE undergoes a cycle of activation that is synchronous with the estrous cycle of the rat [118]. Similarly, the cAMP-PDE that is regulated by FSH increases in vivo during puberty [117] and its induction correlates with the decrease in responsiveness of the Sertoli cell at puberty [119].

Thus, the above data indicate that regulation of high-affinity PDEs occurs during gonadal and gamete development. Further understanding of the role of this PDE regulation in gonadal function could provide some clues to the exact role of cAMP-PDEs in physiologically relevant processes.

10.3.6 Other physiological processes involving a regulation of cAMP-PDEs

A number of other physiological processes have been shown to be affected by second-generation, cAMP-PDE-specific PDE inhibitors [2]. For instance, rolipram, a potent inhibitor of brain PDEs [120], has been shown to markedly stimulate central nervous system functions, including memory formation [121–123]. This obviously must be related to the *Drosophila* 'dunce' mutations in which altered PDE expression affects the fly learning behaviour [105]. This topic will be extensively reviewed elsewhere. Similarly, PDE inhibitors of the rolipram and Ro

20-1724 family have been shown to affect bronchiolar constriction and mast cell histamine release [2,124], but the form of PDE involved is still to be characterized.

ACKNOWLEDGEMENTS

The authors wish to thank Dr R. L. Davis for providing the sequence of ratPDE2 and Dr M. Wigler for making available his manuscript on PDE cloning. The support of Dr Frank French and of the members of the Laboratories for Reproductive Biology during our studies on the characterization of the testicular cAMP-PDEs is also acknowledged. The work from the authors' laboratory was supported by NIH grants HD20788 and 5-P30-HD18968 and by a grant from the Andrew W. Mellon Foundation.

REFERENCES

1. Beavo, J. A. (1988) *Adv. Second Messenger Phosphoprotein Res.*, **22**, pp. 1–38.
2. Weishaar, R. E., Cain, M. H., and Bristol, J. A. (1985) *J. Med. Chem.*, **28**, 537–545.
3. Thompson, W. J., Pratt, M. L., and Strada, S. J. (1984) *Adv. Cyclic Nucleotide Protein Phosphorylation Res.*, **16**, 137–148.
4. Kincaid, R. L., and Manganiello, V. C. (1988) *Methods Enzymol.*, **159**, 457–470.
5. Strada, S. J., and Thompson, W. J. (1984) *Adv.Cyclic Nucleotide Protein Phosphorylation Res.*, **16**, VI.
6. Weber, H. W., and Appleman, M. M. (1982) *J. Biol. Chem.*, **257**, 5339–5341.
7. Hidaka, H., and Endo, T. (1984) *Adv. Cyclic Nucleotide Protein Phosphorylation Res.*, **16**, 245–259.
8. Thompson, W. J., Epstein, P. M., and Strada, S. J. (1979) *Biochemistry*, **18**, 5228–5237.
9. Epstein, P. M., Strada, S. J., Sarada, K., and Thompson, W. J. (1982) *Arch. Biochem. Biophys.*, **218**, 119–133.
10. Moore, J. B., and Schroedter, D. E. (1982) *Arch. Biochem. Biophys.*, **213**, 276–287.
11. Sharma, R. K., Wang, T. H., Wirch, E., and Wang, J. H. (1980) *J. Biol. Chem.*, **255**, 5916–5923.
12. Kincaid, R. L., and Vaughan, M. (1983) *Biochemistry*, **22**, 826–830.
13. Hansen, R. S., and Beavo, J. A. (1986) *J. Biol. Chem.*, **261**, 14636–14645.
14. Harrison, S. A., Reifsnyder, D. H., Gallis, B., Cadd, G. G., and Beavo, J. A. (1986) *Mol. Pharmacol.*, **29**, 506–514.
15. Degerman, E., Belfrage, P., Newman, A. H., Rice, K. C., and Manganiello, V. C. (1987) *J. Biol. Chem.*, **262**, 5797–5807.
16. Francis, S. H., and Corbin, J. D. (1988) *Methods Enzymol.*, **159**, 722–729.
17. Fougier, S., Nemoz, G., Prigent, A. F., Marivet, M., Bourguignon, J. J., Wermuth, C., and Pacheco, H. (1986) *Biochem. Biophys. Res. Commun.*, **138**, 205–214.
18. Conti, M. (1990) *Biochem. J.* (submitted).

19. Thompson, W. J., Shen, C. C., and Strada, S. J. (1988) *Methods Enzymol.*, **159**, 760–766.
20. Weshaar, R. E., Kobylarz-Singer, D. C., and Kaplan, H. R. (1987) *J. Mol. Cell. Cardiol.*, **19**, 1025–1036.
21. Thompson, W. J., Ross, C. P., Pledger, W. J., Strada, S. J., Banner, R. L., and Hersch, E. M. (1976) *J. Biol. Chem.*, **251**, 4922–4929.
22. Epstein, P. M., Mills, J. S., Hersh, E. M., Strada, S. J., and Thompson, W. J. (1980) *Cancer Res.*, **40**, 379–386.
23. Onali, P., Strada, S. J., Chang, L., Epstein, P. M., Hersh, E. M., and Thompson, W. J. (1985) *Cancer Res.*, **45**, 1384–1391.
24. Reeves, M. L., Leigh, B. K., and England, P. J. (1987) *Biochem. J.*, **241**, 535–541.
25. De Mazancourt, P., and Guidicelli, Y. (1988) *Methods Enzymol.*, **159**, 766–772.
26. Davis, C. W. (1984) *Biochim. Biophys. Acta*, **797**, 354–362.
27. Marchmont, R. J., Ayad, S. R., and Houslay, M. D. (1981) *Biochem. J.*, **195**, 645–652.
28. Pyne, N., Cooper, M. E., and Houslay M. D. (1987) *Biochem. J.*, **242**, 33–42.
29. Onali, P., Schwartz, J. P., Hanbauer, I., and Costa, E. (1981) Biochem. Biophys. Acta, **657**, 285–292.
30. Brothers, V. M., Walker, N., and Bourne, H. R. (1982) *J. Biol. Chem.*, **257**, 9349–9355.
31. Groppi, V. E., Steinberg, F., Kaslow, H. R., Walker, N., and Bourne, H. R. (1983) *J. Biol. Chem.*, **258**, 9717–9723.
32. Kithas, P. A., Artman, M., Thompson, W. J., and Strada, S. J. (1988) *Circulation Res.*, **62**, 782–789.
33. Chen, C. N., Denome, S., and Davis, R. L. (1986) *Proc. Natl Acad. Sci. USA*, **83**, 9313–9317.
34. Sass, P., Field, J., Nikawa, J., Toda, T., and Wigler, M. (1986) *Proc. Natl Acad. Sci. USA*, **83**, 9308–9312.
35. Charbonneau, H., Beier, N., Walsh, K. A., and Beavo, J. A. (1986) *Proc. Natl Acad. Sci. USA*, **83**, 9308–9312.
36. Davis, R. L., Takayasu, H., Eberwine, M., and Myres, J. (1989) *Proc. Natl Acad. Sci. USA*, **86**, 3604–3608.
37. Swinnen, J. V., Joseph, D. R., and Conti, M. (1989) *Proc. Natl Acad. Sci. USA*, **86**, 5325–5329.
38. Swinnen, J. V., Joseph, D. R., and Conti, M. (1989) *Proc. Natl Acad. Sci. USA*, **86**, 8197–8201.
39. Nielsen, L. D., Monard, D., and Rickenberg, H. V. (1973) *J. Bacteriol.*, **116**, 857–866.
40. Okabashi, T., and Ide, M. (1970) *Biochim. Biophys. Acta*, **220**, 116–123.
41. Colicelli, J., Birchmeier, C., Michaeli, T., O'Neill, K., Riggs, M., and Wigler, M. (1989) *Proc. Natl Acad. Sci. USA*, **86**, 3599–3603.
42. Epstein, P. M., and Hachisu, R. (1984) *Adv. Cyclic Nucleotide Protein Phosphorylation Res.*, **16**, 303–323.
43. Klee, C. B., Crouch, T. H., and Krinks, M. H. (1979) *Biochemistry*, **18**, 722–729.
44. Ovchinnikov, Yu. A., Gubanov, V. V., Khramtsov, N. V., Ischenko, K. A., Zagranichny, V. E., Muradov, K. G., Shuvaeva, T. M., and Lipkin, V. M. (1987) *FEBS Lett.*, **223**, 169–173.
45. Dayhoff, M. O., Schwartz, R. M., and Orcutt, B. C. (1978) In *Atlas of Protein Sequence and Structure*, Vol. 5, Suppl. 3, National Biochemical Research Foundation, pp. 235–352.

46. Lacombe, M., Podgorski, G. J., Franke, J., and Kessin, R. H. (1986) *J. Biol. Chem.*, **261**, 16811–16817.

47. Scott, J. D., Glaccum, M. B., Zoller, M. J., Uhler, M. D., Helfman, D. M., McKnight, G. S., and Krebs, E. G. (1987) *Proc. Natl Acad. Sci. USA*, **84**, 5192–5196.

48. Charbonneau, H. C., Novack, J., MacFarland, R., Walsh, K., and Beavo, J. A. (1987) In *Fourth International Symposium on Calcium and Calcium Binding Proteins*, Academic Press, pp. 151–176.

49. Erhardt, P. W., Hagedorn, A. A., III, and Sabio, M. (1988) *Mol. Pharmacol.*, **33**, 1–13.

50. McKay, D. B., and Stetz, T. A. (1981) *Nature*, **290**, 744–746.

51. McKay, D. B., Weber, I. T., and Steitz, T. A. (1982) *J. Biol. Chem.*, **257**, 9518–9522.

52. Bubis, J., Neitzel, J. J., Saraswat, L. D., and Taylor, S. S. (1988) *J. Biol. Chem.*, **263**, 9668–9673.

53. Weber, I. T., Steitz, T. A., Bubis, J., and Taylor, S. S. (1987) *Biochemistry*, **26**, 343–351.

54. Kelly, S. J., and Butler, L. G. (1977) *Biochemistry*, **16**, 1102–1104.

55. Culp, J. S., Blytt, H. J., Hermodson, M., and Butler, L. G. (1985) *J. Biol. Chem.*, **260**, 8320–8324.

56. Plagemann, P. G. W., and Erbe, J. (1977) *J. Biol. Chem.*, **252**, 2010–2016.

57. Brunton, L. L., and Mayer, S. E. (1979) *J. Biol. Chem.*, **254**, 9714–9720.

58. Van Sande J., and Dumont, J. E. (1975) *Mol. Cell. Endocrinol.*, **2**, 289–301.

59. Su, Y. F., Cubeddu-Ximenes, L., and Perkins, J. P. (1976) *J. Cyclic Nucleotide Res.*, **2**, 257–270.

60. Barber, R., Clark, R. B., Kelly, L. A., and Butcher, R. W. (1978) *Adv. Cyclic Nucleotide Res.*, **9**, 507–516.

61. Butcher, R. W. (1984) *Adv. Cyclic Nucleotide Protein Phosphorylation Res.*, **16**, 1–12.

62. Erneux, C., Boeynaems, J.-M., and Dumont, E. J. (1980) *Biochem. J.*, **192**, 241–246.

63. Goldberg, N. D., Walseth, T. F., Eide, S. J., Krick, T. P., Kuehn, B. L., and Gander, J. E. (1984) *Adv. Cyclic Nucleotide Protein Phosphorylation Res.*, **16**, 363–379.

64. Rodbard, D. (1974) *Endocrinology*, **94**, 1427–1437.

65. Boeynaems, J. M., Van Sande, J., Pochet, R., and Dumont, J. E. (1974) *Mol. Cell. Endocrinol.*, **1**, 139–155.

66. Rodbard, D., Moyle, W., and Ramanchandran, J. *Current Topics Mol. Endocrinol.*, **1**, 79–87.

67. Barber, R., Goka, T. J., and Butcher, R. W. (1988) *Mol. Pharmacol.*, **32**, 753–759.

68. Goldberg, N. D., Ames, A., Gander, J. E., and Walseth, T. F. (1983) *J. Biol. Chem.*, **258**, 9213–9219.

69. Leiser, M., Fleisher, N., and Erlichman, J. (1986) *J. Biol. Chem.*, **261**, 15486–15490.

70. Alkon, D. L., and Rasmussen, H. (1988) *Science*, **239**, 998–1005.

71. Manganiello, V., and Vaughan, M. (1972) *Proc. Natl Acad. Sci. USA*, **69**, 269–273.

72. D'Armiento, M., Johnson, G. S., and Pastan, I. (1972) *Proc. Natl Acad. Sci. USA*, **69**, 459–462.

73. Bourne, H. R., Tomkins, G. M., and Dion, S. (1973) *Science*, **181**, 952–954.

74. Uzunov, P., Shein, H. M., and Weiss, B. (1973) *Science*, **180**, 304–306.

75. Schwartz, J. P., and Passonneau, J. V. (1974) *Proc. Natl Acad. Sci. USA*, **71**, 3844–3848.

76. Browning, E. T., Brostrom, C. O., and Groppi, V. E. (1976) *Mol. Pharmacol.*, **12**, 32–40.

77. Oleshansky, M. A., and Neff, N. H. (1975) *Mol. Pharmacol.*, **11**, 552–557.
78. Conti, M., Geremia, R., Adamo, S., Stefanini, M. (1981) *Biochem Biophys. Res. Commun.*, **98**, 1044–1050.
79. Verhoeven, G., Cailleau, J. and de Moor, P. (1981) *Mol. Cell. Endocrinol.*, **24**, 41–52.
80. Conti, M., Toscano, M. V., Petrelli, L., Geremia, R., and Stefanini, M. (1982) *Endocrinology*, **110**, 1189–1196.
81. Conti, M., Kasson, B. G., and Hsueh, A. J. W. (1984) *Endocrinology*, **114**, 2361–2368.
82. Conti, M., Toscano, M. V., Geremia, R., and Stefanini, M. (1983) *Mol. Cell. Endocrinol.*, **29**, 79–89.
83. Zinman, B., and Hollemberg, C. H. (1974) *J. Biol. Chem.*, **249**, 2182–2187.
84. Loten, E. G., Assimacopoulos-Jeannet, F. D., Exton, J. H., and Park, C. R. (1978) *J. Biol. Chem.*, **253**, 746–757.
85. Makin, H., and Kono, T. (1980) *J. Biol. Chem.*, **255**, 7850–7854.
86. Alvarez, R., Taylor, A., Fazzari, J. J., and Jacobs, J. R. (1981) *Mol. Pharmacol.*, **20**, 302–309.
87. Macphee, C. H., Reifsnyder, D. H., Moore, T. A., Lerea, K. M., and Beavo, J. A. (1988) *J. Biol. Chem.*, **263**, 10353–10358.
88. Gettys, T. W., Vine, A. J., Simonds, M. F., and Corbin, J. D. (1988) *J. Biol. Chem.*, **263**, 10359–10363.
89. Grant, P. G., Mannarino, A. F., and Colman, R. W. (1988) *Proc. Natl Acad. Sci. USA*, **85**, 9071–9075.
90. Marchnmont, R. J., and Houslay, M. D. (1980) *Nature*, **286**, 904–906.
91. Schwartz, J. P., and Onali, P. (1984) *Adv. Cyclic Nucleotide Protein Phosphorylation Res*, **16**, 195–203.
92. Harden, T. K. (1983) *Pharmacol. Rev.*, **35**, 5–32.
93. Sibley, D. R., Benovic, J. L., Caron, M. G., and Lefkowitz, R. J. (1987) *Cell*, **48**, 913–922.
94. De-Vellis, J., and Brooker, G. (1974) *Science*, **186**, 1221–1223.
95. Su, Y. F., Johnson, G. L., Cubeddu-Ximenes, L., Leichtling, B. H., Ortmann, R., and Perkins, J. P. (1976) *J. Cyclic Nucleotide Res.*, **2**, 271–285.
96. Browning, E. T., Schwartz, J. P., and Breckenridge, B. M. (1974) *Mol. Pharmacol.*, **10**, 162–174.
97. Conti, M., Toscano, M. V., Petrelli, L., Geremia, R., and Stefanini, M. (1983) *Endocrinology*, **113**, 1845–1853.
98. Conti, M., Monaco, L., Geremia, R., and Stefanini, M. (1986) *Endocrinology*, **118**, 901–908.
99. Loeb, J. N., and Strickland, S. (1987) *Mol. Endocrinol.*, **1**, 75–82.
100. Catt, K. J., and Dufau, M. L. (1973) *Nature*, **244**, 219–221.
101. Moyle, W. R., King, Y. C., and Ramachandran, J. (1973) *J. Biol. Chem.*, **248**, 2409–2417.
102. Rozengurt, E. (1986) *Science*, **234**, 161–166.
103. Pledger, W. J., Thompson, W. J., and Strada, S. J. (1975) *Nature*, **256**, 729–731.
104. Pledger, W. J., Thompson, W. J., Epstein, P. M., and Strada, S. J. (1979) *J. Cell. Physiol.*, **100**, 497–509.
105. Saltz, H. K., Davis, R. L., and Kiger, J. A. (1982) *Genetics*, **100**, 587–596.
106. Speaker, M. G., and Butcher, F. R. (1977) *Nature*, **267**, 848–849.
107. Magnusson, C., and Hillensjo, T. (1977) *J. Exp. Zool.*, **201**, 139–147.

108. Cho, W. K., Stern, S., and Biggers, J. D. (1974) *J. Exp. Zool.*, **187**, 383–386.
109. Dekel, N., and Beers, W. M. (1978) *Proc. Natl Acad. Sci. USA*, **75**, 4369–4373.
110. Sadler, S. E., and Maller, J. L. (1981) *J. Biol. Chem.*, **256**, 6367–6373.
111. Finidori-Lepicard, J., Schorderet-Slatkine, S., Hanoune, J., and Baulieau, E. E. (1981) *Nature*, **292**, 255–257.
112. Miot, F., and Erneaux, C. (1982) *Biochim. Biophys. Acta*, **701**, 253–259.
113. Maller, J. L. (1983) *Adv. Cyclic Nucleotide Res.*, **15**, 295–336.
114. Sadler, S. E., and Maller, J. L. (1987) *J. Biol. Chem.*, **262**, 10644–10650.
115. Sadler, S. E., and Maller, J. L. (1989) *J. Biol. Chem.*, **264**, 856–861.
116. Birchmeier, C., Broek, D., and Wigler, M. (1985) *Cell*, **43**, 615–621.
117. Geremia, R., Rossi, P., Pezzotti, R., and Conti, M. (1982) *Mol. Cell. Endocrinol.*, **28**, 37–53.
118. Schmidtke, J., Meyer, H., and Epplen, J. T. (1980) *Acta Embryol. Exp.*, **95**, 404–413.
119. Steinberger, A., Hintz, M., and Heindel, J. J. (1978) *Biol. Reprod.*, **19**, 566–577.
120. Schwabe, U., Miyake, M., Ohga, Y., and Daly, J. W. (1976) *Mol. Pharmacol.*, **12**, 900–910.
121. Mizokawa, T., Kimura, K., Ikoma, Y., Hara, K., Oshino, N., Yamamoto, T., and Ueki, S. (1988) *Japan. J. Pharmacol.*, **48**, 357–364.
122. Horowski, R., and Sastre-Y-Hernandez, M. (1985) *Current Ther. Res.*, **38**, 23–29.
124. Frossard, N., Landry, Y., Pauli, G., and Ruckstuhl, M. (1981) *Br. J. Pharmacol.*, **73**, 933–938.
125. Van Haastert, P. M. J., Dijkgraaf, P. A. M., Konijn, T. M., Abbad, E. G., Petridis, G., and Jastorff, B. (1983) *Eur. J. Biochem.*, **131**, 659–666.

11

STRUCTURE–FUNCTION RELATIONSHIPS AMONG CYCLIC NUCLEOTIDE PHOSPHODIESTERASES

Harry Charbonneau

Department of Biochemistry, SJ-70, University of Washington, Seattle, WA 98195, USA

11.1 INTRODUCTION

Multiple intracellular enzymes account for the total cyclic nucleotide phospho-diesterase (PDE) activity measured in animal cell extracts. Over the past 10 years, much effort in this field has been focused on identifying and distinguishing between these isoenzymes [1–4]. Using a combination of conventional biochemical methods, immunological techniques, and pharmacologic reagents it has been possible to distinguish among these enzymes and to categorize them into several different multienzyme families [1,5]. More recently, molecular genetic approaches have been used effectively to identify several isoenzymes [6–11,12]. The PDE isoenzymes identified thus far can be distinguished by their size, substrate preference, sensitivity to drugs, amino acid sequence, and mode of regulation [1–5].

One remarkable feature of the PDE isoenzymes is the multiplicity of regulatory mechanisms that have evolved to modulate this hydrolysis reaction [1–4]. The activity of cGMP-stimulated PDE isoenzymes is modulated by allosteric site(s) for cyclic nucleotides [13–18]. The photoreceptor PDEs [19,20] and the cGMP-binding PDEs [21–24] have binding sites with physiologically relevant affinities for cGMP, and even though no direct effects on activity have been demonstrated to date, it is likely that cGMP binding serves a regulatory function. Several isoenzymes are regulated by the binding of small intracellular proteins (e.g. calmodulin (CaM), transducin). The binding of the Ca^{2+}–CaM complex activates multiple CaM-dependent isoenzymes [25–29], whereas the interaction of the photoreceptor PDEs with the α subunit of transducin [30], with bound GTP, provides a mechanism whereby the hydrolysis of cGMP is increased

Cyclic Nucleotide Phosphodiesterases: Structure, Regulation and Drug Action
Edited by J. Beavo and M. D. Houslay   © 1990 John Wiley and Sons Ltd

Table 11.1 Cyclic nucleotide PDE sequences

Isoenzyme family	Source	Molecular weight[a]	Sequencing method	Extent of sequence	Isoenzyme class[b]	Reference
CaM-dependent						
61 kDa CaM-PDE	Bovine brain	61 716	Protein	Complete	Class I	[42,43]
59 kDa CaM-PDE	Bovine heart	~59 000	Protein	Partial	Class I	[50,51]
63 kDa CaM-PDE	Bovine brain	~63 000	Protein	Partial	Class I	[50]
Photoreceptor						
α subunit	Bovine rod	99 120	DNA	Complete	Class I	[45]
β subunit	Bovine rod	~84 000	DNA	Partial	Class I	[45]
γ subunit	Bovine rod	9 700	DNA	Complete	Class I	[52]
α' subunit	Bovine cone	98 850	DNA/Protein	Complete	Class I	[46,47]
cGMP-stimulated						
cGS-PDE	Bovine heart	102 920	Protein/DNA	Complete	Class I	[42,44]
Low-K_m, cAMP-specific						
dunce gene product						
dunce homolog 1[c]	Drosophila	64 780	DNA	Complete	Class I	[6]
dunce homolog 2	Rat	68 172	DNA	Complete	Class I	[9]
dunce homolog 3	Rat	59 460	DNA	Complete	Class I	[10]
dunce homolog 4	Rat	>47 000	DNA	Partial	Class I	[11,49]
	Rat	67 000	DNA	Partial	Class I	[11,49]
Yeast PDE						
PDE1	S. cerevisiae	42 056	DNA	Complete	Class II	[7]
PDE2	S. cerevisiae	60 900	DNA	Complete	Class I	[8]
Extracellular PDE	Dictyostelium discoideum	51 078	DNA	Complete	Class II	[48]

[a] If the sequence is complete, the monomeric molecular weight is calculated from the amino acid sequence; if incomplete, the value given is based upon SDS-PAGE, gel filtration analyses, or other estimates.

[b] PDE sequences can be divided into two classes, I and II, on the basis of the structural relationship among them. Proteins of the same class are homologous to one another (see text).

[c] Genes encoding four distinct rat dunce homologs have been reported thus far and no consistent nomenclature has been developed for distinguishing among them. Homologs 1 and 2 refer to the rat dnc-1 and DPD cDNAs cloned by Davis et al. [9] and Colicelli

when light is absorbed by rhodopsin. The large subunits of the photoreceptor PDEs also bind smaller regulatory subunits [20,31,32], e.g. an 11 kDa inhibitory γ subunit of the rod PDE. Several isoenzymes are phosphorylated in vitro by various protein kinases [33–38]; the cGMP-inhibited PDE of bovine platelets is phosphorylated in vivo [39–41].

Some of these isoenzymes have been purified and characterized as described in other chapters in this book. In several cases, their amino acid sequences have been obtained by direct protein sequencing or deduced from the nucleotide sequence of cDNA clones. To date, there are partial or complete amino acid sequences for 15 distinct catalytic subunits of PDE isoenzymes (Table 11.1). The availability of these sequence data, the purified proteins, and efficient systems to express the corresponding genes now make it possible to address questions regarding structural relationships among the isoenzymes, to gain insights into their evolutionary history, and to analyze structure–function relationships.

The relatively large size of these isoenzymes and the complexity of their regulatory features suggest that most PDEs are composed of multiple domains, as discussed later in this chapter. The multidomain character and the diversity of regulatory features provide intriguing questions at the molecular level and at the level of control of intracellular cyclic nucleotide metabolism. Thus this family of isoenzymes affords an excellent opportunity to learn more about the structure and organization of multidomain proteins, to investigate the interactions between catalytic and regulatory domains, and to obtain insights into the evolution of multidomain proteins.

This chapter describes the structural relationships between PDE isoenzymes and summarizes the progress that has been made toward understanding their structure, function, and regulation. Recent advances include the identification and delineation of several functional domains. This has been achieved both through clues derived from the comparison of PDE sequences, which provided valuable hints as to the location and function of domains, and by limited proteolysis, an established technique for identifying domains. Much remains to be done in efforts to probe individual domains in detail and to identify specific residues involved in domain function.

11.2 STRUCTURAL RELATIONSHIPS AMONG PDE ISOENZYMES

The homology among different PDE isoenzymes was first demonstrated by comparisons of partial amino acid sequence data from two mammalian isoenzymes [42], the cGMP-stimulated PDE (cGS-PDE) and the 61 kDa CaM-dependent PDE (CaM-PDE), and the predicted amino acid sequences encoded by the yeast [7], *S. cerevisiae*, PDE2 and the *Drosophila* dunce gene [6]. All four proteins were structurally related and had partial sequence similarities that were restricted to a single segment (250–270 residues) of each sequence [42]. Recently, the amino acid sequence of the 61 kDa CaM-PDE [43] has been completed and all but a few residues at the N-terminus of the cGS-PDE have been determined by combinations of analyses of the protein or corresponding partial cDNA clone [44]. These complete sequences corroborate and extend the structural relationships that were originally proposed with partial sequences [42].

Since these studies, the sequences of several different PDE isoenzymes have been deduced from the nucleotide sequence of cDNA clones (see Table 11.1); these include the α subunit of photoreceptor PDE (ROS-PDE) from bovine rods [45], the α' subunit of the photoreceptor PDE of bovine cones [46,47], an excreted PDE from *Dictyostelium discoideum* [48], and four homologs of the dunce gene from rat tissues [9–11,49]. A second PDE gene from yeast (PDE1) has been isolated and sequenced [8]. In addition, partial amino acid sequence data are now available for the 59 kDa CaM-PDE of bovine heart [50,51], 63 kDa CaM-PDE of bovine brain [50], and the β subunit of the ROS-PDE [45].

Comparisons utilizing these sequences (Table 11.1) show that it is now possible to divide PDE isoenzymes into two separate groups or families on the basis of their structural relationships. One group of isoenzymes is composed of those PDEs bearing the 250 residue conserved segment first detected in the dunce PDE, yeast PDE2, 61 kDA CaM-PDE, and cGS-PDE [42]. This group of isoenzymes, designated herein as class I for the purpose of this discussion, includes all those listed in Table 11.1 with the exception of the yeast PDE1 gene product [8] and the *Dictyostelium* PDE [48]. The presence of this homologous segment suggests an evolutionary relationship with at least a portion of each molecule arising from a common progenitor.

The second group of isoenzymes, class II, includes the yeast PDE1 gene product and the *Dictyostelium* PDE which are homologous to each other but are unrelated to the class I PDE isoenzymes [8]. If chemically conserved substitutions are counted, the extent of similarity between the enzymes is 40%. Unlike the class I isoenzymes, the sequence similarity between the class II enzymes extends throughout both molecules. The absence of homology between class I and II PDEs strongly suggests but does not prove separate evolutionary origins, since it is possible that the two groups have diverged rapidly to a point where a structural relationship is no longer detectable. However, the absence of significant sequence similarity indicates that the two classes of isoenzymes may fold to form different tertiary structures and therefore may employ fundamentally distinct mechanisms for binding and hydrolyzing cyclic nucleotides. One mechanistic distinction between class I and II enzymes may involve the role of metals in catalysis, since the yeast PDE1 gene encodes a high K_m, enzyme that is a metalloprotein, containing two atoms of zinc per subunit [53]. At present, there is insufficient evidence to conclude that the presence of metals is a characteristic feature of class II PDEs. However, the possible involvement of metals in the *Dictyostelium* PDE as well as in the class I isoenzymes is an issue that merits further study. There may also be general differences in the cellular localization of the class I and II enzymes. The class II, *Dictyostelium* enzyme is secreted into the medium or attached to the outer surface of the cell, while all class I enzymes appear to be located intracellularly with no evidence that any are secreted. Although it is not known whether the yeast PDE1 gene is also secreted, it is possible that the class II PDEs are secreted forms that function extracellularly and possess unique structural features (e.g. metal cofactors). Thus far, no vertebrate PDE sequences fall into the class II family.

The complex structural relationship among the class I PDE isoenzymes is illustrated in Figure 11.1. The common conserved feature present in all these enzymes is the aforementioned 250 residue segment which is located within the C-terminal half of each molecule. This homology is demonstrated by the alignment of 10 different class I isoenzymes shown in Figure 11.2. The statistical

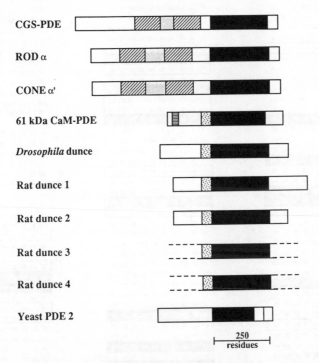

Figure 11.1 Diagram illustrating the structural relationship among class I PDE isoenzymes. The homologous, catalytic domain, present in each isoenzyme, is shown as a solid black box. Boxes with diagonal lines designate the two internally homologous subdomains of the conserved cGMP-binding domain, whereas the intervening shaded boxes are homologous to one another but not to the two internal repeats. The location of the CaM binding site within the 61 kDa CaM-PDE is designated by a box with horizontal lines. The stippled area shows the location of residues from the catalytic domains of the CaM-PDEs, *Drosophila* dunce PDE, and rat dunce homologs that are also homologous to one another but not to the other isoenzymes. Broken lines indicate regions where the length of segments are not known since complete sequences are unavailable

Positions (first block): (252) (219) (221) (281) (592) (592) (296)

Positions (second block): (311) (278) (280) (340) (650) (650) (354)

Positions (third block): (356) (334) (336) (396) (710) (710) (391)

Row labels: CAM: RD1: RD2: RD3: RD4: DUN: ROS: CON: CGS: YE2:

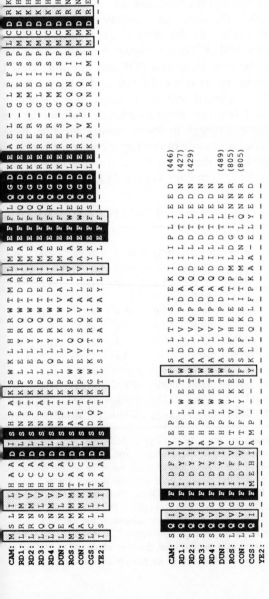

Figure 11.2 Alignment of the conserved catalytic domains from class I PDE isoenzymes. Residues identical in all isoenzymes aligned are enclosed in black boxes; shaded boxes designate positions where all residues are chemically conserved. Gaps (hyphens) have been introduced to optimize alignments. About 50 residues of the yeast PDE2 sequence fail to show significant similarity to corresponding regions of the other PDEs and are not included (see text); hyphens have replaced these residues, which are located at the C-terminal end of the alignment. Abbreviations for the isoenzymes shown and the location of their domains are defined in the legend to Table 11.2

significance of the sequence similarities among the conserved segments of these
proteins is documented by the alignment scores [54] and sequence identities
listed in Table 11.2. This conserved region appears to be a unique feature of class
I PDEs, since searches of sequence databases reveal no other proteins with
significant similarities [42,43]. As mentioned above, the presence of this conserved region indicates that all class I isoenzymes have diverged from a common
ancestor.

Recently, Davis et al. [9] have updated the predicted amino acid sequence for
the *Drosophila* dunce gene product. The dunce gene contains an open reading
frame that is much larger than that originally described [6] and predicts a protein
product with 222 additional amino acids at the N-terminus. A reevaluation of the
structural relationship between the new *Drosophila* PDE and the other class I
isoenzymes has shown that within the 61 kDa CaM-PDE and the mammalian
dunce homologs the size of the conserved segment is larger than previously
described. As shown in Figure 11.1, there are about 50 additional residues at the
N-terminal end of the conserved segment in these isoenzymes that have
significant sequence similarity. The significance of this larger conserved region is
not fully understood at present but it should be noted that the core of the
conserved domains from the rat dunce homologs and the 61 kDa CaM-PDE are

Table 11.2 Homology among the catalytic domains of class I PDE isoenzymes

Isoenzyme (PDE)	CaM	DUN	RD1	RD2	RD3	RD4	cGS	ROS	CON	YE2
CaM		25.9	25.8	27.8	22.0	25.1	16.0	15.8	14.6	8.5
DUN	(42%)		52.1	50.0	54.5	51.9	16.7	14.5	13.4	9.9
RD1	(42%)	(76%)		61.0	57.3	57.9	16.9	17.1	17.0	8.6
RD2	(42%)	(78%)	(92%)		59.5	52.9	18.4	17.1	17.0	9.1
RD3	(41%)	(73%)	(86%)	(84%)		55.9	18.4	17.7	16.2	8.3
RD4	(42%)	(76%)	(91%)	(91%)	(86%)		18.1	18.6	17.1	9.4
cGS	(28%)	(33%)	(31%)	(31%)	(33%)	(31%)		27.0	24.6	5.0
ROS	(28%)	(31%)	(34%)	(31%)	(32%)	(33%)	(35%)		58.6	7.3
CON	(27%)	(29%)	(33%)	(31%)	(31%)	(32%)	(35%)	(78%)		6.0
YE2	(26%)	(28%)	(29%)	(29%)	(27%)	(29%)	(23%)	(27%)	(27%)	

Homology among the catalytic domains of the class I PDEs is indicated with percentage
absolute sequence identity (in parentheses, below the diagonal) and with alignment scores
generated by the ALIGN program [54] of the Protein Identification Resource (above diagonal).
Abbreviations for the PDE isoenzymes and the residues used for each domain are as follows:
CaM, 61 kDa calmodulin-dependent PDE [43] (193–446); **DUN**, *Drosophila* dunce PDE [6]
(223–489); **RD1**, rat dnc-1 [9] (161–427); **RD2**, rat DPD [10] (163–429); **RD3**, rat PDE1 [11];
RD4, rat PDE3 [11]; **cGS**, cGMP-stimulated PDE [44]; **ROS**, α′ subunit of the rod photoreceptor PDE [45] (535–803); **CON**, α′ subunit of the cone photoreceptor PDE [46] (535–802); **YE2**,
S. cerevisiae PDE2 [7] (241–478). Precise residue numbers for domains of the **RD3** [11], **RD4**
[11], and **cGS** PDEs [44] are not given, since total sequences are not yet available.

more similar to one another (42% sequence identity, Table 11.2) than to the segments of other mammalian PDEs (28–34% sequence identity; Table 11.2).

The conserved segment from the yeast PDE2 gene product appears to be significantly smaller than that of the other class I isoenzymes [42], as indicated in Figure 11.1. About 50 residues from the C-terminal end of the yeast PDE2 conserved segment fail to give statistically significant alignments with corresponding sequences from the conserved regions of other class I PDEs. The significance of this finding is unclear, but it is likely that the shortened homologous segment simply reflects the relatively large time span since the divergence of yeast from insects and mammals. The general limits of the conserved regions described here were selected with computer-generated alignments and corresponding statistical alignment scores [54] as a guide; however, the assignment of the specific residues at boundaries was a subjective process.

The pattern of homology (Figure 11.1) among class I isoenzymes can be most easily interpreted by considering these PDEs as chimeric, multidomain proteins. It is now widely accepted that many large proteins are chimeric in nature and are composed of several different domains, each of which is associated with a particular function [55]. These chimeric proteins appear to have been formed by genetic mechanisms that result in the fusion of independent domains of separate evolutionary origin. The fusion of multiple domains, each with a unique function, can generate a single polypeptide chain bearing multiple catalytic sites (e.g. fatty acid synthetase) or a family of enzymes with the same activity linked to distinct regulatory domains (e.g. protein kinases). Each domain is assumed to comprise a large segment of a polypeptide chain that folds independently and has at least one specific function (e.g. catalysis, ligand binding). In many cases, domains are separated spatially by 'hinge' regions that allow a domain to be removed from the parent protein by limited proteolysis [55]. The chimeric character of the class I PDEs is suggested by the presence of a single homologous domain in each enzyme that in many cases is linked to N-terminal segments of different lengths and sequence (Figure 11.1). As discussed in detail below, limited proteolysis data confirm that these PDEs are multidomain proteins.

The large segment of conserved structure in the class I isoenzymes can be viewed as delineating a common independent folding domain. The level of sequence identity (Table 11.2) among these conserved regions is remarkable when one considers the diversity in size, kinetic parameters, regulatory properties, and phylogenetic origins of the class I isoenzymes. Hence there may have been strong evolutionary pressure to maintain this 270 residue segment. Since the common feature of these isoenzymes is their capacity to hydrolyze cyclic nucleotides, it seems reasonable that the function of the conserved domain is catalysis [42]. Experimental evidence demonstrating the existence of the homologous, catalytically active domains is outlined in a later section, after considering the large segments of these proteins that appear to lack a direct role in catalysis.

As illustrated in Figure 11.1, the class I PDE isoenzymes have 200–600 residues that are located outside of an N-terminal to the catalytic domain. The structural diversity in these segments is thought to account for the variety of regulatory mechanisms employed by the class I isoenzymes [42] and to provide the binding site(s) for ligands that modulate activity (e.g. CaM, transducin, cyclic nucleotides). Class I PDEs can be divided into four distinct groups on the basis of sequence similarities among their noncatalytic N-terminal segments: (1) CaM-PDEs (2) *Drosophila* dunce gene product and rat dunce homologs, (3) yeast PDE2 gene product, and (4) α, β, and α' subunits of photoreceptor PDEs and cGS-PDE.

The 61 and 59 kDa CaM-PDEs are identical with the exception of 18 residues at their N-termini, while partial sequences from corresponding regions of the 63 kDa CaM-PDE are similar but not identical. A CaM binding site has been identified near the N-terminal of the 61 and 59 kDa isoenzymes (Figure 11.1) and partial sequence data suggest that a similar site exists at a corresponding position in the 63 kDa isoenzyme [43,50,56] (see below). While these sites do not appear to be homologous, they both have the capability of forming amphiphilic α-helices with a net positive charge, a feature noted for CaM binding sites from other enzymes. These findings suggest that the N-terminal segments from this category of enzymes bind CaM and participate in the CaM activation mechanism.

As might be expected, parts of the N-terminal noncatalytic sequences from the *Drosophila* dunce PDE [6] and two rat dunce homologs [9,10] show a high degree of sequence similarity to each other. The related sequences begin at the N-terminal end of the extended catalytic domain and extend towards the N-terminus of the molecule in contiguous segments of 100 and 80 residues from the *Drosophila* PDE and the two rat dunce homologs, respectively. Within these regions, the two rat PDEs have 72% sequence identity with one another whereas the *Drosophila* PDE shows 53% sequence identity with each of the rat enzymes. Although the two unique rat PDE homolog sequences reported by Swinnen et al. [11] are not complete and appear to be missing N-terminal sequences, the same degree of sequence similarity is indicated by the data that are available. It is not known whether these homologous N-terminal segments constitute an independent domain that can be separated from the catalytic domain by limited proteolysis. However, the homology within these regions suggests that they have the same function, possibly providing the site for interaction with regulatory molecules. At present, little information is available regarding the regulatory properties of these isoenzymes. There are about 25 residues at the N-terminus of the rat homologs and about 70 residues from the *Drosophila* PDE that are not structurally related. Perhaps these sequences confer properties that give each enzyme a unique function.

The N-terminal sequences of the yeast PDE2 gene product have no structural relationship to corresponding sequences from the other class I PDEs. This lack of related structure suggests a function for this part of the molecule that is unlike that of any other class I isoenzyme studied thus far. The regulatory properties of this enzyme are unknown.

In addition to the conserved catalytic domain, the α and α' subunit of the photoreceptor PDEs and cGS-PDE have significant sequence similarity with a segment of about 340 residues [47]. The sequence identity between the cGS-PDE and either the α or α' subunits is 27% and 30%, respectively, whereas the similarity between the two photoreceptor subunits was much greater at 67% sequence identity. Alignment of the segments requires a 30 residue gap near the center of the cGS-PDE segment. The identification of these conserved regions suggest the presence of a second functional domain. This second conserved segment shows internal homology with two repeats consisting of about 120 residues each [47]. The sequence identity between the internal repeats within the cGS-PDE and α subunit are 19% and 29%, respectively. The two internal repeats are separated by an intervening sequence of about 100 residues which is not similar to either of the repeats. However, the intervening sequence from all three proteins are homologous. The presence of internal repeats suggests that this conserved segment was formed by an evolutionary process involving a gene duplication event.

It has been suggested that the extensive N-terminal sequence similarity between the photoreceptor large subunits and cGS-PDEs reflects the ability of these enzymes to bind cGMP with high affinity at noncatalytic sites [47]. As discussed in detail below, these homologous sequences are located within a structural domain that can be separated from the catalytic domain by limited proteolysis [47,57]. Significantly, photolabeling studies show that cGMP-specific binding sites are located within this second conserved domain [47,57].

The overall structural organization of the PDE isoenzymes (Figure 11.1) is reminiscent of that for the protein kinase family [58–60]. Like the PDE isoenzymes, the protein kinases all catalyze the same fundamental reaction but with differing substrate specificities and differing susceptibility to regulating ligands. In fact, the PDE isoenzymes and kinases respond to many of the same regulatory ligands, CaM and cyclic nucleotides. Most of the kinases contain a homologous catalytic domain that is linked with one or more regulatory domains, those that bind CaM, cyclic nucleotides, or span the membrane to form extracellular receptors [58–60]. In at least two cases, it is clear that segments lying outside of the catalytic domain of the PDEs have regulatory functions, as is the case with the protein kinases. As described above, the homologies among the PDE isoenzymes and their general structural motif are consistent with the concept that the PDE isoenzymes are a diverse group of chimeric molecules that may have evolved in a manner analogous to that of the kinases.

In several cases, those PDE domains that are involved in similar functions (e.g. catalysis, cGMP binding) also possess significant sequence similarity. Thus the structural relationships among the PDEs can provide clues regarding the location and probable function of domains. Moreover, this implies that the formidable task of understanding the structure, function, and regulation of this large family of enzymes may be simpler than expected, since it now appears that much of what is learned about the structure and function of one isoenzyme can be applied to other forms.

11.3   EVIDENCE THAT THE CATALYTIC SITE IS LOCATED WITHIN A DOMAIN CONSERVED IN ALL CLASS I PDEs

Thus far, limited proteolysis studies have provided the greatest insight into the function of the conserved segment found in class I PDE isoenzymes (Figure 11.1). The limited proteolysis of three different PDEs yields fragments of similar size (∼36 kDa) that include the residues of the conserved segment. These findings alone demonstrate that the contiguous residues of the conserved segment constitute the core of a globular domain that is separated from the remainder of the molecule by 'hinge' regions. Data from two isoenzymes (cGS-PDE and 61 kDa CaM-PDE) show that one major function of this conserved domain is catalysis.

Stroop et al. [57] have investigated the domain structure of the cGS-PDE of bovine heart using limited chymotryptic cleavage in conjunction with direct photolabeling of cyclic nucleotide binding sites with [^{32}P]cGMP. The cGS-PDE can hydrolyze both cAMP and cGMP with nearly equal efficiency, but micromolar concentrations of cGMP induce a 10-fold stimulation of cAMP hydrolysis. Studies of the kinetic and cyclic nucleotide binding properties of this isoenzyme show that the enzyme has an allosteric site with specificity for cGMP [13–17]. The properties of the cGS-PDE are discussed in detail by Manganiello et al. [18].

Stroop et al. [57] found that limited chymotryptic cleavage separates the allosteric and catalytic domains, since proteolysis produced an increase in the basal activity of the enzyme and a concomitant loss of the stimulatory effects of cGMP on cAMP hydrolysis. This cleavage produced five major fragments with molecular masses of 60 kDa, 57 kDa, 36 kDa, 21 kDa, and 17 kDa as estimated by SDS-PAGE. N-terminal sequence analysis of the fragments allowed their positions to be mapped within the recently completed cGS-PDE sequence [44]. As shown in Figure 11.3, the 36 kDa fragment is generated by cleavage at an exposed 'hinge' region located about 40 residues N-terminal to the conserved domain. The 36 kDa fragment spans most if not all of the conserved domain and can be degraded further by cleavage at a secondary site near its midpoint to produce the 17 kDa and 21 kDa fragments (Figure 11.3). The 60 kDa and 57 kDa fragments have blocked N-termini and are related molecules that must be derived from the N-terminus of the native protein, since it is also blocked (Figure 11.3).

A solid-phase monoclonal antibody that recognizes the native enzyme bound the 60 kDa and 57 kDa fragments but did not remove enzymatic activity or the

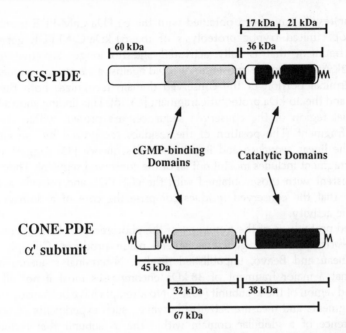

Figure 11.3 Functional domain organization of cGMP-stimulated PDE (cGS-PDE) and the α′ subunit of the photoreceptor PDE from cones (CONE-PDE). The catalytic and cGMP-binding domains are represented by boxes and are connected by narrow lines indicating interdomain 'hinge regions' that are protease-sensitive. The positions of the major proteolytic fragments (identified by their SDS-PAGE derived molecular masses) are shown by the horizontal lines. The position of fragments were mapped by N-terminal sequencing; however, the C-terminal ends of fragments were not directly determined and are estimated from apparent molecular masses. The solid black and shaded areas indicate the position of conserved sequences within the catalytic and noncatalytic cGMP-binding domains, respectively. The domain organization depicted was deduced from mapping the location of fragments and direct photolabelling experiments as described [47,57]

36 kDa fragment, suggesting an association between enzymatic activity and the 36 kDa fragment [57]. Purification of the proteolytic products by ion exchange HPLC chromatography produced a fraction that contained significant activity and was composed of the 36 kDa, 21 kDa, and 17 kDa fragments which are derived from the conserved domain. Furthermore, photolabeling experiments [57] with [^{32}P]cGMP showed that the 36 kDa fragment has a nucleotide binding site with an apparent affinity ($K_d = 30 \ \mu M$) and specificity that are consistent with its being the catalytic site [13,14,57]. Thus several lines of evidence suggest that the 36 kDa fragment of the cGS-PDE is a conserved domain that bears the active site of the enzyme. The smaller 17 kDa and 21 kDa fragments form structural subdomains of unknown significance.

Similar results have been obtained with the 61 kDa CaM-PDE from bovine brain [56]. Limited tryptic proteolysis of the 61 kDa CaM-PDE generates a 36 kDa fragment that is fully activated and no longer sensitive to CaM [56,61,62]. An antipeptide antiserum directed against a 12 residue segment from the C-terminal portion of the conserved domain recognizes both the native enzyme and the 36 kDa proteolytic fragment [43,56]. This finding shows that the C-terminal region of the conserved sequences are present within the active 36 kDa fragment. The position of the residues recognized by the antiserum within the linear sequence and its N-terminal sequence [43] suggest that the 36 kDa fragment includes most if not all of the conserved segment. These results are consistent with those obtained with the cGS-PDE and provide additional evidence that the conserved residues comprise the core of a domain having enzymatic activity.

Limited proteolysis of the α' subunit of the photoreceptor PDE of cones with *Achromobacter* protease I generates several major products (Gillespie, Prusti, Charbonneau and Beavo, unpublished results). N-terminal sequence analysis shows that a major fragment of 38 kDa encompasses most if not all of the conserved region of the α' subunit [46,47]. No attempts have been made to purify these fragments and measure activity; however, such experiments demonstrate the existence of a globular domain within the α' subunit that includes the conserved segment (Figure 11.3).

Conti and his colleagues [11,49] have expressed active, rat dunce PDE homologs in bacteria by transformation with the corresponding cDNA clones. Active PDEs are not produced when truncated cDNA clones that lack parts of the conserved domain are expressed in bacteria [49]. While this observation is consistent with the results obtained by limited proteolysis, it must be interpreted with caution. There is always a concern that the lack of activity in a deletion mutant of this type is an indirect effect that occurs because of a global change in protein conformation resulting from the deletion of residues that may be important for maintenance of native structure. The converse experiments in which activity is detected upon expression of cDNA constructs encoding only those residues surrounding the conserved domain would provide more direct evidence.

No data are available regarding the precise location of the active site or the identity of residues that are essential for catalysis. Thus far there have been no studies to specifically label or modify active site residues. However, a careful examination of the alignment of the catalytic domain from 10 different isoenzymes may provide some clues (Figure 11.2) about critical residues. As shown in Figure 11.2, there are 20 positions where all 10 isoenzymes have identical residues. If the non-homologous C-terminal residues of yeast PDE2 are excluded, there are 27 positions that are totally conserved. It is noteworthy that of these 20, five positions are occupied by histidine residues. The relatively low abundance of this chemically active residue in proteins adds to the significance of this finding. It is likely that one or more of the conserved histidines is involved in catalysis.

Active site studies with the bovine intestinal 5'-nucleotide PDE which hydrolyzes several different types of phosphodiester bonds, including those of cyclic nucleotides, may provide insights into the active site of class I PDEs [63]. Although this enzyme appears to be structurally unrelated to the class I PDEs [42], it is possible that the basic mechanism of hydrolysis of phosphodiester bonds may be similar. Culp et al. [63] have shown that a stable phosphothreonine intermediate is formed during catalysis. In this regard, it is interesting to note that one Thr and two Ser residues are totally conserved within the catalytic domains of all 10 isoenzymes (Figure 11.2).

Within the alignment of Figure 11.2, there is a single region (≈ 20–30 residues) where the overall similarity among the 10 isoenzymes is particularly low. This region is a site where relatively large gaps are required to align some isoenzymes, suggesting that either deletions or insertions have occurred here during evolutionary divergence from a common progenitor domain. It is tempting to speculate that this variable region may somehow govern substrate specificity. This is supported by the observation that those isoenzymes having the greatest similarity in both sequence and size of gaps from this region also have the same specificity.

11.4 IDENTIFICATION OF A PUTATIVE NONCATALYTIC, cGMP BINDING DOMAIN

The cGMP-stimulated PDE (cGS-PDE), the α subunit of the rod photoreceptor PDE, and the α' subunit of the cone photoreceptor PDE all possess, in addition to a homologous catalytic domain, a second conserved segment (~ 340 residues) located in the N-terminal half of the molecule [47]. This second conserved site will be designated here as conserved segment B, to distinguish it from the catalytic domain. In addition to their capacity to hydrolzye cyclic nucleotides, one function shared by all three isoenzymes is the ability to specifically bind cGMP at noncatalytic sites. Cyclic GMP binds specifically to an allosteric site on the cGS-PDE and positively stimulates the hydrolysis of cAMP [13–18]. Gillespie and Beavo [19,20] have shown that both rod and cone PDEs possess two noncatalytic, cGMP binding sites with dissociation constants of 10 nM or less. A possible function for the conserved segment B is to provide cGMP binding sites, since it is found only in those isoenzymes that bind cGMP at noncatalytic sites. Results from limited proteolysis and direct photolabeling with [^{32}P]cGMP support this concept [57].

As mentioned above, limited chymotryptic cleavage of the cGS-PDE produces N-terminally blocked 60 kDa and 57 kDa fragments that are derived from the N-terminus of the native protein [57]. If the molecular weights determined by SDS-PAGE are accurate, both fragments are of sufficient length to span most of the conserved segment B. As indicated in Figure 11.3, the site of cleavage producing the 36 kDa catalytic domain is located near the C-terminal end of the conserved segment B and is consistent with the presence of a 'hinge' region separating two domains, one of which contains most of segment B [57]. Since the 60 kDa and 57 kDa segments are much larger, segment B may not correspond directly to a single domain but may simply be a conserved element within a much bigger domain.

Stroop et al. [57] have photolabeled a cGMP-specific site on the cGS-PDE with an apparent dissociation constant of 1 μM and characteristics consistent with it being the allosteric site. The 57 kDa and 60 kDa limited proteolytic fragments that encompass most of the conserved segment B (Figure 11.3) are photolabeled and must possess the cGMP-specific allosteric site [57]. In addition, photolabeling has localized the allosteric site to a large, CNBr fragment (28 kDa) which is totally encompassed within the conserved segment B. Thus the noncatalytic, cGMP binding site(s) of the cGS-PDE must be formed from residues located within the conserved segment B.

Limited proteolysis of the CONE-PDE with *Achromobacter* protease I (Gillespie, Prusti, Charbonneau and Beavo, unpublished results) produces three major fragments of 32 kDa, 45 kDa and 67 kDa that have been mapped to positions that span most if not all of segment B [46,47]. As shown in Figure 11.3, the termini of the 32 kDa fragment correspond very closely to the boundaries of segment B [47]. These findings indicate that residues within segment B are capable of folding to form a globular domain that is separated by limited proteolysis. When the α' subunit is photolabeled at noncatalytic, cGMP binding sites under conditions similar to those used for the cGS-PDE, radiolabeled polypeptides with SDS-PAGE mobilities similar to those of the 32 kDa and 45 kDa fragments were detected. These data suggest that the noncatalytic, cGMP binding sites on the α' subunit are also located within segment B.

Combined photolabeling and limited proteolysis studies performed with the cGS-PDE and the α' subunit indicate that the conserved segment B forms the nucleus of a distinct domain, with at least one of its major functions being cGMP binding. However, more definitive experiments are needed to support this hypothesis. For example, it will be important to map the precise location of residues involved in cGMP binding and to isolate fully functional domains so that their cGMP binding properties can be studied in detail. As is the case with the catalytic domain, in vitro mutagenesis experiments should provide a powerful tool for analyzing the functional properties of these domains. In the case of the cGS-PDE, the domain appears to be associated with an allosteric site. At present, the functional role of the homologous domains in the photoreceptor PDEs is not fully understood and it is not clear what effect cGMP binding may have on enzyme function. These findings emphasize the importance of carefully investigating the role of cGMP in modulating photoreceptor PDE function. The conserved cGMP binding domains from the cGS-PDE and photoreceptor PDEs may be viewed as prototypes for similar domains that may be present in other isoenzymes having noncatalytic sites for cGMP such as the cGMP binding PDEs [21–24] or the low K_m, cGMP-inhibited PDE [37,64]. Thus all of the PDEs that bind cGMP at noncatalytic sites may be chimeric proteins that evolved by the fusion of a catalytic domain with a regulatory, cGMP binding domain.

11.5 IDENTIFICATION OF THE CaM BINDING SITE FROM THE 59 kDa AND 61 kDa CaM-PDEs

The binding of CaM to the CaM-PDEs produces a 4–15-fold stimulation of the rate of hydrolysis of either cAMP or cGMP. The sequencing of the 61 kDa CaM-PDE enzyme has been completed [43] and more than 90% of the sequence of the 59 kDa isoenzyme from heart has been determined [50] (Table 11.1). This structural information forms the basis for ongoing studies of the mechanism by which CaM binds to and activates the enzyme. Identification of the CaM binding site is an important element of such mechanistic studies. Recently, CaM binding sites of the 61 kDa and 59 kDa isoenzymes have been located by experiments employing synthetic peptide analogs of PDE sequences [43,50,51,56].

Both naturally occurring and synthetic peptides that bind CaM in a Ca^{2+}-dependent manner with nanomolar affinities have the tendency to form basic, amphiphilic α helices [65,66]. Using these criteria as a guide, the 61 kDa CaM-PDE sequence was screened for potential CaM binding segments. A screening procedure similar to that described by Erickson-Viitanen and Degrado [67] identified only one region (residues 23–44) with appropriate properties (Figure 11.4). A 22 residue synthetic peptide analog of this region was prepared and its CaM binding properties were evaluated [43]. This peptide analog inhibits the CaM-dependent activation with an IC_{50} of 27 nM. Studies using the intrinsic Trp fluorescence of the peptide analog have shown that it binds CaM with high affinity in a Ca^2-dependent manner. The peptide mimics the CaM binding properties of the native enzyme, providing evidence that the basic, amphiphilic segment including residues 23–44 comprises most if not all of the CaM binding site of the 61 kDa CaM-PDE [43].

This location for the CaM binding site is entirely consistent with results from limited tryptic proteolysis experiments [43,56]. A 36 kDa CaM-independent fragment bearing the putative catalytic domain does not bind CaM and is believed to have lost the CaM binding site. The exact position of the 36 kDa fragment has been mapped by N-terminal sequencing [43]. The fragment is generated by the loss of residues 1–135 at the N-terminus and about 80 residues from the C-terminus.

Clearly, additional studies will be needed to precisely define those regions of the molecule involved in CaM binding. Although the dissociation constant has not been directly determined, the data from inhibition studies [43,50] suggest that

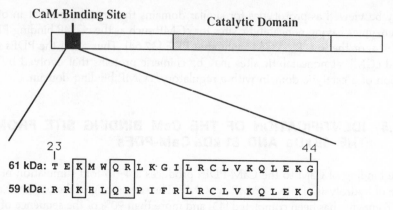

Figure 11.4   Linear model depicting the domain organization of the 59 kDa and 61 kDa CaM-dependent PDEs. The shaded and black boxes indicate the position of the conserved catalytic domain and the CaM binding site, respectively. The sequences of the proposed CaM binding sites from both isoenzymes are given; boxes enclose residues that are identical. All available sequences for the 59 kDa CaM-PDE that are located on the C-terminal side of the CaM binding site are identical to corresponding residues of the 61 kDa isoenzyme [50,51]

the peptide analogs have an affinity that is approximately 10-fold lower than the affinity of the native enzyme for CaM. This suggests that other key elements of the native CaM binding site are not present in the synthetic peptide analogs studied. The characterization of additional peptide analogs of different lengths and derived from distinct locations will reveal whether residues adjacent to this site also participate in binding.

Novack et al. [50] have obtained partial sequence data covering 92% of the expected residues from the 59 kDa CaM-PDE of bovine heart. The 59 kDa and 61 kDa isoenzymes have similar substrate specificities but differ in apparent size and affinity for CaM [29]. The 10–20-fold greater affinity measured for the 59 kDa isoenzyme [29] suggests that the two isoenzymes must have significant structural differences at or near the CaM binding site. The 59 kDa and 61 kDa isoenzymes are identical at all positions where sequence data are available, except for an 18 residue segment corresponding to a portion of the CaM binding site of the 61 kDa isoenzyme [50] (Figure 11.4). Synthetic peptides from this region of the 59 kDa isoenzyme also bind directly to CaM with high affinity and inhibit the CaM activation of CaM-PDE [50,51]. These results confirm the identification of the CaM binding site for the 61 kDa isoenzyme. Figure 11.4 illustrates the position of the CaM binding site and catalytic domain within the 61 kDa and 59 kDa CaM-PDEs.

Partial sequence data for the 63 kDa CaM-PDE of bovine brain have been obtained from the analysis of peptides obtained by cleavage at lysine [50]. The sequences of most peptides were unambiguously aligned with the 61 kDa CaM-PDE sequence. This alignment clearly shows that the two proteins have similar but non-identical sequences and indicates that the two PDEs are products of distinct genes. This result is in accord with the peptide mapping results reported earlier by Sharma et al [26]. It will be important to determine whether other CaM-PDE isoenzymes, such as the 67 kDa form from mouse testis [28] or the 58 kDa enzyme from bovine lung [27], are derived from unique genes or whether they are alternatively processed products of genes encoding either the 61 kDa or 63 kDa enzymes.

Most CaM-dependent target enzymes can be irreversibly activated by limited proteolysis in a manner similar to that described for the CaM-PDEs. With the plasma membrane Ca^{2+}-ATPase [68,69], smooth and skeletal muscle myosin light chain kinase [70,71], CaM-kinase II [72], and calcineurin [73] there is evidence that one effect of proteolysis is the removal of a distinct inhibitory domain. With these enzymes, it is postulated that the inhibitory domains suppress activity in the absence of CaM and that this suppression of activity is somehow relieved upon the binding of CaM. With the myosin light chain kinases, the CaM binding domain resembles kinase substrates and can serve as the inhibitory domain. With the ATPase and calcineurin, the inhibitory and CaM binding sites are distinct [68,69,73].

Kincaid et al. [61] have reported limited proteolysis data that suggest the presence of a distinct inhibitory domain within the CaM-PDE. Of particular interest is the 45 kDa chymotryptic fragment that has retained its ability to bind CaM, but is fully and irreversibly activated [61]. The formation of such a fragment is consistent with the presence of a distinct inhibitory domain. The size of this fragment and the location of the CaM binding site near the N-terminus suggest that this putative inhibitory domain must be located near the C-terminus. Establishing whether or not a separate inhibitory domain exists is critical for elucidating the mechanism of activation by CaM. Additional studies focusing on the characterization of limited proteolytic fragments should provide important evidence about the existence of inhibitory domains. If a distinct inhibitory domain exists within the CaM-PDEs, it may have properties, including primary sequence, resembling those of the inhibitory subunits of the photoreceptor enzymes (see below). Hence, a knowledge of the photoreceptor PDE activation mechanism may provide clues about the activation of PDEs by CaM.

11.6 STRUCTURE–FUNCTION ANALYSES OF THE INHIBITORY γ SUBUNIT OF THE ROD PHOTORECEPTOR PDE

Photoreceptor PDEs play a key role in the visual transduction pathway, since they produce a light-dependent reduction in the cGMP concentration which results in the closure of cGMP-sensitive Na^+ channels [30,74,75]. Channel closure hyperpolarizes the cell and eventually generates a neural response. A photoreceptor-specific G-protein, transducin, couples the PDE to the light receptor, rhodopsin [30]. The GTP complex of the transducin α subunit ($T\alpha$–GTP) is formed upon photoactivation of rhodopsin [30] and is capable of stimulating the PDE by displacing the inhibitory γ subunit [76,77]. The PDE holoenzyme is believed to be composed of one α, one β, and two γ subunits [78].

The bovine brain γ subunit has been cloned and the protein sequence deduced from the cDNA sequence [52]. Recently, a synthetic gene for the γ subunit has been prepared and used to produce a functional protein by expression in *E. coli* [79]. These developments have allowed considerable progress in understanding how the γ subunit inhibits the α and β subunits of the PDE. Lipkin et al. [80] found that removal of seven residues from the C-terminus of the γ subunit resulted in loss of inhibitory capacity. Using the synthetic γ subunit gene, Brown and Stryer [79] found that a truncated gene encoding all but 13 residues at the C-terminus (residues 1–74) produces a mutant protein that is incapable of inhibiting trypsin-activated PDE but is capable of binding the $\alpha\beta$ complex. These studies demonstrate that residues 1–74 are involved in anchoring the γ subunit to the $\alpha\beta$ subunits while residues 75–87 contain the residues that are essential for inhibition. Lipkin et al. [80] provided evidence that residues 24–45 of the γ subunit are necessary for interaction with $T\alpha$–GTP. Thus the $\alpha\beta$ subunits and $T\alpha$–GTP complex appear to compete for similar if not identical sites within the highly positively charged, N-terminal segment of the γ subunit. At present, little is known about the location of the γ subunit binding site on the α and β subunits. The positively charged character of the interacting region of the γ subunit indicates that the sites on the $\alpha\beta$ subunits may be negatively charged.

More extensive investigations of the γ–$\alpha\beta$ subunit interactions will not only help to elucidate the mechanism for the light-dependent activation of photoreceptor PDEs but may also provide insights into the regulation of other PDEs. CaM-PDEs are postulated to have an inhibitory domain that suppresses the activity of the enzyme in the absence of CaM. It is possible that even though it is part of the same polypeptide chain, such an inhibitory domain may act in a manner analogous to that of the γ subunit so that the activation of the photoreceptor PDE may be mechanistically similar to the activation of CaM-PDE by CaM. The putative inhibitory domain of the CaM-PDE may even resemble that of the γ subunit.

11.7 EVIDENCE THAT DIVERSITY AMONG PDE ISOENZYMES IS PARTIALLY GENERATED BY ALTERNATIVE mRNA SPLICING

Alternative mRNA splicing allows eukaryotic cells to generate multiple isoenzymes in a cell-type-specific and developmentally regulated manner [81]. This differential exon usage leads to the production of several protein variants with structural alterations that are restricted to one particular region of the molecule. These variants all have the same basic structural motif and primary function but have been tailored to function in specific cellular environments or at specific stages of development. The available sequence data suggest that alternative splicing mechanisms may account for some of the diversity among the rat dunce homologs [9,12,49] and bovine CaM-PDEs [51,50].

Fifteen residues within a single 18 residue segment at the N-terminus of the 61 kDa CaM-PDE isoenzyme are different from the corresponding residues of the 59 kDa PDE [50] (Figure 11.4) All known sequence differences between the two isoenzymes are restricted to this segment. Such an arrangement is the basis for the suggestion that the two PDEs are encoded by a single gene and are products of alternative RNA splicing [50,51]. The differential usage of exons encoding sequences that are part of the CaM binding site is a mechanism that could generate isoenzymes having different affinities for CaM. This mechanism allows the expression of an isoenzyme having a CaM binding affinity that matches the CaM levels or range of Ca^{2+} concentrations that are characteristic of its intracellular environment. Similar evidence for alternative RNA splicing has been obtained from cDNA clones encoding rat dunce homologs [9,49]. In these cases the possible biological significance of the postulated splicing mechanism is not understood. At this time, the evidence for alternative splicing with either the dunce homologs or CaM-PDEs is indirect and is inferred from sequence data. Definitive proof of differential exon usage will require the isolation and characterization of appropriate genomic sequences. It is likely that other examples of alternativly spliced PDE genes will become evident as additional isoenzymes are investigated.

11.8 POSSIBLE STRUCTURAL RELATIONSHIP BETWEEN PDEs AND OTHER CYCLIC NUCLEOTIDE BINDING PROTEINS

One intriguing question arising from the analysis of PDE sequences is whether they are structurally related to other proteins that bind cyclic nucleotides. One family with related cyclic nucleotide binding domains includes the regulatory subunits of the cAMP-dependent protein kinases, the cGMP-dependent protein kinase, and the E. coli catabolite gene activator protein (CAP) [58–60]. A careful comparison of class I and class II PDE sequences with those of this protein family reveal no statistically significant structural relationship. Neither the catalytic nor the noncatalytic domains of the PDEs has extended sequence similarities with the CAP family of proteins. This suggests the independent evolution of at least three distinct types of cyclic nucleotide binding sites. However, it is possible that these proteins share a common progenitor but have diverged so long ago or at such a rate as to obscure any relationship at the primary sequence level.

Chen et al. [6] described a seven residue segment from the dunce gene product that is identical to a corresponding segment within the cAMP binding segment of the regulatory subunit of the cAMP-dependent protein kinase. Such an observation must be interpreted with caution, since it is difficult to establish a statistically meaningful relationship between two segments of such a short length [82,83]. Furthermore, corresponding segments from other PDE isoenzymes do not show the same level of sequence similarity with the regulatory subunits of the kinases.

The lack of extensive sequence similarity suggests, but does not prove, that the PDEs and the CAP family of proteins bind cyclic nucleotides in a different mode and have different tertiary structures at their cyclic nucleotide binding sites. Studies with cyclic nucleotide analogs also indicate distinct structural differences between the binding sites of the CAP family [84] of proteins and the PDEs [24, 85]. For example, the 2'-OH of the ribose moiety and the oxygens of the cyclic phosphate group, which interact directly with residues in the binding site of proteins of the CAP family [59,84,86,87], do not appear to be critical for binding at the catalytic sites of the PDEs [24,85]. Additional analog studies may give further insight into the nature of the noncatalytic and catalytic cycle nucleotide binding sites on the PDEs. Obviously, definitive answers to questions regarding differences in the tertiary structure of the cyclic nucleotide binding sites from the PDEs and proteins of the CAP family will require three-dimensional analyses. The crystal structure of CAP has been determined [87] and the three-dimensional structure of binding sites from PDEs may be obtained in the future if sufficient quantities of the native proteins are available or if smaller subfragments retaining catalytic and/or binding functions can be prepared.

11.9   COMMENTS ON THE EVOLUTION OF CLASS I PDE ISOENZYMES

All class I PDE isoenzymes have retained homologous catalytic domains, yet evolved to perform a variety of specialized functions. Gene duplication is probably a primary mechanism generating this diversity among PDE isoenzymes. For example, gene duplication and subsequent divergence must have given rise to the homologous α and α' subunits of the rod and cone photoreceptor PDEs which are uniquely adapted to their respective cells. With other isoenzymes, gene duplication cannot easily explain how a catalytic domain became linked to a non-homologous segment within the same polypeptide chain (see above). Gene fusion mechanisms provide the simplest explanations for such molecular chimeras.

Figure 11.5 illustrates a general and hypothetical scheme that may account for the evolution of this enzyme family. According to this model, an ancestral cyclic nucleotide binding domain was formed by tandem duplication of a primordial

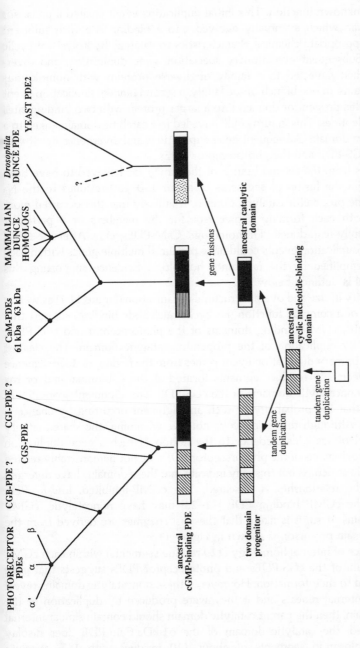

Figure 11.5 Hypothetical scheme for the evolution of class I PDE isoenzymes. This diagram is not intended to represent a rigorously derived evolutionary tree; branch lengths and node positions do not accurately reflect evolutionary distance. Circles denote those branch points where gene duplications give rise to distinct PDE isoenzymes. Abbreviations for PDE isoenzymes depicted are defined as follows: CGS-PDE, cGMP-stimulated PDE; CGI-PDE, low-K_m, cGMP-inhibited PDE [37,64], CGB-PDE, cGMP-binding PDE [21–23]; CaM-PDE, CaM-dependent PDE; and DUNCE PDE, *Drosophila* dunce gene product

segment of unknown function. This initial duplication event created a precursor binding domain which eventually evolved into a binding site with sufficient affinity and appropriate chemical characteristics to catalyze hydrolysis of cyclic nucleotides. Subsequent evolutionary speciation, gene duplication, and divergence may then have led to a family of diverse proteins with homologous catalytic domains. In one branch of the family, a second tandem duplication event transformed this progenitor domain into a larger protein with two binding sites for cyclic nucleotides. This in turn could have led to a catalytic domain linked to a cGMP binding domain. Subsequent gene duplications and divergence eventually formed the cGS-PDE and the photoreceptor PDEs.

Isoenzymes from the second branch of the family are believed to have been generated from the fusion of segments of about 150–250 residues to the N-terminus of the progenitor catalytic domain. It is likely that this occurred more than once, with each forming a precursor for the members of a particular isoenzyme family (e.g. dunce PDE homologs, CaM-PDEs, etc.). After the initial fusions, gene duplication events could have produced multiple genes within each family as exemplified by the rat dunce homologs. Evidence supporting this general model is outlined below.

The similarity in size and overall structural organization (Figure 11.1) as well as the existence of a common function (i.e. cyclic nucleotide binding) suggest that the noncatalytic, cGMP binding domains of the photoreceptor and cGS-PDEs were formed by duplication of the progenitor catalytic domain. The clearest evidence of a primary duplication event comes from the finding of 18% sequence identity between 220 residue segments located at the C-terminal end of the catalytic and noncatalytic domains of the cGS-PDE. The alignment score [54] of 3.5 indicates that the probability of such an alignment occurring by chance is $< 2 \times 10^{-4}$. Although this evidence is not overwhelming, the shared affinity for cyclic nucleotides adds credence to the notion of such a gene duplication event. The two domains of the photoreceptor PDEs are not demonstrably related to each other in structure, but this may indicate that these domains have diverged so far that the relationship is obscured. The cGMP-inhibited, low-K_m PDE [37,64] and the cGMP binding PDE [21–23] may have noncatalytic, cGMP binding domains; if so, it is likely that these isoenzymes are derived from the same two-domain precursor as shown in Figure 11.5.

The presence of internal homology (120 residue segments) within each cGMP binding domain of the cGS-PDEs and photoreceptor PDEs suggests that gene duplication led to their formation. However, if these noncatalytic domains reveal evidence of internal repeats and if they were produced by duplication of the catalytic domain, then the parent catalytic domain should contain similar internal repeats. Indeed, the catalytic domain of the 61 kDa Cam-PDE does display internal homology in segments of about 120 residues with 16% absolute sequence identity (32% with chemically conserved residues included) and an alignment score of 4.1. Internal repeats of similar size are suggested in the catalytic domains of the cGS-PDE, *Drosophila* dunce PDE, and rat dunce homologs, but the evidence is weaker, with sequence identities ranging from 17% to 20% and corresponding alignment scores from 1.4 to 2.8. As proposed here, the initial duplication producing an ancestral catalytic domain is the earliest step in PDE evolution and it is not surprising that this internal homology is now difficult to detect. The catalytic domain may have diverged more rapidly than the noncatalytic domain, where retention of cGMP specificity and regulatory capacity required more constraints than retention of catalytic capacity.

An examination of Table 11.2 leads to tentative conclusions regarding the relative similarity and point of divergence of the catalytic domains. As expected from the time of divergence of fungi from animals, the yeast PDE2 catalytic domain is least similar to the other domains (23–29% sequence identity). The catalytic domains of the 61 kDa CaM-PDE and the dunce homologs resemble each other more than those of other mammalian enzymes, suggesting that these two isoenzyme types diverged from a common ancestor sometime after the PDEs bearing cGMP binding domains separated. Surprisingly, as Davis et al. [9] pointed out, the rat and *Drosophila* dunce homologs are strikingly similar (76–78% sequence identity within the conserved domain) despite the 600 million year divergence of vertebrates and invertebrates. Davis et al. [9] also noted that this PDE was significantly more conserved than other *Drosophila* proteins that are involved in intracellular signaling, suggesting a specific and essential physiological role. Taken together these observations indicate that the various genes for PDE isoenzymes may have distinctly different rates of change, making it difficult to estimate the timing of their divergence or of putative gene fusions. All but two of the known PDE sequences are from mammals. To acquire a more rigorous and comprehensive view of PDE evolution, it will be necessary to assess the distribution of the various PDE isoenzymes among a wider range of organisms and to compare their sequences.

11.10   SUMMARY

The elucidation of structure–function relationships and the mechanisms underlying the complex regulation of multiple PDE isoenzymes is clearly a challenging problem. The acquisition of the primary sequences for several isoenzymes over the past three years has made it possible to begin an analysis of these problems. Even though this is a relatively new area of endeavor, considerable progress has been made on three fronts. A clear picture of the structural relationships among the PDE isoenzymes is emerging and the overall domain organization of several isoenzymes has been determined. Furthermore, significant insights into their evolution have been gained. Of course, these advances are only the first steps toward structure–function investigations, and much remains to be done.

A comparison of amino acid sequences has revealed the existence of a family of PDE isoenzymes from both vertebrate and invertebrate species that are structurally related despite having considerable differences in substrate preference, kinetic parameters, and mode of regulation. All of these isoenzymes possess a conserved domain composed of about 250 residues. Several limited proteolysis studies have provided substantial evidence that this conserved domain includes the catalytic site. The observed conservation of primary structure suggests that all of these isoenzymes employ the same fundamental mechanism for cyclic nucleotide hydrolysis. However, a second category of isoenzymes, including a second PDE from *S. cerevisiae* and one from *Dictyostelium discoideum*, are homologous to one another but do not have this conserved catalytic domain. It is likely that they have different biological roles and employ a distinct mechanism of catalysis.

The existing structural data clearly demonstrate the existence of multiple PDE genes within one organism and tend to eliminate earlier concerns that the multiplicity of PDE isoenzymes might have arisen from artifactual modification or proteolysis occurring during extraction or purification. Several of these genes appear to be expressed in specific cell types (e.g. α' subunit in cones). However, recent data suggest that isoenzyme diversity among some CaM-PDEs and dunce homologs has been achieved by alternative mRNA processing of a single gene. In the case of the CaM-PDEs, it appears that alternative splicing gives rise to two molecules with nearly identical kinetic characteristics but distinct affinities for CaM.

N-terminal sequences lying outside of the conserved catalytic domain of the class I isoenzymes appear to have primarily regulatory functions, as indicated by the finding of significant sequence similarity among those isoenzymes regulated by the same molecules. For example, a putative cGMP binding domain of about 350 residues has been identified within the N-terminal sequences of the bovine, cGMP-stimulated PDE and the α and α' subunits of bovine photoreceptor PDEs. The binding of cGMP at a site(s) in this domain stimulates the activity of the cGS-PDE by an allosteric mechanism, but the precise role of the corresponding domains in the photoreceptor PDEs is not fully understood. It is likely that a homologous domain may also be located within the cGMP-inhibited PDE or cGMP binding PDEs. The CaM binding sites of the 59 and 61 kDa CaM-PDEs have been localized to 22 residue segments near the N-termini of these enzymes. Thus far, no functions have been identified for the N-terminal sequences of the yeast PDE2 enzyme or the dunce homologs.

Now that the location and boundaries of domains from several isoenzymes are known, the stage has been set for more detailed studies designed to probe each domain individually so that those residues critical to proper domain function can be identified. Then questions can be asked regarding the ligand–domain, domain–domain interactions that must play an important role in their regulation. These more detailed structure–function questions can be addressed using conventional approaches such as residue-specific chemical modification, affinity reagents, substrate analogs, inhibitors, and covalently attached spectroscopic probes. Since several isoenzymes have been cloned, recombinant DNA techniques for efficient expression and mutagenesis should provide extremely powerful tools capable of producing a wealth of information on the structure and function of PDE isoenzymes. For example, the large-scale expression of truncated cDNA clones encoding small but functional domains could make it possible to obtain three-dimensional structures for PDE domains by X-ray crystallography or NMR spectroscopy. Such studies would be extremely difficult if not impossible without recombinant DNA techniques, since PDEs are relatively large proteins that are present in trace amounts in most cells. The next several years should be exciting as new PDE sequences become available and as the techniques of modern molecular biology are brought to bear on these structure–function problems. A knowledge of the structure–function relationships of the PDE isoenzymes and the regulatory mechanisms that govern their activities will be important pharmacologically to facilitate the design of therapeutic agents capable of modulating cyclic nucleotide levels and will also provide valuable insights toward understanding the diverse physiological roles of these isoenzymes.

ACKNOWLEDGEMENTS

The author acknowledges the support, advice, and encouragement given by Joseph A. Beavo and Kenneth A. Walsh. The author's studies on PDE structure and function were carried out in their laboratories and was supported by National Institute of Health Grants DK 21723 and EY 08197 to Dr Beavo and GM 15731 to Dr Walsh. I am grateful to Hans Neurath for his guidance and encouragement. I thank Norbert Beier, Steven D. Stroop, Hai Le Trong, Rabi K. Prusti, Santosh Kumar, and Jeff Novack for their collaboration and for many stimulating discussions.

REFERENCES

1. Beavo, J. A. (1988) *Adv. Second Messengers Phosphoprotein Res.*, **22**, 1–38.
2. Beavo, J. A., Hansen, R. S., Harrison, S. A., Hurwitz, R. L., Martins, T. J., and Mumby, M. C. (1982) *Mol. Cell. Endocrinol.*, **28**, 387–410.
3. Strada, S. J., Martin, M. W., and Thompson, W. J. (1984) *Adv. Cyclic Nucleotide Res.*, **16**, 13–29.
4. Wells, J. N., and Hardman, J. G. (1977) *Adv. Cyclic Nucleotide Res.*, **8**, 119–143.
5. Beavo, J. A. (1990) In *Cyclic Nucleotide Phosphodiesterases: Structure, Regulation and Drug Action*, (Beavo, J. and Houslay, M. D., eds), John Wiley & Sons, Chichester, pp. 3–17.
6. Chen, C. N., Denome, S., and Davis. R. L. (1986) *Proc. Natl Acad. Sci. USA.*, **83**, 9313–9317.

7. Sass, P., Field, J., Nikawa, J., Toda, T., and Wigler, M. (1986) *Proc. Natl Acad. Sci. USA*, **83**, 9303–9307.
8. Nikawa, J.-I., Sass, P., and Wigler, M. (1987) *Mol. Cell. Biol.*, **7**, 3629–3636.
9. Davis, R. L., Takayasu, H., Eberwine, M., and Myres, J. (1989) *Proc. Natl Acad. Sci. USA*, **86**, 3604–3608.
10. Colicelli, J., Birchmeier, C., Michaeli, T., O' Neill, K., Riggs, M., and Wigler, M. (1989) *Proc. Natl Acad. Sci. USA*, **86**, 3599–3603.
11. Swinnen, J. V., Joseph, D. R., and Conti, M. (1989) *Proc. Natl Acad. Sci. USA*, **86**, 5325–5329.
12. Davis, R. L. (1990) In *Cyclic Nucleotide Phosphodiesterases: Structure, Regulation and Drug Action*, (Beavo, J. and Houslay, M. D., eds), John Wiley & Sons, Chichester, pp. 227–241.
13. Martins, T. J., Mumby, M. C., and Beavo, J. A. (1982) *J. Biol. Chem.*, **257**, 1973–1979.
14. Yamamoto, T., Manganiello, V. C., and Vaughan, M. (1983) *J. Biol. Chem.*, **258**, 12526–12533.
15. Erneux, C., Couchie, D., Dumont, J. E., Baraniak, J., Stec, W. J., Garcia Abbad, E., Petridis, G., and Jastorff, B. (1981) *Eur. J. Biochem.*, **115**, 503–510.
16. Miot, F., Van Haastert, P. J. M., and Erneux, C. (1985) *Eur. J. Biochem.*, **149**, 59–65.
17. Wada, H., Manganiello, V. C., and Osborne, J. C. (1987) *J. Biol. Chem.*, **262**, 13938–13945.
18. Manganiello, V. C., Tanaka, T., and Murashima, S. (1990) In *Cyclic Nucleotide Phosphodiesterases: Structure, Regulation and Drug Action*, (Beavo, J. and Houslay, M. D., eds), John Wiley & Sons, Chichester, pp. 61–85.
19. Gillespie, P. G., and J. A. Beavo. (1989) *Proc. Natl Acad. Sci. USA*, **86**, 4311–4315.
20. Gillespie, P. G., and J. A. Beavo (1988) *J. Biol. Chem.*, **263**, 8133–8141.
21. Francis, S. H., Lincoln, T. M., and Corbin, J. D. (1980) *J. Biol. Chem.*, **255**, 620–626.
22. Hamet, P., and Tremblay, J. (1988) *Methods Enzymol.*, **159**, 710–722.
23. Francis, S. H., and Corbin, J. D. (1988) *Methods Enzymol.*, **159**, 722–729.
24. Francis, S. H., Thomas, M. K., and Corbin, J. D. (1990) In *Cyclic Nucleotide Phosphodiesterases: Structure, Regulation and Drug Action*, (Beavo, J. and Houslay, M. D., eds), John Wiley & Sons, Chichester, pp. 117–140.
25. Hansen, R. S., and Beavo, J. A. (1982) *Proc. Natl Acad. Sci. USA*, **79**, 2788–2798.
26. Sharma, R. K., Adachi, A.-M., Adachi, K., and Wang, J. H. (1984) *J. Biol. Chem.*, **259**, 9248–9254.
27. Sharma, R. K., and Wang, J. H. (1986) *J. Biol. Chem.*, **261**, 14160–14166.
28. Rossi, P., Giorgi, M., Geremia, R., and Kincaid, R. L. (1988) *J. Biol. Chem.*, **263**, 15521–15527.
29. Hansen, R. S., and Beavo, J. A. (1986) *J. Biol. Chem.*, **261**, 14636–14645.
30. Hurley, J. B. (1987) *Annu. Rev. Physiol.*, **49**, 793–812.
31. Hurley, J. B., and Stryer, L. (1982) *J. Biol. Chem.*, **257**, 11094–11099.
32. Gillespie, P. G., Prusti, R. K., Apel, E. D., and Beavo, J. A. (1989) *J. Biol. Chem.*, **264**, 12187–12193.
33. Sharma, R. K., Wang, T. H., Wirch, E., and Wang, J. H. (1980) *J. Biol. Chem.*, **255**, 5916–5923.
34. Hashimoto, Y., Sharma, R. K., and Soderling, T. R. (1989) *J. Biol. Chem.*, **264**, 10884–10887.
35. Sharma, R. K., and Wang, J. H. (1986) *J. Biol. Chem.*, **261**, 1322–1328.

36. Sharma, R. K., and Wang, J. H. (1985) *Proc. Natl Acad. Sci. USA*, **82**, 2603–2607.
37. Harrison, S. A., Reifsnyder, D. H., Gallis, B., Cadd, G. G., and Beavo, J. A. (1986) *Mol. Pharmacol.*, **29**, 506–514.
38. Houslay, M. D., Wallace, A. V., Marchmont, R. J., Martin, B. R., and Heyworth, C. M. (1984) *Adv. Cyclic Nucleotide Res.*, **16**, 159–176.
39. Macphee, C. H., Reifsnyder, D. H., Moore, T. A., Lerea, K. M., and Beavo, J. A. (1988) *J. Biol. Chem.*, **263**, 10353–10358.
40. Macphee, C. H., Reifsnyder, D. H., Moore, T. A., and Beavo, J. A. (1987) *J. Cyclic Nucleotide Protein Phosphorylation Res.*, **11**, 487–496.
41. Grant, P. G., Mannarino, A. F., and Colman, R. W. (1988) *Proc. Natl Acad. Sci. USA*, **85**, 9071–9075.
42. Charbonneau, H., Beier, N., Walsh, K. A., and Beavo, J. A., (1986) *Proc. Natl Acad. Sci. USA*, **83**, 9308–9312.
43. Charbonneau, H., Kumar, S., Novack, J. P., Blumenthal, D. K., Griffen, P. R., Shabanowitz, J., Hunt, D. F., Beavo, J. A., and Walsh, K. A. (1990) (submitted).
44. Le Trong, H., Beier, N., Sonnenburg, W. K., Stroop, S. D., Beavo, J., Walsh, K. A., and Charbonneau, H. (1990) (in preparation).
45. Ovchinnikov, Y. A., Gubanov, V. V., Khramtsov, N. V., Ischenko, K. A., Zagranichny, V. E., Muradov, K. G., Shuvaeva, T. M., and Lipkin, V. M. (1987) *FEBS Lett.*, **223**, 169–173.
46. Li, T., Volpp, K., and Applebury, M. L. (1989) *Proc. Natl Acad. Sci. USA*, **87**, 293–297.
47. Charbonneau, H., Prusti, K., LeTrong, H., Sonnenburg, W. K., Mullaney, P. J., Walsh, K. A., and Beavo, J. A. (1989) *Proc. Natl Acad. Sci. USA*, **87**, 288–292.
48. Lacombe, M.-L., Podgorski, G. J., Franke, J., and Kessin, R. H. (1986) *J. Biol. Chem.*, **261**, 16811–16817.
49. Conti, M., and Swinnen, J. V. (1990) In *Cyclic Nucleotide Phosphodiesterases: Structure, Regulation and Drug Action*, (Beavo, J. and Houslay, M. D., eds), John Wiley & Sons, Chichester, pp. 243–266.
50. Novack, J. P., Charbonneau, H., Walsh, K. A., and Beavo, J. A. (1990) (submitted).
51. Novack, J. P., Charbonneau, H., Blumenthal, D. K., Walsh, K. A., and Beavo, J. A. (1989) In *Calcium Protein Signaling* (Hidaka, H., ed.), Plenum Press, New York, pp. 387–395.
52. Ovchinnikov, Y. A., Lipkin, V. M., Kumarev, V. P., Gubanov, V. V., Khramtsov, N. V., Akhmedov, N. B., Zagranichny, V. E., and Muradov, K. G. (1986) *FEBS Lett.*, **204**, 288–292.
53. Londesborough, J., and Suoranta, K. (1983) *J. Biol. Chem.*, **258**, 2966–2972.
54. Dayhoff, M. O., Barker, W. C., and Hunt, L. T. (1983) *Methods Enzymol.*, **91**, 524–545.
55. Coggins, J. R., and Hardie, D. G. (1986) In *Multidomain Proteins—Structure and Evolution* (Hardie, D. G. and Coggins, J. R., eds), Elsevier, New York, pp. 1–12.
56. Charbonneau, H., Novack, J. P., MacFarland, R. T., Walsh, K. A., and Beavo, J. A. (1987) In *Calcium-Binding Proteins in Health and Disease* (Norman, A. W., Vanaman, T. C. and Means, A. R., eds), Academic Press, Orlando, Fl, pp. 505–517.
57. Stroop, S. D., Charbonneau, H., and Beavo, J. A. (1989) *J. Biol. Chem.*, **264**, 13718–13725.
58. Takio, K., Wade, R. D., Smith, S. B., Krebs, E. G., Walsh, K. A., and Titani, K. (1984) *Biochemistry*, **23**, 4207–4218.

59. Taylor, S. S. (1989) *J. Biol. Chem.*, **264**, 8443–8446.
60. Walsh, K. A. (1987) In *Signal Transduction and Protein Phosphorylation* (Heilmeyer, L. M. G., ed.), Plenum, New York, pp. 185–193.
61. Kincaid, R. L., Stith-Coleman, I. E., and Vaughan, M. (1985) *J. Biol. Chem.*, **260**, 9009–9015.
62. Krinks, M. H., Haiech, J., Rhoads, A., and Klee, C. B. (1984) *Adv. Cyclic Nucleotide Res.*, **16**, 31–47.
63. Culp, J. S., Blytt, H. J., Hermodson, M., and Butler, L. G. (1985) *J. Biol. Chem.*, **260**, 8320–8324.
64. Degerman, E., Belfrage, P., Newman, A. H., Rice, K. C., and Manganiello, V. C. (1986) *J. Biol. Chem.*, **262**, 5797–5807.
65. DeGrado, W. F. (1988) *Adv. Protein Chem.*, **39**, 51–124.
66. Cox, J. A., Comte, M., Fitton, J. E., and DeGrado, W. F. (1985) *J. Biol. Chem.*, **260**, 2527–2534.
67. Erickson-Viitanen, S., and DeGrado, W. F. (1987) *Methods Enzymol.*, **139**, 455–478.
68. Brandt, P., Zurini, M., Neve, R. L., Rhoads, R. E., and Vanaman, T. C. (1988) *Proc. Natl Acad. Sci. USA*, **85**, 2914–2918.
69. Benaim, G., Zurini, M., and Carafoli, E. (1984) *J. Biol. Chem.*, **259**, 8471–8477.
70. Kennelly, P. J., Edelman, A. M., Blumenthal, D. K., and Krebs, E. G. (1987) *J. Biol. Chem.*, **262**, 11958–11963.
71. Kemp, B. E., Pearson, R. B., Guerriero, V., Bagchi, I. C., and Means, A. R. (1987) *J. Biol. Chem.*, **262**, 2542–2548.
72. Colbran, R. J., Fong, Y.-L., Schworer, C. M., and Soderling, T. R. (1988) *J. Biol. Chem.*, **263**, 18145–18151.
73. Hubbard, M. J., and Klee, C. B. (1989) *Biochemistry*, **28**, 1868–1874.
74. Stryer, L. (1986) *Annu. Rev. Neurosci.*, **9**, 87–119.
75. Owen, W. G. (1987) *Annu. Rev. Physiol.*, **49**, 743–764.
76. Deterre, P., Bigay, J., Robert, M., Pfister, C., Kuhn, H., and Chabre, M. (1986) *Proteins*, **1**, 188–193.
77. Wensel, T. G., and Stryer, L. (1986) *Proteins*, **1**, 90–99.
78. Deterre, P., Bigay, J., Forquet, F., Robert, M., and Chabre, M. (1988) *Proc. Natl Acad. Sci. USA*, **85**, 2424–2428.
79. Brown, R. L., and Stryer, L. (1989) *Proc. Natl Acad. Sci. USA*, **86**, 4922–4926.
80. Lipkin, V. M., Dumler, I. L., Muradov, N. O., Artemyev, N. O., and Etingof, R. N. (1988) *FEBS Lett.*, **234**, 287–290.
81. Breitbart, R. E., Andreadis, A., and Nadal-Ginard, B. (1987) *Annu. Rev. Biochem.*, **56**, 467–495.
82. Doolittle, R. F. (1986) *Of URFS and ORFS*, University Science Books, Mill Valley, Ca., pp. 10–17.
83. Argos, P., and Leberman, R. (1985) *Eur. J. Biochem.*, **152**, 651–656.
84. deWit, R. J. W., Hoppe, J., Stec, W. J., Baraniak, J., and Jastorff, B. (1982) *Eur. J. Biochem.*, **122**, 95–99.
85. Couchie, D., Petridis, G., Jastorff, B., and Erneaux, C. (1983) *Eur. J. Biochem.*, **136**, 571–575.
86. Weber, I. T., Steitz, T. A., Bubis, J., and Taylor, S. S. (1987) *Biochemistry*, **26**, 343–351.
87. McKay, D. B., Weber, I. T., and Steitz, T. A. (1982) *J. Biol. Chem.*, **257**, 9518–9524.

DEVELOPMENT OF SELECTIVE INHIBITORS OF CYCLIC NUCLEOTIDE PHOSPHODIESTERASES AS THERAPEUTIC AGENTS

PART E

12

CARDIAC PHOSPHODIESTERASES AND THE FUNCTIONAL EFFECTS OF SELECTIVE INHIBITION

Martin L. Reeves and Paul J. England

Department of Cellular Pharmacology, SmithKline Beecham Pharmaceuticals, The Frythe, Welwyn, Herts AL6 9AR, UK

12.1 INTRODUCTION

Ever since the realisation that isoenzymes of cyclic nucleotide phosphodiesterase (PDE) exist, there has been considerable interest in developing the tools for selectively perturbing the activity of a single PDE isoenzyme to probe its functional role in any particular tissue. This interest was recently heightened when it was realised that such selective compounds may have major therapeutic utility [1], in, for example, modulating the cardiovascular system [2] or altering behaviour [3]. The heart was one of the first tissues in which more than one isoenzyme of PDE was detected [4]. Using non-selective PDE inhibitors (e.g. IBMX (3-isobutyl-1-methylxanthine), theophylline), it was recognised that PDE inhibition led to a positive inotropic response in heart [5], but whether any specific PDE was more important in regulating contractility could not be addressed in the absence of isoenzyme-selective agents. The premise that PDE inhibitors were potential regulators of cardiac contractility was founded on the evidence that cAMP was involved in mediating the inotropic response to β-adrenergic receptor agonists. The mechanisms for the hormonal stimulation of cAMP production and how increases in cAMP lead to an increase in contraction have been reviewed many times, and will not be discussed further here.

Clinically, although β-adrenergic receptor agonists and non-selective PDE inhibitors had been identified as inotropes, their usefulness as cardiotonic agents was limited by their side-effects [6]. Subsequently, however, a group of positive inotropes were discovered which lacked many of the undesirable effects of β-adrenergic receptor agonists and non-selective PDE inhibitors [7]. Although initial studies suggested that these agents were not PDE inhibitors [8], later work

Cyclic Nucleotide Phosphodiesterases: Structure, Regulation and Drug Action
Edited by J. Beavo and M. D. Houslay   © 1990 John Wiley and Sons Ltd

indicated that their mechanism of action was inhibition of cardiac PDE [2,9,10] and that inhibition of only one of the isoenzymes of PDE found in heart was important in producing an inotropic response.

This chapter will describe the various isoforms of PDE found in heart, and review the biochemical and functional effects of the selective inhibitors currently available. Unfortunately, much of the earlier work designed to assess the selectivity of PDE inhibitors was carried out on poorly separated and character-ised enzyme preparations, and a direct comparison of the data published from different laboratories and from different tissues is not always possible. The evidence that inhibition of only one of the forms of PDE in heart is responsible for the inotropic action of PDE inhibitors will also be discussed. This will include references to recent work which demonstrates that these PDE inhibitors alter the phosphorylation of key proteins involved in regulating contractility, thus confirming the involvement of PDE inhibition and cAMP elevation in the response. However, research into the mechanisms of action of these new agents has revealed an increasing complexity within the cAMP system in heart and its role in regulating contraction. In particular, there is evidence for the compartmen-tation of cAMP, with different compartments being coupled to different 'pools' of adenylate cyclase and different PDE isoenzymes.

12.2 STATUS OF CURRENT KNOWLEDGE OF CARDIAC PDEs

The existence of multiple forms of PDE in many tissues including heart is well documented [11] (Chapter 1). Inexplicably, the complexity of the isoforms of PDE appears to have been recognised only recently by many groups, which is puzzling when so much fine work was done over a decade ago. One excellent example of the demonstration of the presence of several isoenzymes of PDE in a single tissue was reported by Bergstrand and Lundquist [12]; they also described the properties of two high-affinity (or 'low-K_m') cAMP-specific PDE isoenzymes, one of which was very sensitive to inhibition by submicromolar concentrations of cGMP.

At the time of the discovery of the newer inotropic agents/PDE inhibitors, at least three forms of PDE were known to be present in heart [13,14], including human heart [15]. These forms were: a Ca^{2+} calmodulin-stimulated PDE (CaM-PDE), a cGMP-stimulated PDE (cGS-PDE), and an activity demonstrating substrate selectivity and high affinity for cAMP, the so-called 'low-K_m' PDE. Despite the recognition of these various isoforms of PDE, inhibition by the newer positive inotropes continued to be studied in heart homogenates, making it particularly difficult to interpret data if selective PDE inhibitors were being tested. Later it was assumed that one-step chromatography on ion exchange resins would yield kinetically 'clean' preparations of PDE. It is now clear that this was an incorrect assumption which has led to many of the problems in relating data on the properties of the PDE prepared in different laboratories, and in comparing the selectivity of PDE inhibitors. Moreover, it has recently become apparent that the heart contains a more complex array of PDE activities than the three described above, with the result that preparations of a 'low-K_m' PDE from heart used in many studies almost certainly contained at least two PDE activities, both of which were selective for cAMP as substrate but possessing very different inhibitor sensitivities [16]. Preparations were often used which showed non-linear Eadie–Hofstee plots, and only partial inhibition by high concentrations of inhibitors, a clear indication that the preparations contained multiple isoenzymes. This has led to a number of confusing reports, in which inhibitors of very different properties have been claimed to inhibit the 'low-K_m' PDE, whereas subsequent work has shown that the inhibitors actually inhibit different isoenzymes present in the enzyme preparations used.

Currently, a consensus is emerging on the PDE activities present in cardiac tissue, with only slight variations depending on the species. These isoenzymes are not exclusive to heart, since PDE activities with almost identical properties have been described in many other tissues, including non-muscle sources. Four major types of PDE have been described in cardiac muscle, as summarised in Table 12.1. For a detailed description of the properties of these isoenzymes, the reader is referred to the relevant chapters of this book. The nomenclature used to describe the various isoenzymes of PDE and favoured most by the authors is that introduced by Beavo [11], which we believe to be most definitive. Other commonly used classifications for the various isoenzymes found in heart are shown in parentheses.

Table 12.1   Summary of the properties of cardiac PDE

PDE	Source	Molecular mass	$K_m(\mu M)$ cAMP	$K_m(\mu M)$ cGMP	Ratio V_{max} cAMP/cGMP	Reference
CaM-PDE	Bovine	59 kDa	36	5.1	3.3	Unpublished
	Guinea pig	—	0.75	1.0	~1	[16]
cGS-PDE	Bovine	105 kDa	30[a]	10[a]	1	[30]
cGI-PDE	Bovine	110 kDa	0.2	0.1	10	[39]
Rolipram sensitive-PDE	Guinea pig	45 kDa	2.1	> 50	ND	[16]

[a] Hill coefficients are 1.9 and 1.3. for cAMP and cGMP respectively.
ND = not determined.

12.2.1 Ca^{2+}-calmodulin-stimulated phosphodiesterase (type I, PDE I)

The Ca^{2+}–CaM-stimulated enzyme is one of the most extensively studied PDE isoenzymes. Several forms of this PDE are known to exist, differing in substrate specificity, characteristics of the interaction with Ca^{2+}–CaM complex, and molecular weight. The enzyme from bovine heart has been studied in considerable detail [17–19].

Table 12.1 summarises the characteristics of bovine heart CaM-PDE. More recently, it has been recognised that a second type of CaM-PDE is found in heart, distinguishable on the basis of substrate specificity [16,20]. This activity, found for example in hearts of guinea pig, rat and man, has not been purified, but does appear similar to the activity found in immature rat testis [21].

CaM-PDE can show considerable activation (56-fold) by Ca^{2+}–CaM [22], but reports on the level of activation vary from 2- to 56-fold, particularly for non-purified preparations. Some of this variability could be due to incomplete resolution of CaM-PDE from other PDE activities, or to the fact that different forms of CaM-PDE have different degrees of activation by Ca^{2+}–CaM. In addition, limited proteolysis activates CaM-PDE by abolishing its requirement for Ca^{2+}–CaM binding [23] and so it is possible that differences in the method of preparation of CaM-PDE could affect the ratio of proteolysed/Ca^{2+}–CaM-dependent species and thus alter the level of activation by Ca^{2+}–CaM.

In view of the role of cyclic nucleotides in regulating muscle contraction via effects on Ca^{2+} movements, and the Ca^{2+} dependency of the activity of CaM-PDE, it is possible that CaM-PDE has an important regulatory function in controlling muscle contraction. In support of this proposal was the observation in smooth muscle that a correlation between contractile state and CaM-PDE activity could be measured [24,25]. The technique employed for estimating the fraction of CaM-PDE that was in an EGTA-sensitive state, i.e. with Ca^{2+}–CaM bound to it in the muscle, was to homogenise tissues in the presence and absence of trifluorperazine. This was used to stop association of free Ca^{2+}–CaM and unactivated CaM-PDE upon homogenisation, while dissociation of bound Ca^{2+}–CaM from CaM-PDE was inhibited by homogenisation and assay of the PDE activity at low temperatures. The results of these experiments indicated that Ca^{2+}–CaM and CaM-PDE did associate in intact cells, and that the degree of association correlated with the contractile state of the tissue, and hence by inference cytoplasmic Ca^{2+}. Other reports of a Ca^{2+}–CaM-dependent regulation of cAMP levels in other tissues include astrocytoma cells and thyroid [26,27]. A second potential mechanism for the regulation of cardiac CaM-PDE activity is via phosphorylation by protein kinase(s) (see Chapter 2). It has been shown that a brain CaM-PDE can be phosphorylated in vitro by cAMP-dependent protein kinase, and that the phosphorylation decreases the affinity of the enzyme for Ca^{2+}–CaM [28]. In other experiments the phosphorylation of brain CaM-PDE was catalysed by a Ca^{2+}–CaM-dependent protein kinase [29]. To date, however, there are no reports of phosphorylation of heart CaM-PDE. Thus, it would appear that multiple forms of the CaM-PDE exist and that these are selectively expressed in different tissues (Chapter 2).

12.2.2 Cyclic GMP-stimulated phosphodiesterase (type II, PDE II)

The second PDE isoenzyme found in heart, the cGMP-stimulated PDE (cGS-PDE), would appear to be the same activity in the majority of studies of heart PDE. The main characteristic of this PDE is that cAMP hydrolysis is markedly stimulated by micromolar concentrations of cGMP (see Chapter 3). The mechanism of this activation is believed to result from the association of cGMP with an allosteric binding site on the enzyme, leading to a conformational change at the separate hydrolytic binding site. This allosteric regulation is also responsible for the marked co-operativity in the kinetics of cyclic nucleotide hydrolysis observed when studying this enzyme. The most well characterised cGS-PDE is again that from bovine heart [30]. Table 12.1 summarises the main characteristics of this PDE. The observation that cAMP hydrolysis is stimulated by cGMP has tempted many to propose that this is a mechanism by which cGMP could control the level of cAMP in a cell, and thus explain some of the apparent 'antagonistic' effects of cGMP on cAMP-induced cellular responses, the so-called 'Yin-Yang' hypothesis. In frog and guinea pig heart there is evidence that the cGMP-induced inhibition of a cAMP-stimulated Ca^{2+} current is dependent on an increased hydrolysis of cAMP [31,32]. However, there are also reports that suggest that cGMP levels can be increased in heart without altering the level of cAMP, e.g. [33]. As for the CaM-PDE, much has to be done to understand the physiological relevance of cGS-PDE in heart.

12.2.3 Rolipram-sensitive phosphodiesterase (type IV, PDE IV)

The rolipram-sensitive PDE is so named by virtue of the fact that rolipram, an agent developed as an antidepressant, selectively and potently inhibits this activity. That rolipram sensitive- and cGMP-inhibited PDE isoenzymes were separate activities was first indicated by Yamamoto et al. [34] in a study of calf liver PDE. Since then the presence of this isoenzyme in heart has been reported in many species [16,35–37]. Ro 20-1724 also acts as a selective inhibitor of this enzyme. An important characteristic of this isoenzyme is that it is very specific for the hydrolysis of cAMP ($K_m \sim 1 \mu M$, cf. K_m cGMP $> 100 \mu M$). Of the PDE isoenzymes described in heart, least is known about this enzyme since there are no reports of its purification from heart or any other source. More recently, it was reported that a PDE activity from skeletal muscle with properties very similar to those of the rolipram-sensitive PDE from heart could be activated by cAMP-dependent protein kinase, presumably as a result of phosphorylation of the enzyme [38]. This observation indicates that phosphorylation and activation of rolipram-sensitive PDE could be a mechanism for limiting the effects of cAMP elevation following stimulation by an agonist, as has been suggested for other PDE isoenzymes (see Chapter 8).

Rolipram-sensitive PDE was alluded to earlier as one of the activities that was probably present in the 'low-K_m' PDE preparations used for inhibitor studies with PDE inhibitors that served as inotropic agents [16]. As discussed below, these crude preparations contributed to much of the misleading data in the literature regarding the inhibitory characteristics of compounds selective for the fourth PDE activity in heart, the cGMP-inhibited PDE.

12.2.4   Cyclic GMP-inhibited phosphodiesterase (type IV, PDE III)

The cGI-PDE is perhaps the most relevant enzyme in the context of this chapter, as it would seem to be the most important PDE activity for controlling the 'compartment' of cAMP involved in regulating contractility in heart. In many tissues, including heart, cGI-PDE appears to be a membrane-bound enzyme, although membrane association may vary with species (see below). For many years the kinetic characteristics reported for this activity from heart and other tissues varied considerably. As previously mentioned, some of this variation was probably due to the incomplete resolution of cGI-PDE from other PDE activities. However, more recently, following the isolation of antibodies raised against purified cGI-PDE from bovine heart [39], it is apparent that this isoenzyme is exquisitely sensitive to proteolysis, generating active fragments of a variety of molecular weights, each with slightly different kinetic characteristics. The bovine cardiac enzyme appears identical to the bovine and human platelet enzyme kinetically, immunologically, and by mobility on SDS gels (110 kDa) [40]. The presence of cGI-PDE in platelets and the potent anti-aggregatory effects of cGI-PDE inhibitors suggests that in such cells, as for heart, inhibition of cGI-PDE underlies the functional effects of these agents. As with the rolipram-sensitive PDE, cAMP is the favoured substrate ($K_m = 0.2\ \mu$M, V_{max} cAMP/V_{max} cGMP $= 10$) of cGI-PDE, though cGMP also binds to the active site with high affinity ($K_m = 0.1\ \mu$M). The net result of this binding is that cGMP is a very effective inhibitor of cAMP hydrolysis, hence its description as the cGMP-inhibited PDE [41]. This property has led to the suggestion that cGMP may increase cAMP levels in heart by inhibiting cAMP hydrolysis by cGI-PDE. However, to date there is no evidence for such a regulation.

cGI-PDE has also reported to be associated with membrane fractions in heart, more specifically the sarcoplasmic reticulum (SR) [37,42]. In order to isolate membrane-associated activity from heart, the choice of homogenisation conditions is important, i.e. isotonic buffers and gentle homogenisation are required [43]. It has been suggested that the potency of the cGI-PDE inhibitors as inotropes in any species correlates with the degree of association of cGI-PDE with membranes. This is because in species where cGI-PDE is apparently not found associated with the membrane fraction, the cGI-PDE inhibitors are found to act only as weak, positive inotropic agents [43,44]. This is particularly interesting because of the presence in cardiac membranes of proteins whose phosphorylation by cAMP-dependent protein kinase is of primary importance in the regulation of the inotropic state of the heart. Two of particular relevance are phospholamban (present in the SR), the phosphorylation of which results in an increased rate of uptake of Ca^{2+} into the SR, and the voltage-operated Ca^{2+} channel, whose phosphorylation produces an increased influx of extracellular Ca^{2+} during the action potential. The localisation of these proteins in the same membrane(s) with the cGI-PDE could explain why inhibition of this isoenzyme results in such potent positive inotropic effects with only relatively small increases in total tissue cAMP (see below).

From studies in platelets it is apparent that the cGI-PDE activity may be regulated by cAMP-dependent phosphorylation [45,46]. In these studies phosphorylation and activation of cGI-PDE was demonstrated in response to agents that increased cAMP, although the degree of activation was somewhat variable. This observation suggests that phosphorylation of cGI-PDE may be a mechanism for the regulation of cAMP levels in response to an agonist. Indeed, phosphorylation of cardiac cGI-PDE has been reported, though in this study no change in activity was observed [39]. Further evidence for regulation of cardiac cGI-PDE by a cAMP-dependent mechanism was provided by the report that cAMP analogues (which would activate A-kinase) lowered cAMP levels in rat heart cells. The inference from this is that such a fall in cAMP was due to phosphorylation and activation of the cGI-PDE [47].

This brief review of the properties of PDE isoenzymes found in heart now leads us to a discussion of the various inhibitors available to selectively inhibit heart PDE and how these have been used to probe the functional role of such activities in the myocardium.

12.3 THE ROLE OF PDEs IN CONTROLLING HEART FUNCTION

The functional significance of multiple forms of PDE in heart will be discussed from three aspects: the data supporting the theory that PDE inhibition and concomitant rises in cAMP mediate the inotropic actions of the newer inotropic agents, the functional consequences of inhibiting the other forms of PDE present in heart and a brief analysis of the proposal that compartments of cAMP may exist in heart.

12.3.1 Evidence that cGI-PDE inhibitors exert their positive inotropic actions through altering cAMP concentrations

The earliest indications that the newer inotropic agents such as amrinone and milrinone exerted their positive inotropic effects by inhibition of cGI-PDE in heart came from correlations of positive inotropic potency in heart with inhibitory activity against cGI-PDE [48]. These correlations were found to be very good and suggested a strong link between these two effects [42,49]. Support for PDE inhibition being the mechanism of action of these inotropes came from the many observations that the levels of cAMP in heart were increased following exposure to these cGI-PDE inhibitors [2,50–54]. In other studies where the time-course of cAMP accumulation and inotropy were compared, there appeared to be an increase in contraction before a significant increase in cAMP when cGI-PDE inhibitors were used, but not when non-selective PDE inhibitors were studied [55]. The lack of correlation in these time-courses could be explained if inotropic responses were elicited by small increases in cAMP which would be below the level of statistical significance. In general, the increases in cAMP in the heart in response to cGI-PDE inhibitors are relatively small when compared to those produced by β-adrenergic agonists, and this makes it difficult to detect initial increases. In addition, evidence is accumulating that there may be compartmentation of cAMP and/or PDE isoenzymes in heart. Thus, localised small increases in cAMP in the relevant compartment may be all that is needed to elicit an inotropic response. The concept of compartmentation of cAMP in cardiac muscle has been proposed previously, based on experiments using a variety of agonists of the cAMP system (PGE_1, β_1- and β_2-adrenergic agonists, forskolin). This is discussed in more detail below, with particular reference to the use of isoenzyme-selective PDE inhibitors.

The only proven mediator of the actions of cAMP in mammalian cells is the cAMP-dependent protein kinase (A-kinase) [56]. Therefore, several attempts have been made to demonstrate that exposure of hearts to cGI-PDE inhibitors could activate A-kinase. It would be expected that if these agents produce only small changes in intracellular cAMP then the increases in the activation of A-kinase would also be small. Indeed, data are sparse in the literature reporting such an activation of A-kinase [37,57,58], though this is probably due as much to the technical difficulties of the assay used to measure the activation of A-kinase as to the magnitude of the activation itself. In the authors' laboratory, efforts have been concentrated on developing a more reliable assay of A-kinase activation using a synthetic peptide substrate for the phosphorylation reaction, instead of the more widely used artificial substrate, histone [59]. Using this improved method we have been able to demonstrate clearly an activation of A-kinase in guinea pig hearts perfused with the selective cGI-PDE inhibitor SK&F 94120 even at low concentrations of SK&F 94120 (Table 12.2) [60].

Table 12.2 Time-course of the effects of the cGI-PDE inhibitor SK&F 94120 on perfused guinea pig hearts

Perfusion time (s)	Cyclic AMP (pmol/mg protein)	Soluble A-kinase (% active)	Developed tension
0	9.7 ± 0.9	23.5 ± 1.7	100
40	12.7 ± 1.4	31.7 ± 2.4^a	114 ± 1.9^b
70	17.1 ± 1.2^b	35.0 ± 1.0^b	120 ± 4.2^b
120	12.8 ± 1.2	33.0 ± 0.4^b	126 ± 6.5^b
300	18.0 ± 2.7^a	36.2 ± 1.6^b	127 ± 6.7^a

Values are means $\pm$ SEM. $^aP < 0.05$, $^bP < 0.01$ compared with controls.

Further strong evidence for a cAMP-dependent mechanism mediating the inotropic responses of the cGI-PDE inhibitors would be the demonstration that protein substrates known to be phosphorylated in response to β-adrenergic receptor agonist were also phosphorylated in response to cGI-PDE inhibitors. Such a study has been reported [57] using amrinone and, consistent with cAMP mediating the inotropic response to this agent, increases in the phosphorylation of several proteins was observed. Similarly, in the authors' laboratory, using the cGI-PDE inhibitor SK&F 94120, increases in the phosphorylation of a number of proteins thought to be involved in regulating cardiac contraction was observed, with the profile of proteins phosphorylated being similar to that induced by the β-adrenergic receptor agonist isoprenaline [61]. In contrast, however, Rapundalo et al. [62] recently reported that no significant changes in the phosphorylation of any protein were observed in response to the perfusion of hearts with milrinone, another cGI-PDE inhibitor.

Other evidence consistent with cGI-PDE inhibitors acting via stimulation of A-kinase and protein phosphorylation is their effect in producing carbachol-inhibitable increases in intracellular Ca^{2+} transients. These show an increased inward Ca^{2+} current during systole [53] and increased $^{45}Ca^{2+}$ influx [54]. Such effects are consistent with phosphorylation of the Ca^{2+} channel, which has been demonstrated to result in increased Ca^{2+} channel opening during depolarisation ([63] for review).

In using cGI-PDE inhibitors to study the functional role of cGI-PDE in heart, it should always be remembered that it is possible that, with some of these agents, mechanisms in addition to the inhibition of PDE isoenzymes may be partly responsible for their inotropic action. For example, increase in Ca^{2+} sensitivity of myofibrils [64], inhibition of G_i (the inhibitory GTP-dependent receptor coupling protein), or antagonism of A_1-adenosine receptors [65] have been observed to be elicted by certain cGI-PDE inhibitors. Care must therefore be exercised in using these compounds to study the pharmacological consequences of inhibition of cGI-PDE.

In summary, cGI-PDE inhibitors have been shown to be positive inotropic agents in the heart. There is an increasing amount of evidence, typified by increases in cAMP, activation of A-kinase, and phosphorylation of appropriate proteins, that inhibition of cGI-PDE and resultant elevation of cAMP mediates the inotropic actions of these inhibitors. In the next sections we will see that these inhibitors could prove very useful in investigating whether compartmentation of the cAMP system exists in the heart cell, as suggested by Hayes et al. [64].

12.3.2   Effects of other PDE inhibitors on cardiac function

Although the evidence is clear that inhibition of cGI-PDE results in increased cardiac contractility, the evidence for effects of inhibition of the other PDE isoenzymes on functional parameters of the heart is much weaker. Until recently the study of the functional role of PDE in any tissue has relied on the use of non-selective inhibitors of PDE, e.g. IBMX, theophylline. Obvious difficulties in interpretation of experimental data arise from the fact that more than one isoenzyme may be affected by these non-selective inhibitors (see Table 12.3). In the absence of alternative mechanisms for selectively altering the activity of PDE, the use of selective inhibitors of the various PDE isoenzymes in heart remains the only mechanism for investigating the role of PDEs in regulating contractility and other cellular functions. With the advent of a more general appreciation of the existence and understanding of the multiple isoenzymes of PDE, other selective inhibitors have been identified and, in some cases, agents once thought of as selective inhibitors of one isoenzyme of PDE have been found to inhibit other PDE isoenzymes to a significant extent. Table 12.3 summarises the effects of the better known selective PDE inhibitors on cardiac PDE isoenzymes.

Exposure of hearts to appropriate concentrations of these selective inhibitors can permit investigation of the specific physiological processes regulated by the relevant PDE isoenzyme. Indeed, a few experiments of this nature have been published, in which a variety of cardiac muscle preparations were incubated or perfused with these and other selective PDE inhibitors. Table 12.4 summarises the results of some of these studies.

Earlier it was speculated that the activity of CaM-PDE might be closely coupled with the contractile state of the heart, since both are regulated by Ca^{2+} and this indicates a key role for CaM-PDE in regulating cardiac contractility. For an investigation of the importance of cardiac CaM-PDE in regulating heart contractility, zaprinast is the only compound available to date which can inhibit it selectively. However, zaprinast is also a potent inhibitor of the cGMP-specific PDE found in vascular smooth muscle [66], which is probably the same enzyme as the cGMP-binding PDE described in lung [67] and platelets [68]. This finding demonstrates that, in the assessment of the 'PDE selectivity' of a compound, it is important to recognise that different PDE isoenzymes may exist in tissues other than the one under investigation. In those experiments where hearts were

Table 12.3 Selectivity of a variety of PDE inhibitors for cardiac PDE

Compound	Guinea pig CaM-PDE (μM)	Bovine CaM-PDE (μM)	cGS-PDE (μM)	cGI-PDE (μM)	Rolipram-sensitive PDE (μM)
Amrinone	> 1000[c]	880[a]	400[a]	7.9[a]	280[e]
Milrinone	310[c]	48[a]	180[a]	0.26[a]	11[e]
SK&F 94120	NI[b]	—	NI[b]	0.8[b]	NI[b]
Adibendan (BM 14478)	> 100[f]	—	> 100[f]	0.52[f]	> 100[f]
Imazodan (CI 914)	> 1000[c]	—	750[c]	8.2[c]	> 180[d]
Fenoximone (MDL 17043)	> 1000[c]	240[a]	430[a]	15[e]	NI[f]
Cilostamide (OPC 3689)	—	—	—	0.02[d]	31[d]
Sulmazole (ARL 115 BS)	> 1000[c]	—	240[c]	40[c]	155[e]
Zaprinast (M&B 22948)	13[f]	6.8[f]	> 100[f]	> 100[f]	> 100[f]
Dipyridamole	54[c]	—	29[f]	174[f]	6.1[f]
Rolipram (ZK 62711)	NI[b]	—	NI[b]	90[d]	0.4[f]
Ro 20-1724	NI[b]	240[a]	300[a]	62[a]	3.1[b]
Denbufylline	~ 140[f]	—	> 100[f]	> 100[f]	0.83[f]
IBMX	16[f]	6.7[a]	14[a]	1.3[a]	8.0[f]
Papaverine	24[c]	12[a]	4.5[a]	0.66[a]	—
Theophylline	230[c]	—	210[c]	340[c]	310[d]
MY 5445	~ 135[f]	—	114[f]	NI[f]	90[f]

NI = No inhibition observed up to 100 μM.
[a] K_i values obtained using bovine heart PDE. cGS-PDE assayed in the presence of 1 μM cGMP [39].
[b] K_i values obtained using human heart PDE [16].
[c] IC$_{50}$ values obtained using guinea pig heart and determined using 1 μM cAMP as substrate [20].
[d] IC$_{50}$ values obtained using guinea pig heart cGI-PDE and canine heart rolipram-sensitive PDE and determined using 1 μM cAMP as substrate [43].
[e] K_i values obtained using rabbit heart PDE [37].
[f] IC$_{50}$ values obtained using guinea-pig heart PDE and determined using 1 μM cyclic nucleotide (authors' unpublished data).

exposed to zaprinast, a doubling of the concentration of cGMP resulted with only a slight negative inotropic response being recorded. However, in view of the lack of totally selective CaM-PDE inhibitors and the limited studies that have been conducted with zaprinast, no conclusions regarding the functional role of CaM-PDE in heart can be made.

It should also be noted that there exists another group of agents that will inhibit CaM-PDE, the so called 'calmodulin antagonists'. These agents inhibit the CaM-PDE by interfering with the activation of this PDE activity by the Ca^{2+}–CaM complex [69]. However, their usefulness for investigating the role of CaM-PDE in heart is rather limited by the fact that these antagonists inhibit Ca^{2+}–CaM activation of many CaM-dependent processes of direct relevance to

Table 12.4 Effect of isoenzyme-selective PDE inhibitors on cardiac contractility and cyclic nucleotide levels

PDE inhibited	Inhibitor (dose)	Test system	Increase cAMP	Increase cGMP	Inotropic effect	Reference
CaM-PDE	Zaprinast					
	(30 μM)	Cat ventricle	11%[a]	69%[a]	[a]	[72]
	(10 μM)	Guinea pig atria	10%	ND	[a]	[35]
cGS-PDE	Dipyridamole					
	(10 μM)	Guinea pig atria	ND	ND	Negative	[35]
Rolipram-	Rolipram					
sensitive	(30 μM)	Cat ventricle	27%	5%[a]	[a]	[72]
PDE	(32 μM)	Guinea pig ventricle	147%	ND	[a]	[73]
	Ro 20-1724					
	(0.1 mg/kg)	Anaesthetised dog	ND	ND	[a]	[35]
cGI-PDE	Milrinone					
	(10 μM)	Guinea pig papillary	25%	16%[a]	Positive	[58]
	SK&F 94120					
	(100 μM)	Guinea pig heart	76%	ND	Positive	[60]
	Imazodan					
	(32 μM)	Guinea pig atria	62%	-1%[a]	Positive	[81]

ND = not determined.
[a] Not significantly different from controls.

cardiac cell function, and also may be inhibitors of protein kinase-C [69]. They also do not inhibit the 'basal' state of the CaM-PDE, i.e. when Ca^{2+} concentrations are insufficiently high to activate this enzyme.

The screening of compounds for activity as cGS-PDE inhibitors is complicated by the presence of two potential sites of interaction on cGS-PDE, the hydrolytic site and the allosteric binding site [70]. Binding of an agent at the active site would normally result in inhibition of hydrolysis of cyclic nucleotide. However, an interaction at the allosteric site could result in either activation or inhibition of hydrolysis of the cyclic nucleotide substrate if the compound is an agonist or antagonist, respectively, at the allosteric site. Therefore, screening of potential cGS-PDE effectors should take this into account, and assays using cAMP as the substrate should be conducted in the presence and absence of an activating concentration of cGMP. To date no selective inhibitors or activators of cGS-PDE have been identified. Some investigators have exposed cardiac preparations to the agent dipyridamole in an attempt to selectively inhibit cGS-PDE [35]. However, the selectivity of dipyridamole for cGS-PDE when compared to other PDEs is not good (Table 12.3), and this lack of selectivity is further complicated by its effects on other cell processes, making the interpretation of the results very difficult. Furthermore, the fact that cGMP is a marked activator of cAMP hydrolysis by this isoenzyme will influence the interpretation of experiments where selective inhibitors of other PDEs in heart produce alterations in cGMP levels, since this could effect the activity of cGS-PDE. Demand for truly selective inhibitors of cGS-PDE is high, for the obvious reason that the regulation of cAMP levels by cGMP has been implicated in several important biological systems, but as yet biochemical probes for such a regulation by cGS-PDE are not available.

Rolipram and Ro 20-1724 are selective inhibitors of the rolipram-sensitive PDE. The finding that heart contains this PDE activity was made by virtue of the fact that heart contained a 'low-K_m' cAMP-selective PDE that was sensitive to inhibition by rolipram, but not the cGI-PDE inhibitors. In this case the identification of a particular PDE in a heart was dependent solely on inhibitor sensitivity, and the lack of inhibition of cAMP hydrolysis by cGMP. Much work is still to be done to characterise this activity more fully in heart and other tissues. Again the data available on the effects of exposure of heart to agents such as rolipram are sparse. Of interest, though, was that rolipram caused a 2–3-fold increase in cAMP and activation of A-kinase [71] with little positive inotropic effect (except at very high concentrations, where inhibition of the cGI-PDE could be significant). This result is reminiscent of that obtained by perfusion with PGE_1, both suggesting that compartments of cyclic nucleotides may exist in heart.

It would appear from the very limited studies carried out to date that only agents that inhibit cGI-PDE have an inotropic effect in heart, even though inhibition of other PDEs may increase the level of cAMP and/or cGMP. Other explanations could be invoked, however, one being that since such a limited range of compounds has been used in these experiments (this is not true for cGI-PDE inhibitors), other effects of the inhibitors may mask the positive inotropic effect. Indeed, in studies where hearts were exposed to combinations of inhibitors (e.g. zaprinast and SK&F 94120), there are reports of a slight attenuation of the inotropic response to the cGI-PDE inhibitor [72]. In contrast, it has been reported that rolipram acted synergistically with SK&F 94120 to increase the force of contraction of isolated guinea pig ventricle strips [73].

Although inhibitors of CaM-PDE and rolipram-sensitive PDE isoenzymes have little or no functional effect in cardiac muscle, it should be noted that these compounds can have major effects in other tissues. Zaprinast, although a selective inhibitor of CaM-PDE in heart, causes marked relaxation of smooth muscle by inhibition of a cGMP-specific PDE isoenzyme [74,75] not present in cardiac muscle. Rolipram and other inhibitors of the same isoenzyme not only have major effects in the central nervous system [3], but are also smooth muscle relaxants [74,75] and potent stimulators of gastric acid secretion [76]. Thus the relative importance of the various PDE isoenzymes in the control of cell function shows considerable variation between different tissues. The presence of an isoenzyme by itself is not indicative of a role in a particular cell function, and it is necessary to show also that inhibition of the isoenzyme results in a change in the relevant response.

Since only a few experiments using selective inhibitors of enzymes other than cGI-PDE have been reported, no clear answer as to the role of these isoenzymes in controlling cardiac contractility or any other cellular function has emerged. In the case of cGI-PDE there are indications in several species that this isoenzyme controls a pool of cAMP important in regulating contraction, at least acutely. Whether other PDEs are involved in regulating cardiac contractility in the long term, perhaps by influencing the metabolic potential of the heart muscle, or even its structural characteristics, is yet to be addressed.

12.3.3   Compartmentation of cAMP in the heart

The suggestion that compartments of cAMP exist in heart cells was first invoked to explain the anomalous observations that although both isoprenaline and PGE_1 caused similar increases in cardiac cAMP, only isoprenaline gave an increased inotropic response [77]. Hayes et al. [78] extended these observations by showing that the increase in cAMP induced by PGE_1 did not result in the phosphorylation of troponin-I or phosphorylase in isolated cardiac myocytes, indicating that such compartments existed in the same cell (see also [79]). In addition, there is other evidence for a lack of correlation between levels of cAMP, inotropic response, and contractile protein phosphorylation [80], including that discussed above using selective inhibitors of PDE isoenzymes.

The identity or localisation of these 'compartments' is presently unclear. They could represent spatially localised regions in which there is limited diffusion, or the compartmentation could be of a more functional nature, in which the enzymes producing, removing and interacting with cAMP are in close proximity. In support of this second hypothesis is the observation discussed earlier that cGI-PDE in several species is associated with cell membrane fractions which also have been shown to contain A-kinase and known substrate proteins. The lack of any positive inotropic effect of the cGI-PDE inhibitors in rat and hamster could then be explained by the observation that cGI-PDE in these species is not associated with the membrane fraction, whereas in responding species it is. A pattern is now apparently emerging which implicates positive inotropic responses with a compartmentalisation of cAMP related to heart cell particulate structures. However, much further work is necessary to define the complexities of such proposed multiple 'compartments' of cAMP within the myocyte.

12.4   CONCLUSIONS

Although the heart contains several distinct PDE isoenzymes, their involvement in the control of cardiac function is far from clear. The isoenzyme for which the evidence for a major regulatory role is best established is the cGI-PDE. A range of agents of widely different chemical structures which inhibit this isoenzyme are positive inotropes, suggesting that their main mechanism of action in the intact tissue is via inhibition of cGI-PDE. Evidence is also accumulating that inhibition of cGI-PDE results in an increased positive inotropic response through activation of cAMP-dependent protein kinase and phosphorylation of relevant contractile and membrane proteins.

Concerning the role of the other cAMP-hydrolysing PDE isoenzymes in the regulation of cardiac function, our knowledge is very unclear. Inhibition of the rolipram-sensitive PDE results in large increases in cAMP with no apparent effect on function. This compares with the activation of adenylate cyclase by PGE_1. This raises the question of how an increase in cAMP and related activation of A-kinase can occur without phosphorylation of proteins. In addition, the issue arises of whether this PDE isoenzyme, which has a significant catalytic activity in heart in several species, has any physiological role in controlling cardiac function.

The importance of the cGS-PDE is even less well understood. Hearts of most species contain a high specific activity of this enzyme compared to other PDE isoenzymes, although because of its kinetics much of this will only be expressed in the presence of elevated concentrations of cGMP. However, because of the lack of high-affinity, specific inhibitors of this isoenzyme, it has been impossible to investigate its function using the approaches which have proved successful with the other isoenzymes.

Inhibition of the CaM-PDE in the heart results in an increase in cGMP, indicating that this enzyme is important in controlling the hydrolysis of this nucleotide in heart. Cyclic GMP in heart normally has been associated with negative inotropism, although often there is not a clear inverse relationship between concentrations of cGMP and the inotropic state, and the potential exists for considerable interplay between cGMP and cAMP in the heart. The CaM-PDE (at least in some species) can hydrolyse both nucleotides equally; the hydrolysis of cAMP by the cGS-PDE is increased many-fold by cGMP, while the activity of the cGI-PDE is inhibited by cGMP. Although some of these regulatory mechanisms appear to be contradictory, if 'compartments' of cyclic nucleotides do exist in heart there is the potential for specific regulation by different isoenzymes of various 'pools' of cyclic nucleotides coupled to specific cell functions. There is also evidence that similar compartmentation of cAMP can exist in tissues other than heart. These observations open up the possibility of developing isoenzyme-specific PDE inhibitors which not only selectively affect different cell functions, but also can be targetted to a range of therapeutic utilities.

REFERENCES

1. Weishaar, R. E., Cain, M. H., and Bristol, J. A. (1985) *J. Med. Chem.*, **28**, 537–545.
2. Honerjager, P., Schafer-Korting, M., and Reiter, M. (1981) *Naunyn-Schmiedeberg's Arch. Pharmacol.*, **318**, 112–120.

3. Wachtel, H. (1983) *Neuropharmacology*, **22**, 267–272.
4. Beavo, J. A., Hardman, J. G., and Sutherland, E. W. (1970) *J. Biol. Chem.*, **245**, 5649–5655.
5. Korth, M (1978) *Naunyn-Schmiedeberg's Arch. Pharmacol.*, **302**, 77–86.
6. Smith, T. W., and Braunwald, E. (1984) *Heart Disease: A Textbook of Cardiovascular Medicine* (Braunwald, E. ed.) 2nd edn, W B Saunders, Philadelphia, pp. 503–590.
7. Alousi, A., Farah, A. E., Lesher, G., and Opalka, C. J. Jr (1978) *Fed. Proc.*, **37**, 914.
8. Alousi, A. A., Farah, A. E., Lesher, G. Y., and Opalka, C. J. Jr (1979) *Circ. Res.*, **45**, 666–677.
9. Endoh, M., Satoh, K., and Yamashita, S. (1980) *Eur. J. Pharmacol.*, **66**, 43–52.
10. Kariya, T., Wille, L. J., and Dage, R. C. (1982) *J. Cardiovasc. Pharmacol.*, **4**, 509–514.
11. Beavo, J. A., Hansen, R. S., Harrison, S. A., Hurwitz, R. L., Martins, T. J., and Mumby M. C. (1982) *Mol. Cell. Endocrinol.*, **28**, 387–410.
12. Bergstrand, H., and Lundquist, B. (1976) *Biochemistry*, **15**, 1727–1735.
13. Wells, J. N., and Hardman, J. G. (1977) *Adv. Cyclic Nucleotide Res.*, **8**, 119–143.
14. Thompson, W. J., Terasaki, W. L., Epstein, P. M., and Strada, S. J. (1979) *Adv. Cyclic Nucleotide Res.*, **10**, 69–92.
15. Hidaka, H., Yamaki, T., Ochiai, Y., Asano, T., and Yamabe, H. (1977) *Biochim. Biophys. Acta.*, **484**, 398–407.
16. Reeves, M. L., Leigh, B. K., and England, P. J. (1987) *Biochem. J.* **241**, 535–541.
17. Ho, H. C., Wirch, E., Stevens, F. C., and Wang, J. H. (1977) *J. Biol. Chem.*, **252**, 43–50.
18. LaPorte, D. C., Toscano, W. A. Jr., and Storm, D. R. (1979) *Biochemistry*, **18**, 2820–2825.
19. Hansen, R. S., and Beavo, J. A. (1982) *Proc. Natl Acad. Sci. USA*, **79**, 2788–2792.
20. Weishaar, R. E., Burrows, S. D., Kobylarz, D. C., Quade, M. M., and Evans, D. B. (1986) *Biochem. Pharmacol.*, **35**, 787–800.
21. Purvis, K., Olsen, A., and Hansson, V. (1981) *J. Biol. Chem.*, **256**, 11434–11441.
22. Tucker, M. M., Robinson, J. B. Jr, and Stellwagen, E. (1981) *J. Biol. Chem.*, **256**, 9051–9058.
23. Kincaid, R. L., Stith-Coleman, I. E., and Vaughan, M. (1985) *J. Biol. Chem.*, **260**, 9009–9015.
24. Saitoh, Y., Hardman, J. G., and Wells, J. N. (1985) *Biochemistry*, **24**, 1613–1618.
25. Miller, J. R., and Wells, J. N. (1987) *Biochem. Pharmacol.*, **36**, 1819–1824.
26. Tanner, L. T., Harden, T. K., Wells, J. N., and Martin, M. W. (1986) *Mol. Pharmacol.*, **29**, 455–460.
27. Erneux, C., Van Sande, J., Miot, F., Cochaux, P., Decoster, C., and Dumont, J. E. (1985) *Mol. Cell. Endocrinol.*, **43**, 123–134.
28. Sharma, R. K., and Wang, J. H. (1985) *Proc. Natl Acad. Sci. USA*, **82**, 2603–2607.
29. Sharma, R. K., and Wang, J. H. (1986) *J. Biol. Chem.*, **261**, 1322–1328.
30. Martins, T. J., Mumby, M. C., and Beavo, J. A. (1982) *J. Biol. Chem.*, **257**, 1973–1979.
31. Hartzell, H. C., and Fischmeister, R. (1986) *Nature*, **323**, 273–275.
32. Fischmeister, R., and Hartzell, H. C. (1987) *J. Physiol.*, **387**, 453–472.
33. England, P. J. (1976) *Biochem. J.*, **160**, 295–304.
34. Yamamoto, T., Lieberman, F., Osborne, J. C. Jr, Manganiello, V. C., Vaughan, M., and Hidaka, H. (1984) *Biochemistry*, **23**, 670–675.
35. Wieshaar, R. E., Kobylarz-Singer, D. C., Quade, M. M., Steffen, R. P., and Kaplan, H. R. (1987) *J. Cyclic Nucleotide Protein Phosphorylation Res.*, **11**, 513–527.

36. Tenor, H., Bartel, S., and Krause, E.-G. (1987) *Biomed. Biochim. Acta.*, **46**, S749–S753.
37. Kithas, P. A., Artman, M., Thompson, W. J., and Strada, S. J. (1988) *Circ. Res.*, **62**, 782–789.
38. Cordle, S. R., and Corbin, J. D. (1988) *Faseb J.*, **2**, A595.
39. Harrison, S. A., Reifsnyder, D. H., Gallis, B., Cadd, G. G., and Beavo, J. A. (1986) *Mol. Pharmacol.*, **29**, 506–514.
40. Macphee, C. H., Harrison, S. A., and Beavo, J. A. (1986) *Proc. Natl Acad. Sci.*, **83**, 6660–6663.
41. Harrison, S. A., Chang, M. L., and Beavo, J. A. (1986) *Circulation*, suppl. 3, 109–116.
42. Kauffman, R. F., Crowe, V. G., Utterback, B. G., and Robertson, D. W. (1986) *Mol. Pharmacol.*, **30**, 609–616.
43. Weishaar, R. E., Kobylarz-Singer, R. P., and Kaplan, H. R. (1987) *J. Mol. Cell. Cardiol.*, **19**, 1025–1036.
44. Manganiello, V. C. (1987) *J. Mol. Cell. Cardiol.*, **19**, 1037–1040.
45. Macphee, C. H., Reifsnyder, D. H., Moore, T. A., Lerea, K. M., and Beavo, J. A. (1988) *J. Biol. Chem.*, **263**, 10353–10358.
46. Grant, P. G., Mannarino, A. F., and Colman, R. W. (1988) *Proc. Natl Acad. Sci.*, **85**, 9071–9075.
47. Gettys, T. W., Blackmore, P. F., Redmon, J. B., Beebe, S. J., and Corbin, J. D. (1987) *J. Biol. Chem.*, **262**, 333–339.
48. Bristol, J. A., Sircar, I., Moos, W. H., Evans, D. B., and Weishaar, R. E. (1984) *J. Med. Chem.*, **27**, 1099–1101.
49. Sircar, I., Weishaar, R. E., Kobylarz, D., Moos, W. H., and Bristol, J. A. (1987) *J. Med. Chem.*, **30**, 1955–1962.
50. Endoh, M., Yanagisawa, T., Taira, N., and Blinks, J. R. (1986) *Circulation*, **73** (suppl. 3), 117–113.
51. Weishaar, R. E., Quade, M., Schenden, J. A., Boyd, D. K., and Evans, D. B. (1985) *Eur. J. Pharmacol.*, **119**, 205–215.
52. Ahn, H. S., Eardley, D., Watkins, R., and Prioli, N. (1986) *Biochem. Pharmacol.*, **35**, 1113–1121.
53. Gristwood, R. W., English, T. A. H., Wallwork, J., Sampford, K. A., and Owen, D. A. A. (1987) *J. Cardiovasc. Pharmacol.*, **9**, 719–727.
54. Olson, E. M., Kim, D., Smith, T. W., and Marsh, J. D. (1987) *J. Mol. Cell. Cardiol.*, **19**, 95–104.
55. Hsieh, C. P., Kariya, T., Dage, R. C., and Ruberg, S. J. (1987) *J. Cardiovasc. Pharmacol.*, **9**, 230–236.
56. Krebs, E. G., and Beavo, J. A. (1979) *Annu. Rev. Biochem.*, **48**, 923–959.
57. Hayes, J. S., Bowling, N., Boder, G. B., and Kauffman, R. (1984) *J. Pharmacol. Exp. Ther.*, **230**, 124–132.
58. Silver, P. J., Harris, A. L., Canniff, P. C., Lepore, R. E., Bentley, R. G., Hamel, L. T., and Evans, D. B. (1989) *J. Cardiovasc. Pharmacol.*, **13**, 530–540.
59. Murray, K. J., England, P. J., Lynahm, J. A., Mills, D., and Reeves, M. L. (1988) *Biochem. Soc. Trans*, **16**, 355.
60. Murray, K. J., England, P. J., and Reeves, M. L. (1987) *Br. J. Pharmacol.*, **92**, 755.
61. Reeves, M. L., England, P. J., and Murray, K. J. (1989) *Biochem. Soc. Trans*, **17**, 169.
62. Rapundalo, S. T., Solaro, R. J., and Kranias, E. G. (1989) *Circ. Res.*, **64**, 104–111.

63. Sperelakis, N. (1988) *J. Mol. Cell. Cardiol.*, **20** (suppl. 2), 75–105.
64. Hayes, J. S., Brunton, L. L., and Mayer, S. E. (1980) *J. Biol. Chem.*, **255**, 5113–5119
65. Parsons, W. J., Ramkumar, V., and Stiles, G. L. (1988) *Mol. Pharmacol.*, **33**, 441–448
66. Lugnier, C., Schoeffter, P., Le Bec, A., Stouthou, E., and Stoclet, J. C. (1986) *Biochem Pharmacol.*, **35**, 1743–1751.
67. Francis, S. H., Lincoln, T. M., and Corbin, J. D. (1980) *J. Biol. Chem.*, **255**, 620–626
68. Coquil, J. F., Franks, D. J., Wells, J. N., Dupuis, M., and Hamet, P. (1980) *Biochim Biophys. Acta*, **631**, 148–165.
69. Weishaar, R. E., Quade, M. M., Schenden, J. A., and Evans, D. B. (1985) *J. Cyclic Nucleotide Protein Phosphorylation Res.*, **10**, 551–564.
70. Yamamoto, T., Yamamoto, S., Osborne, J. C. Jr, Manganiello, V. C., Vaughan, M. and Hidaka, H. (1983) *J. Biol. Chem.*, **258**, 14173–14177.
71. Reeves, M. L., England, P. J., and Murray, K. J. (1987) *Biochem. Soc. Trans*, **15** 955–956.
72. Gristwood, R. W., Sampford, K. A., and Williams, T. J. (1986) *Br. J. Pharmacol.*, **89** 573.
73. Gristwood, R. W., and Owen, D. A. A. (1986) *Br. J. Pharmacol.*, **87**, 91.
74. Fredholm, B. B., Brodin, K., and Strandberg, K. (1979) *Acta Pharmacol. Toxicol.*, **45** 336–344.
75. Schoeffter, C., Lugnier, C., Demesy-Waeldele, F., and Stoclet, J. C. (1987) *Biochem. Pharmacol.*, **36**, 3965–3972.
76. Puurunen, J., Lucke, C., and Schwabe, U. (1978) *Naunyn-Schmiedeberg's Arch. Pharmacol.*, **304**, 69–75.
77. Kaumann, A. J., and Birnbaumer, L. (1974) *Nature*, **251**, 515–517.
78. Hayes, J. S., Bowling, N., King, L. K., and Boder, G. B. (1982) *Biochim. Biophys. Acta.*, **714**, 136–142.
79. Buxton, I. L. O., and Brunton, L. L. (1982) *J. Biol. Chem.*, **258**, 10233–10239.
80. England, P. J., and Shahid, M. (1987) *Biochem. J.*, **246**, 687–695.
81. Hidaka, H., Inagaki, M., Nishikawa, M., and Tanaka, T. (1988) *Methods Enzymol.*, **159**, 652–660.

13

SECOND-GENERATION PHOSPHODIESTERASE INHIBITORS: STRUCTURE–ACTIVITY RELATIONSHIPS AND RECEPTOR MODELS

Paul W. Erhardt

Berlex Laboratories, Cedar Knolls, NJ 07927, USA

13.1   INTRODUCTION

Since the initial designation of cyclic adenosine 3′,5′-monophosphate (cAMP, **1**) as a second messenger [1], details for the various intracellular actions of the cyclic nucleotides have been characterized extensively [2]. It is now clear that 'cyclic nucleotide catabolism is controlled by (several isozymic) forms' [3] of phosphodiesterase (PDE). Recognition that the various PDEs represented a 'fertile field for the development of new drugs' [4] has led to considerable pharmaceutical research directed toward obtaining selective PDE inhibitors. For example, selective inhibitors of cyclic guanosine 3′-5′-monophosphate (cGMP, **2**) PDE, isolated from human lung tissue, have been sought as potential bronchodilator drugs [3]. Most of these studies have employed structural analogs of the parent nucleotides **1** and **2** [5] or derivatives of natural products such as xanthine [6] and papaverine [7], structures **3** and **4**, respectively. Despite complications arising from the 'heterogeneity of the enzyme preparations and wide variations in (these early) assays' [4], several review articles are available which provide excellent discussions about the structure–activity relationships (SAR) associated with this 'first generation' of PDE inhibitors [4–6].

Differences among the PDE isoenzymes are now well established [8,9]. At least three types have been characterized and are considered to be important in most tissues. Their classification varies in the literature and while it has been

Cyclic Nucleotide Phosphodiesterases: Structure, Regulation and Drug Action
Edited by J. Beavo and M. D. Houslay   © 1990 John Wiley and Sons Ltd

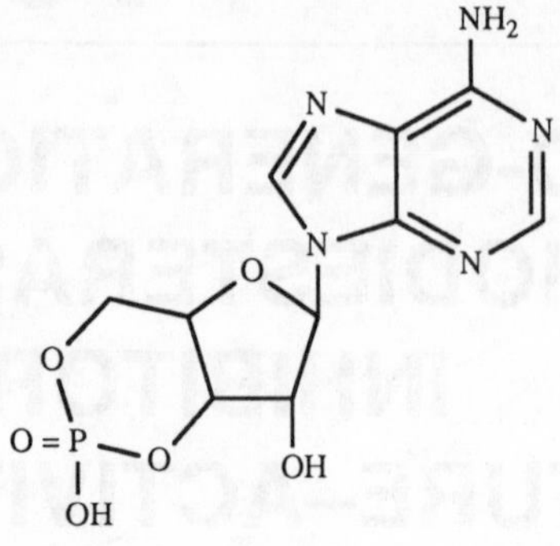

1 c-AMP
anti-conformation

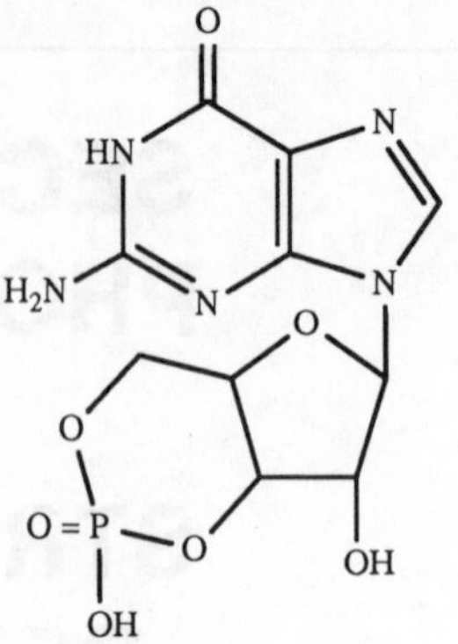

2 c-GMP
syn-conformation

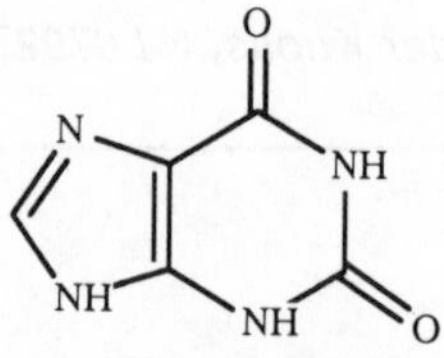

3 Xanthine

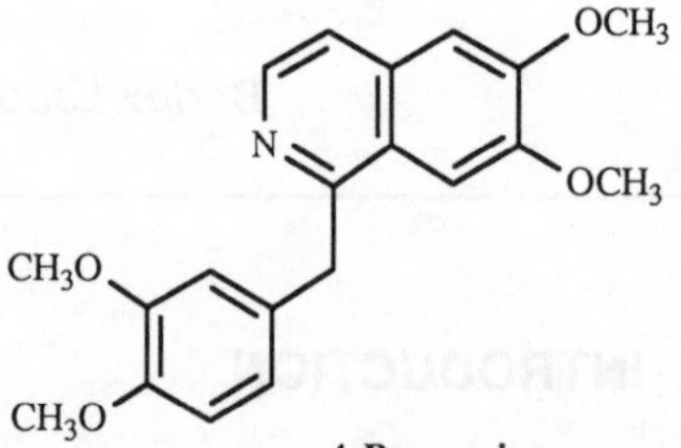

4 Papaverine

recommended to employ a nomenclature based solely on 'substrate preference and regulatory properties' [10], these types are commonly referred to as peak I, II or III, corresponding to their elution times using DEAE–cellulose anion exchange chromatography [11]. Numerous exceptions related to species and tissue differences further complicate a general classification. Nevertheless, for the column-derived nomenclature, peak I PDE demonstrates sensitivity toward activation by Ca^{2+}–calmodulin and generally exhibits a preference toward hydrolysis of cGMP, peak II is sensitive to cGMP and hydrolyzes both cGMP and cAMP, and peak III has a 'low K_m' or 'high affinity' for cAMP as its substrate and is inhibited by cGMP. In the recommended nomenclature [10], the column classification corresponds to types I, II and IV, respectively. The recommended nomenclature will be used for the remainder of this chapter. Prototype structures for a nonspecific PDE inhibitor and inhibitors having specificity for PDE I or PDE IV are represented by 5, 6 and 7, respectively. The concentrations required to effect 50% inhibition (IC_{50}) of PDE I and PDE IV are provided for each compound. These values were obtained by employing crude enzyme preparations [12] from canine aorta and cardiac tissues.

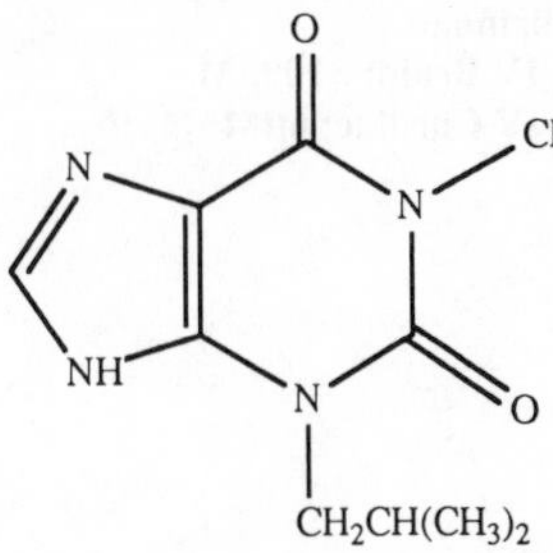

5 IBMX

IC$_{50}$ I: 20μM

IC$_{50}$ IV: 28μM

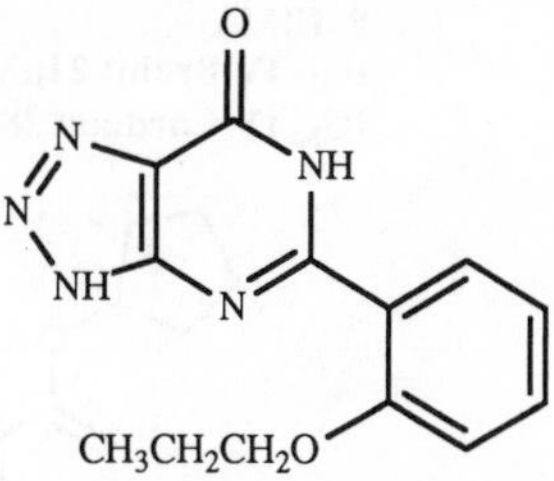

6 Zaprinast

IC$_{50}$ I: 7μM

IC$_{50}$ IV: >100μM

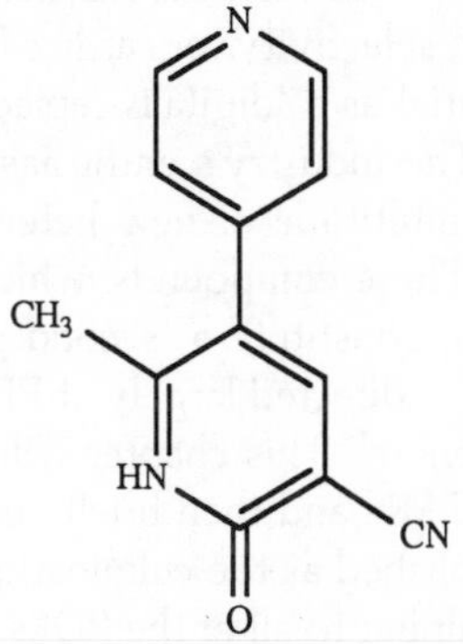

7 Milrinone

IC$_{50}$ I: >100μM

IC$_{50}$ IV: 6μM

Today, efforts are being directed toward even further differentiation of these isoenzymes. Subtypes are being identified for the PDEs when they are present in different tissues and when they can be localized to discrete intracellular organelles [13]. For example, we have recently suggested [12] that there are at least two subtypes of PDE IV in order to explain SAR differences obtained for this enzyme when it is isolated in crude form from brain versus cardiac tissue. The data provided for compounds 8, 9 and 10 illustrate this possibility. It is noteworthy that while our biochemical results are compatible with observed pharmacology (milrinone being a cardiovascular drug with essentially no CNS activity and rolipram an antidepressant drug with essentially no effects on the heart), the enzyme specificity of PDE can, apparently, be altered when column purification of the isoenzymes is performed. In a similar study by Silver et al. [14], reasonable inhibitory potency was observed for rolipram using canine cardiac PDE IV after it had been purified by column chromatography. Thus, it is clear that attempts to further subclassify these isoenzymes will not be straightforward and we will need to proceed with extreme care.

8 IBMX
IC_{50} IV Brain: 21μM
IC_{50} IV Cardiac: 28μM

9 Milrinone
IC_{50} IV Brain: >100μM
IC_{50} IV Cardiac: 6μM

10 Rolipram
IC_{50} IV Brain: 1μM
IC_{50} IV Cardiac: >100μM

Concurrent with these developments has been the realization by the pharmaceutical industry that an inhibitor with selectivity for cardiac PDE IV represents a positive inotropic drug having potential as a 'digitalis replacement' [15] for the therapy of congestive heart failure. The industry's enthusiastic response to this possibility has been to synthesize a multitude of new heterocyclic compounds intended to exploit this mechanism. These compounds, which have proliferated extensively during the last five years, constitute a 'second generation' of PDE inhibitors and, although they have been directed largely at PDE IV, they provide a wealth of SARs for the PDEs in general. This chapter will provide a detailed review of recent, second-generation SARs and then briefly compare the various receptor models which have been published as the culmination of many of these studies. A few general principles pertaining to all of the PDEs will be summarized at the end of the chapter.

13.2 SECOND-GENERATION PDE STRUCTURE–ACTIVITY RELATIONSHIPS

All of the new PDE IV selective inhibitors fit the following very general pharmacophoric relationship: dipole-heteroatom; π or lone-pair electron group; electron-rich moiety. This relationship, as represented by Figure 13.1a, encompasses the more specific relationship: heterocycle–phenyl–imidazole (H–P–I as shown in Figure 13.1b). The latter has been previously suggested by us [16,17] and by others [18] to be a key pharmacophore for positive inotropic activity in cardiac muscle.

Since structural depiction of 'H–P–I,' as in Figure 13.1b, affords a convenient molecular framework, it will be used to further discuss details of the overall SARs associated with the second generation of PDE inhibitor compounds. Structural examples, all of which conform to the representation of Figure 13.1a, are

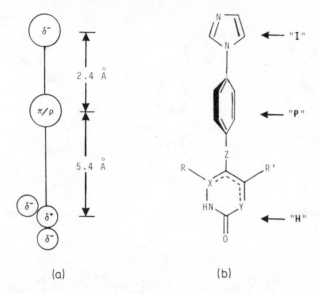

Figure 13.1

provided by compounds **11–22**. Most of these compounds are positive inotropic drugs under development by various pharmaceutical companies for potential use in cardiovascular therapy. References involved with SAR for each compound are given in parentheses. Compounds **14–16** also exemplify the 'H–P–I' relationship.

13.2.1   The heterocyclic portion 'H'

A wide variety of structures are contained in the list of representative compounds. However, this should not be taken to imply that only a very general or random heterocyclic arrangement is required for activity. In fact, a rather strict structural pattern emerges after close analysis of the individual atoms within each

<u>11</u> Amrinone
(19)

<u>12</u> Pelrinone
(20)

<u>13</u> Enoximone
(21,22)

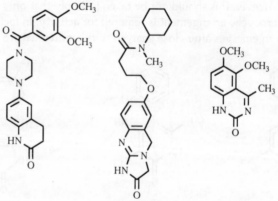

14 Imazodan
(18)

15 CK-2289
(17)

16 Pimobendan
(23)

17 Indolidan
(24)

18 Carbazeran
(25)

19 SKF-94120
(26)

20 Vesnarinone
(27)

21 Lixazinone
(28)

22 Bemarinone
(29,30)

heterocycle. The presence of a dipole and adjacent acidic proton system can be consistently recognized. All of the structures have been aligned to highlight this relationship. It has been noted previously in the literature [18] and is considered to be of primary importance. In addition, early [25] and recent [31] studies have suggested that these atoms probably simulate 'the electrophilic center in the natural phosphate moiety' [25] present in cAMP. However, the pK_a for the latter is approximately 2 and it is, essentially, completely ionized at pH 7.4. Thus, cAMP lacks an acidic proton. To resolve this apparent conflict of pharmaco-phores between substrate and inhibitors, we have suggested [31] that it is the electronic similarity between the carbonyl (or phosphoryl in the case of cAMP) dipoles and their conjugated heteroatoms that is of primary importance and not the presence of an adjacent acidic proton. Indeed, the similarity in electronic structure between the ionized phosphate in cAMP and a typical neutral heterocycle such as the imidazolone system present in **13** and **15** is dramatic. This similarity results in a corresponding similarity in their molecular electrostatic potentials which are shown in Figure 13.2. Nevertheless, the 'acidic proton hypothesis' repeatedly gains popularity because in all cases where the nitrogen atom in the 'H' ring of Figure 13.1b has been further alkylated (thereby removing the acidic proton), activity is significantly diminished (typically by a factor of 100 or more). Unfortunately, as a general rule in SAR studies, even a single, subtle structural change usually causes several variables to change. In this case, the N-alkylation also imparts a significant new steric demand at this locale as well as causing a subtle change in the overall electronics of 'H'. Thus, whether or not the presence of an acidic proton is actually requisite for inhibitory activity should still be regarded as an open SAR question.

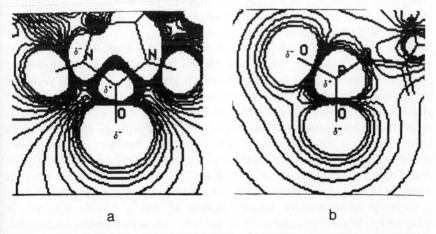

a   b

Figure 13.2   AM1 derived in-plane molecular electrostatic potentials of the imidazo-lone (nonionized) portion of **15** (a) and the phosphate (ionized) portion of cAMP (b) [31]

Additional unsaturation in 'H' in the form of at least one double bond also appears to be a feature generally present in the new structures. Even in the 'opened heterocyclic' version represented by compound **18**, an oxygen atom is also present to contribute its lone-pair electrons, perhaps analogously to the π-electrons present in double bonds. While the presence of additional unsaturation has a dramatic effect upon the electronic nature of 'H,' a specific receptor binding site for the unsaturation has also been postulated [31].

Although there are occasional exceptions (e.g. see [29]), alkyl substitution on 'H' at either R or R' typically enhances potency when the alkyl groups are limited to either methyl or ethyl. Initially, based on an analysis of compounds related to **14**, this was attributed to the presence of a 'small lipophilic pocket [on PDE IV] which would accommodate and bind small alkyl groups' [18]. However, it has been subsequently pointed out [31] that the presence of a single pocket cannot explain the identical results when such substitutions are effected on either side of 'H' (at R or R'). Either two nearly identical pockets must be invoked or the effects on potency are, instead, attributable to some other phenomenon that is common to substitution at either locale, e.g. the influence that such substituents might have on overall molecular conformation (see below). When the alkyl groups are lengthened past ethyl, e.g. propyl, inhibitory potency typically diminishes sharply, suggesting the presence of steric limitations on each side of 'H' and/or a further twisting of the overall conformation past a preferred range of angles between 'H' and 'P' [31].

13.2.2   The phenyl ring portion 'P'

After consideration of SARs related to **14**, it was concluded that the central phenyl ring 'acts mainly as a spacer' [32], thus implying that it is not a functional pharmacophoric unit itself. Consideration of a broader range of structures and their overall SARs has prompted an alternative proposal. As discussed above, it has been suggested [31] that the beneficial effect of the small alkyl groups at either R or R' in Figure 13.1b is to twist the 'P' ring away from exact coplanarity with the 'H' ring. This implies that a specific orientation of 'P' may be preferred for optimal inhibitory activity. The latter, then, suggests that the 'P' ring does play a functional role as an actual pharmacophoric unit. This notion gains strong support upon consideration of how cAMP might bind with the PDE receptor. In cAMP there is a rigid twist away from coplanarity between the two planes established by the adenine ring and the phosphoryl moiety where these groups would then be analogous to the 'P' ring and 'H' ring, respectively. In addition, when the 'P' ring in inhibitor structures is replaced by a simple alkyl chain which can serve as an appropriate 'spacer' between 'H' and 'I,' inhibitory activity is either completely lost or largely attenuated [33]. This further implies a functional, rather than static, role for 'P'. Analysis of this region within all of the structures **11–22** suggests that electron-rich heteroatoms may also serve as surrogates for the 'P' ring pharmacophore. For instance, the piperidine nitrogen in **18** and the methoxy oxygen atoms in **22** have lone-pair electrons which could presumably play this role.

13.2.3 The imidazole portion 'I'

An approximate potency ranking for various substituents that have been placed at the p-position of the phenyl ring as pharmacophores for the 'I' region is provided below. The list represents a composite analysis of independent SAR reports from several different investigators where direct comparisons of at least two of these substituents were made while maintaining constant 'H–P' systems [17,22,33–36]. In general, whenever SAR overlaps occurred, there was close agreement among the different investigators.

Inhibitory potency: lactam (e.g. **17**) $\geq$ alkyl–CONH– $\geq$ imidazoyl = pyridine in place of phenyl with its nitrogen in the analogous 4-position $\geq$ alkyl–S– > simple ethers > halide = amines > imidazolium (inactive). A general pharmacophore, which encompasses all of these groups, has previously been proposed to be that of a 'hydrogen-bond acceptor' [18]. However, there is no compelling evidence that such a specific type of interaction occurs during binding (as Bronsted bases) and it is probably better to consider the representatives on this list as simply electron-rich moieties (as Lewis bases) until such data become available. In support of both proposals, the permanently charged, alkylated imidazolium group is totally inactive such that a positively charged (non-basic) moiety cannot be tolerated in this region of the PDE 'receptor' [17].

Interestingly, for the imidazoyl group, potency and selectivity have been shown to be 'exquisitely dependent upon the nature of its [additional] substituents' [17]. The following 'general trend [for simple alkyl substituents] appears for the rank order of [potency and to a certain extent PDE IV] selectivity: no substitution > mono substitution > disubstitution > trisubstitution' [32]. For example, within a series of structures related to **15** we have found that introduction of a single methyl substituent at the 2-position of the imidazolyl group (Figure 13.3, compound **23**) leads to a dramatic alteration in biological profile. While **15** is a very potent inhibitor of PDE IV (IC_{50} 2 μM) and acts as a potent inotropic and vasodilator drug when tested in vivo, **23** is only a weak to moderately potent inhibitor of PDE IV (IC_{50} 53 μM). Furthermore, while **23** still demonstrates strong inotropic activity in vivo, it has essentially no effects on the vasculature. Figure 13.3 is intended to illustrate the general influence that the presence of small alkyl groups can have on the overall conformation of 'H–P–I' systems using **15** and **23** as examples. For **15**, the ethyl substituent on the 'H' ring tends to turn the 'P' ring approximately 45° away from their coplanarity. For **23**, the additional methyl substituent on the 'I' ring tends to push the 'P' ring even further (to an angle approximately 60°) away from the 'H' ring plane. The magnitude of the latter twist may be such that the proper alignment for maximal

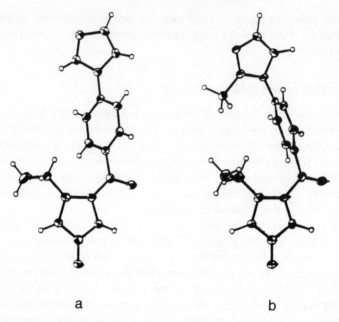

a   b

Figure 13.3 ORTEP drawings of **15** (a) and its 2-methylimidazo-1-yl analogue **23** (b) derived from X-ray analysis of their free-base crystals

binding of this pharmacophore has actually been surpassed, thus decreasing its PDE inhibitory potency relative to **15**.

In addition to a pyridine ring with its nitrogen near the 'I' region and an imidazoyl group, two other substituents also deserve special comment. First, the N-cyclohexyl-N-methyl-4-oxybutyramide side chain, as present in **21**, has been found to be of 'significant value as a steric and/or lipophilic pharmacophore within' 'H–N' systems related to **20** and **21** [37,38]. Similarly, the 3,4-dimethoxybenzoylpiperazine side chain, as present in **20**, probably interacts with special features in this region of PDE IV to enhance its selectivity as an inhibitor [27].

13.2.4 Structure–activity relationship summary

In summary, the general pharmacophoric relationship heterocycle–phenyl–imidazole represents an extremely useful molecular framework from which the second-generation PDE inhibitors can be related and analyzed in terms of their detailed SARs. Furthermore, in a recent retrograde study of this relationship, we have determined that 'while all three [of the 'H', 'P' and 'I'] components contribute significantly toward potent activity, any combination of two components, in approximately the preferred geometry, represents the minimal requirement for weak activity' and that 'no single component appears to be requisite in an absolute sense' [33].

13.3 PDE IV RECEPTOR MODELS

Accompanying the second generation of PDE inhibitors, with its surge of new structures and SARs, has been the development of several cAMP-PDE receptor models. Several earlier papers have also discussed models of the PDE enzyme active sites, particularly with an emphasis toward detailing the process of nucleotide hydrolysis (for examples see [39–42]). Interestingly, the early notion that the phosphate moiety assumes a trigonal bipyramidal transition state during the hydrolytic process complements the planar arrangement (Figure 13.2b) derived from more recent receptor modeling. In fact, when the cyclized 5'-oxygen [41] joins the plane defined by the PO_2^- resonance hybrid (thereby establishing the central triangle which acts as the common base for each of the symmetrical pyramids in the transition state) it assumes a position that is analogous to the additional electron-rich nitrogen region (right side) as depicted in Figure 13.2a. Thus, heterocyclic systems like the ones present in **13**, **15**, **18**, **19** and **22** may actually inhibit PDE IV while binding as transition state analogues, at least with regard to their 'H' regions.

The new models are summarized in Table 13.1, which uses the 'H–P–I' framework to conveniently compare specific features and differences for each model.

13.4 SUMMARY

Most authors seem to agree that the second-generation inhibitors bind at the active site of PDE IV with the heterocyclic amidic moiety (without an absolute requirement for an acidic proton) simulating the phosphate group present in cAMP. Since hydrolysis is an identical function for all of the PDEs, it seems reasonable to assume that molecular conservation has occurred in this region of the various isoenzymes such that these SAR details are likely to be common to all of the PDEs. One can speculate, then, that the same heterocycles will be quite useful when further applied toward new ('third-generation') PDE inhibitors, perhaps as cGMP (type I)-selective agents for use as novel vasodilators or bronchodilators. There is less agreement, however, about the role of small alkyl substituents when they are present on the heterocyclic portion of the inhibitors. Essentially two proposals are in place: first, that the alkyl groups bind with (at least) one small alkyl pocket at the receptor and, further, that the preferred overall molecular conformation for the inhibitors is nearly that of a single plane; or second, that their foremost role is to twist the conformation of the central ring approximately 20° away from that of overall molecular coplanarity. There is

Table 13.1 PDE IV receptor models[a]

No.[b]	Author[c]	Affiliation	Date	Reference	Structure[d]	H Region				P Region		I Region	
						Site[e]	Sel.[f]	H/N[g]	Alkyl[h]	D/S[i]	Twist[j]	PA/ER[k]	Other[l]
1[m]	Wells	Vanderbilt University, USA	1981	6	3	A	cA	—	—	D	30*	ER	A
2	Roberts	Pfizer, UK	1983	25	18	A	cA	N	—	D	20	ER	A
3	Leclerc	Institut de Pharmacologie, France	1986	43	20[n]	U	—	N	C	D	15–45	ER	—
4	Davis	Smith Kline & French, UK	1987	26	19	A	cA	N	P	D	0	ER	A
5	Robertson	Lilly, USA	1987	46	17	U	—	—	P	—	0	ER	A
6	Moos (Bristol)	Warner-Lambert, USA	1987 (1984)	47 (18)	14	A	U	H	P	S	0	ER	U
7	Venuti	Syntex, USA	1988 (1987)	28 (37)	21	A	U	H	—	D	17–22	PA	A
8	Erhardt	Berlex, USA	1988 [1986]	31 [48]	15	A	cA	N	C	D	20	ER	A

[a] List of topographical receptor models which have been designed to accommodate SARs associated with compounds that inhibit PDE IV. In addition to explicitly published models, the list contains 'models' perceived by this author to be clearly present as a major guide or theme during the SAR studies published by the indicated authors. The various models are compared by employing the Heterocycle–Phenyl–Imidazole ('H' region, 'P' region and 'I' region) framework, as shown in Figure 13.1b to categorize their specific features. See Section 13.2 for further explanation of the subdivisions within the 'H', 'P' and 'I' categories. In cases where data were not provided for a given entry, an attempt has been made by this author, whenever appropriate, to derive such data. These entries bear a superscript asterisk. When the latter could not be accomplished with confidence, a dash has been entered.

[b] The models are listed chronologically according to publication date.

[c] Correspondence or first author as listed in indicated reference. In general, this reference is the principcal publication pertaining to the receptor model and its date was used per footnote b. Information listed in parentheses represents a preceding/preliminary publication of essentially the same model. Information listed in square brackets represents a preceding scientific presentation of essentially the same model.

[d] Key structural type for studies involving the receptor model.

[e] Models in which binding of the inhibitors is thought to occur within the active site of the enzyme (substrate site) are listed as 'A'. 'U' indicates that this relationship is unstated or not addressed by the authors.

[f] Active site models which also account for selectivity of cAMP hydrolysis over cGMP are listed as 'cA'. Active site models where cAMP/cGMP selectivity is not addressed are listed as 'U'.

[g] Models which have a binding site for a carbonyl with an adjacent acidic proton are designated as 'H', and those which have a binding site for a carbonyl with a conjugated heteroatom are designated as 'N'.

[h] Models which suggest that the primary influence of substituents at R or R' is to interact with a small lipophilic pocket(s) are designated 'P', and those which suggest that their primary influence is upon overall molecular conformation are designated 'C'.

[i] Models which suggest that the phenyl ring serves as a source of π-electrons which interact with the receptor in a functional manner are designated 'D', and those which suggest that the phenyl ring plays only a static role as a spacer are designated 'S'.

[j] The approximate deviation from coplanarity between the 'H' and 'P' planes is given in degrees. For model 1, this analysis was made between the planes perceived for the phosphate and adenine moieties.

[k] Models in which the key binding site in this region require a proton acceptor moiety to be present in the inhibitor or substrate are designated as 'PA', and those that require only an electron-rich moiety are designated 'ER'.

[l] Models which allude to additional, but less defined, important binding sites in this region are designated 'A', and those that do not are designated 'U'.

[m] While this model actually represents a review of 'first-generation' PDE inhibitors, it is included as the key historical reference and perspective from earlier SAR studies.

[n] Based upon subsequent publications by these authors [44,45].

nearly complete agreement that a central aromatic ring (or π or lone-pair electrons appropriately located in the 'P' region) plays a functional, rather than static, role during binding. At this time, the importance of electron-rich groups in the 'I' region is probably best associated with their electronic character rather than any specific hydrogen bond (as an acceptor) formation. There also seems to be agreement that different arrangements of at least two binding sites in the 'I' region probably account for the selectivities of the various PDEs along with the possibility of *syn* versus *anti* conformations also playing a key role in the cGMP versus cAMP specificity of the enzymes. The latter variable seems ideally suited for studies employing rigid conformer analogs. Since many of the recent structures lend themselves to such manipulation, this variable will probably be better defined in the near future. Hopefully, many of the other SAR controversies will also be resolved as another ('third') generation of PDE inhibitors is pursued.

REFERENCES

1. Sutherland, E. W., and Rall, T. W. (1960) *Pharmacol. Rev.*, **12**, 265.
2. For periodic coverage of this topic see the sequential volumes of *Advances in Cyclic Nucleotide Research* (Greengard, P. et al., eds), Raven Press, New York, beginning with Vol. 1 in 1972.
3. Bergstrand, H., Kristoffersson, J., Lundquist B., and Schurmann, A. (1977) *Mol. Pharmacol.*, **13**, 38.
4. Amer, M. S., and Kreighbaum, W. E. (1975) *J. Pharm. Sci.*, **64**, 1.
5. Drummond, G. I., and Severson, D. L. (1971) In *Annual Reports in Medicinal Chemistry 1970*, Vol. 6 (Cain, C. K., ed.), Academic Press, New York, p. 215.
6. Wells, J. N., Garst, J. E., and Kramer, G. L. (1981) *J. Med. Chem.*, **24**, 954.
7. Hanna, P. E., O'Dea, R. F., and Goldberg, N. D. (1972) *Biochem. Pharmacol.*, **21**, 2266.
8. Hidaka, H., Tanaka, T., and Itoh, H. (1984) *Trends Pharmacol. Sci.*, 237.
9. Weishaar, R. E., Cain, M. H., and Bristol, J. A. (1985) *J. Med. Chem.*, **28**, 537.
10. Strada, S. J., and Thompson, W. J. (1984) *Adv. Cyclic Nucleotide Protein Phosphorylation Res.*, **16**, vi.
11. Thompson, W. J., Terasaki, W. L., Epstein, P. M., and Strada, S. J. (1979) *Adv. Cyclic Nucleotide Res.*, **10**, 69.
12. Pang, D. C., Cantor, E., Hagedorn, A., Erhardt, P., and Wiggins, J. (1988) *Drug Dev. Res.*, **14**, 141.
13. Kauffman, R. F., Crowe, V. G., Utterback, B. G., and Robertson, D. W. (1986) *Mol. Pharmacol.*, **30**, 609.
14. Silver, P. J., Hamel, L. T., Perrone, M. H., Bentley, R. G., Bushover, C. R., and Evans, D. B. (1988) *Eur. J. Pharmacol.*, **150**, 85.
15. Erhardt, P. W. (1987) *J. Med. Chem.*, **30**, 231.
16. Davey, D., Erhardt, P. W., Lumma, W. C. Jr, Wiggins, J., Sullivan, M., Pang, D., and Cantor, E. (1987) *J. Med. Chem.*, **30**, 1337.
17. Hagedorn, A. A. III, Erhardt, P. W., Lumma, W. C. Jr, Wohl, R. A., Cantor, E., Chou, Y.-L., Ingebretsen, W. R., Lampe, J. W., Pang, D., Pease, C. A., and Wiggins, J. (1987) *J. Med. Chem.*, **30**, 1342.

18. Bristol, J. A., Sircar, I., Moos, W. H., Evans, D. B., and Weishaar, R. E. (1984) *J. Med. Chem.*, **27**, 1099.
19. Weishaar, R. E., Quade, M., Boyd, D., Schenden, J., Marks, S., and Kaplan, H. R. (1983) *Drug Dev. Res.*, **3**, 517.
20. Bagli, J., Bogri, T., Palameta, B., Rakhit, S., Peseckis, S., McQuillan, J., and Lee, D. K. H. (1988) *J. Med. Chem.*, **31**, 814.
21. Kariya, T., Wille, L. J., and Dage, R. C. (1982) *J. Cardiovasc. Pharmacol.*, **4**, 509.
22. Schnettler, R. A., Dage, R. C., and Grisar, J. M. (1982) *J. Med. Chem.*, **25**, 1477.
23. Honerjager, P., Heiss, A., Schafer-Korting, J., Schnosteiner, G., and Reiter, M. (1984) *Naunyn-Schmiedeberg's Arch. Pharmacol.*, **325**, 259.
24. Robertson, D. W., Krushinski, J. H., Pollock, G. D., Wilson, H., Kauffman, R. F., and Hayes, J. S. (1987) *J. Med. Chem.*, **30**, 824.
25. Campbell, S. F., Cussans, N. J., Danilewicz, J. C., Evans, A. G., Ham, A. L., Jaxa-Chamiec, A. A., Roberts, D. A., and Stubbs, J. K. (1983) In *Second SCI-RSC Medicinal Chemistry Symposium* (Emmett, J. C., ed.), Special Publication No. 50, Burlington House, London, p. 47.
26. Davis, A., Warrington, B. H., and Vinter, J. G. (1987) *J. Computer-Aided Molecular Design*, **1**, 97.
27. Tiara, N., Endoh, M., Ijima, T., Satoh, K., Yanaqisawa, T., Yamashita, S., Maruyama, M., Kawada, M., Morita, T., and Wada, Y. (1984) *Arzneim.-Forsch.*, **34**, 347.
28. Venuti, M. C., Stephenson, R. A., Alvarez, R., Bruno, J. J., and Strosberg, A. M. (1988) *J. Med. Chem.*, **31**, 2136.
29. Bandurco, V. T., Schwender, C. F., Bell, S. C., Combs, D. W., Kanojia, R. M., Levine, S. D., Mulvey, D. M., Appolina, M. A., Reed, M. S., Malloy, E. A., Falotico, R., Moore, J. B., and Tobia, A. J. (1987) *J. Med. Chem.*, **30**, 1421.
30. Falotico, R., Moore, J. B., Bandurco, V., Bell, S. C., Levine, S. D., and Tobia, A. J. (1988) *Drug Dev. Res.*, **12**, 241.
31. Erhardt, P. W., Hagedorn, A. H. III, and Sabio, M. (1988) *Mol. Pharmacol.*, **33**, 1.
32. Sircar, I., Weishaar, R. E., Kobylarz, D., Moos, W. H., and Bristol, J. A. (1987) *J. Med. Chem.*, **30**, 1955.
33. Erhardt, P. W., Hagedorn, A. A. III, Davey, D., Pease, C. A., Venepalli, B. R., Griffin, C. W., Gomez, R. P., Wiggins, J. R., Ingebretsen, W. R., Pang, D., and Cantor, E. (1989) *J. Med. Chem.*, **32**, 1173.
34. Robertson, D. W., Krushinski, J. H., Beedle, E. E., Wyss, V., Pollock, G. D., Wilson, H., Kauffman, R. F., and Hayes, J. S. (1986) *J. Med. Chem.*, **29**, 1832.
35. Sircar, I., Duell, B. L., Bristol, J. A., Weishaar, R. E., and Evans, D. B. (1987) *J. Med. Chem.*, **30**, 1023.
36. Robertson, D. W., Krushinski, J. H., Pollock, G. D., and Hayes, J. S. (1988) *J. Med. Chem.*, **31**, 461.
37. Venuti, M. C., Jones, G. H., Alvarez, R., and Bruno, J. J. (1987) *J. Med. Chem.*, **30**, 303.
38. Jones, G. H., Venuti, M. C., Alvarez, R., Bruno, J. J., Berks, A. H., and Prince, A. (1987) *J. Med. Chem.*, **30**, 295.
39. Springs, B., and Haake, P. (1977) *Tet. Lett.*, No. 37, 3223.
40. Landt, M., and Butler, L. G. (1978) *Biochemistry*, **17**, 4130.
41. Vanool, P. J. J. M., and Buck, H. M. (1982) *Eur. J. Biochem.*, **121**, 329.
42. Lowe, G. (1983) *Acc. Chem. Res.*, **16**, 244.
43. Rakhit, S., Marciniak, G., Leclerc, G., and Schwartz, J. (1986) *Eur. J. Med. Chem.*, **21**, 511.

44. Leclerc, G., Marciniak, G., Decker, N., and Schwartz, J. (1986) *J. Med. Chem.*, **29**, 2427.
45. Leclerc, G., Marciniak, G., Decker, N., and Schwartz, J. (1986) *J. Med. Chem.*, **29**, 2433.
46. Robertson, D. W., Jones, N. D., Krushinski, J. H., Pollock, G. D., Schwartzendruber, J. K., and Hayes, J. S. (1987) *J. Med. Chem.*, **30**, 623.
47. Moos, W. H., Humblet, C. C., Sircar, I., Rithner, C., Weishaar, R. E., Bristol, J. A., and McPhail, A. T. (1987) *J. Med. Chem.*, **30**, 1963.
48. Erhardt, P. W., Hagedorn, A. A. III, and Lampe, J. In *IXth International Symposium on Medicinal Chemistry, September 1986*, West Berlin, FRG. Abstract Number 111.

ABBREVIATIONS

ADA	adenosine deaminase
A-kinase, cAMP-PrK	cAMP-dependent protein kinase
CaM	calmodulin
CAP	catabolite gene activation protein
cBIMP	benzimadole 3',5'-monophosphate
cG-BPDE	cGMP-binding cGMP-specific phosphodiesterase
cGI-PDE	cGMP-inhibited cAMP phosphodiesterase
cGS-PDE	cGMP-stimulated phosphodiesterase
CIT	N-(2-isothiocyanato)-ethyl derivative of cilostamide
cPMP	purine riboside 3',5'-monophosphate
ER	endoplasmic reticulum
FAB	fast atom bombardment
HTC	cultured rat hepatoma cell
IBMX	3-isobutyl-1-methylxanthine
$MDCK_T$	Maden Darby Canine Kidney Cells transformed by Harvey Murine Sarcoma Virus
MIKES	mass analysed ion kinetic energy spectrum scanning
OPC	Otsuka Pharmaceutical Company
8PCPT-cAMP	8-p-chlorophenylthio-cAMP
8pCPT-cGMP	8-(4-chlorophenylthio)-cGMP
PDE	phosphodiesterase
PET-cGMP	β-phenyl-1-N^2-etheno-cGMP
PGI_2	prostacyclin
PIA	N^6-(phenylisopropyl)-adenosine
Rh*	activated rhodopsin
RIA	radioimmunoassay
ROS	rod outer segment
SAR	structure–activity relationship
SR	sarcoplasmic reticulum
TPA	12-O-tetradecanoyl phorbol 13-acetate

INDEX